THE REGULATORY GENOME

THE REGULATORY GENOME

Gene Regulatory Networks
in Development and Evolution

Eric H. Davidson

Division of Biology
California Institute of Technology
Pasadena, California

ELSEVIER
ACADEMIC
PRESS

Amsterdam Boston Heidelberg London New York Oxford
Paris San Diego San Francisco Sydney Tokyo

Academic Press is an imprint of Elsevier
30 Corporate Drive, Suite 400, Burlington, MA 01803, USA
525 B Street, Suite 1900, San Diego, California 92101-4495, USA
84 Theobald's Road, London WC1X 8RR, UK

This book is printed on acid-free paper. ∞

Library of Congress Cataloging-in-Publication Data
Application submitted

British Library Cataloguing-in-Publication Data
A catalogue record for this book is available from the British Library.

ISBN 13: 978-0-12-088563-3
ISBN 10: 0-12-088563-8

For information on all Academic Press publications
visit our Web site at www.books.elsevier.com

Printed and bound by CPI Group (UK) Ltd, Croydon, CR0 4YY
Transferred to Digital Print 2011

CONTENTS

PREFACE

This book came about in the following way. In the winter of 2004, Jasna Markovac, the publisher of my monograph "Genomic Regulatory Systems" (2001), asked me to think about doing a second edition of that work. But no sooner had I opened that door and begun to survey the landscape than, as in an old legend of magic, everything seemed to have been transformed to another landscape. The conceptual peaks that were the landmarks of "Genomic Regulatory Systems" were still there, but now new mountains towered over them, and the whole domain appeared to be much more brightly illuminated, and to extend farther toward the horizon. It was beyond possibility to "update" a vista so changed.

"Genomic Regulatory Systems" sought as its main objective to crystallize the incontrovertible evidence that causality in development resides ultimately in *cis*-regulatory control of spatial gene expression. But development is the output of regulatory systems comprising large numbers of regulatory genes. Though in that work I often referred to the gene networks that would someday represent developmental programs, the few examples were anecdotal, and their general properties remained entirely obscure. Then in 2002 we published the first real scale gene regulatory network (including about 50 genes) explanatory of a major piece of development, specification of the endomesoderm of the sea urchin embryo. As the ancients used to say, the scales fell from our eyes. A whole field of developmental gene regulatory networks has now sprung forth, that encompasses many different animal systems. The structure/function properties that emerge from the architecture of these networks are a large part of what has transformed the conceptual terrain of this large area of bioscience. There follow more new things: a different way to think comparatively about various forms of development; a different way to think about the process of evolution; hence the subtitle of this book, "Gene Networks in Development and Evolution". Exploration of these new pathways toward scientific explanation of the developmental and evolutionary phenomena of biology is the central object of this volume. As with its predecessor, the approach I have taken here is that of demonstration by example: the points to be made rest upon powerful, exemplary experimental demonstrations, detailed, for those who desire experimental substance, in the figure captions. However, in no way have I attempted to be encyclopedic. So, willy nilly, there is much that could equally well have been included but was not, and my apologies en masse to the authors of these works.

I have not shied away from what are sometimes pejoratively been termed "big ideas," nor taken the view of an anonymous reviewer of a paper of mine who recently amused me with the complaint "But the original ideas in this paper are speculative!" This book includes many diagrams in which concepts are set forth in specific form, just so they can be subjected to precise tests of falsification, and just so they can be used in precise ways to generate predictions I may not have thought of. One such idea, which underlies everything in this book, is the concept of genomically encoded information processing. To return to my metaphor above, this is like the geological basis of the landscape. In my view, *cis*-regulatory information processing, and information processing at the gene regulatory network circuit level, are the real secret of animal development. Probably the appearance of genomic regulatory systems capable of information processing is what made animal evolution possible.

This book begins with an overview of the regulatory genome and the concept of information processing in gene regulation (Chapter 1). It proceeds to an in-depth analysis of modular cis-regulatory designs for generation of spatial patterns of gene expression, and consideration of how they generate regulatory output (Chapter 2); thence to a comparative treatment of developmental pathways in terms of transient regulatory states (Chapter 3); to gene regulatory network theory and the character of diverse real developmental regulatory networks (Chapter 4); and finally to the application of network structure/function relations to some unsolved problems of animal evolution (Chapter 5). The image of a genomically encoded information processing system that throughout the life cycle responds conditionally to incident regulatory inputs can never lie far from the surface of any of these subject areas.

Science is made by scientists, whose creations deeply affect each others' progress. For me there have been certain scientists in each period of my own progress whose work and ideas have particularly illuminated the world: among them I must mention as of particular importance in this present period, and for what is included herein, Mike Levine, Ellen Rothenberg, Doug Erwin, Sorin Istrail, Bill McGinnis, and Lee Hood. This book would not have whatever worth it does were it not for the generosity of these people and also of Paola Oliveri and Joel Smith, postdoctoral colleagues in my laboratory, in reading, criticizing, and improving drafts of various parts, and in some cases all, of the manuscript. I have been extremely fortunate to have had the very expert services of a superb illustrator, Tania Dugatkin. In my own domain Deanna Thomas has provided invaluable assistance with figures, references, and everything else; and my graduate student Pei-yun Lee not only helped with technical research but also with figure attributions. Nor would this project have ever reached fruition were it not for the continued encouragement of Jasna Markovac, and of the careful, obsessive work of the production manager Paul Gottehrer at Academic Press/ Elsevier. I also wish to say that since so much of what follows is linked to our expanding experimental invasion of gene regulatory networks, the support we have had for that research has been indirectly essential for this book as well: mainly this support has come from the National Institute of Child Health and Human Development and from the Genomes to Life Program of DOE, but also from NIGMS, NIRR, NIHGRI, NSF, NASA, Caltech's Beckman Institute, and Applied Biosystems, Inc.

Finally, I would like to dedicate this book to the person who has worked most closely with me on it, good days and bad, and that is Jane Rigg. She has been my editor, judge, administrator, research aide, and advisor throughout, as also on three other books I have written in the more than 35 years that we have worked together. Only my first book, "Gene Activity in Early Development" (1968) preceded the Jane Rigg era, but that was a very long time ago indeed.

Eric Davidson
April 2006

CHAPTER 1

The "Regulatory Genome" for Animal Development

THE FRAMEWORK

Animal body plans, their structures and the functions with which their morphology endows them, are the integrals over time and space of their successive developmental processes. In abstract terms the mechanism of development has many layers, expanding in the diversity of its parts the farther removed from its core. At the outside, development is mediated by the spatial and temporal regulation of expression of thousands and thousands of genes that encode the diverse proteins of the organism, and that catalyze the creation of its nonprotein constituents. Deeper in is a dynamic progression of regulatory states, defined by the presence and state of activity in the cell nuclei of particular sets of DNA-recognizing regulatory proteins (transcription factors), which determine gene expression. At the core is the genomic apparatus that encodes the interpretation of these regulatory states. Physically, the core apparatus consists of the sum of the modular DNA sequence elements that interact with transcription factors. These regulatory sequences "read" the information conveyed by the regulatory state of the cell, "process" that information, and enable it to be transduced into instructions that can be utilized by the biochemical machines for expressing genes that all cells possess. The sequence content, arrangement, and other aspects of the organization of these modular control elements are the heritage of each species. They contain the sequence-specific code for development; and they determine the particular outcome of developmental processes, and thus the form of the animal produced by every embryo. In evolution, the alteration of body plans is caused by changes in the organization of this core genomic code for developmental gene regulation.

This book is about the system level organization of the core genomic regulatory apparatus, and how this is the locus of causality underlying the twin phenomena of animal development and animal evolution. Because the sequence of the DNA regulatory elements is the same in every cell of each organism, the regulatory genome can be thought of as hardwired, and genomic sequence may be the only thing in the cell that is. Indeed that is a required property of gene regulatory elements, for they must endow each gene with the information-receiving capacity that enables it to respond properly to every conditional regulatory state to which it might be exposed during all phases of the life cycle, and in all cell types. For development, and therefore for major aspects of evolution, the most important part of the core control system is that which determines the spatial and temporal expression of regulatory genes. As used here, "regulatory genes" are those encoding the transcription factors that interact with the specific DNA sequence elements of the genomic control apparatus. The reason that the regulation of genes encoding transcription factors is central to the whole core system is, of course, that these genes generate the determinant regulatory states of development.

There follow several important and general principles of organization of the developmental regulatory apparatus, that is, of the control machinery directing expression of the regulatory genes themselves. First, signaling affects regulatory gene expression: The intercellular signals upon which spatial patterning of gene expression commonly depends in development must affect transcription of regulatory genes, or else they could not affect regulatory state. Therefore, the transcriptional termini of the intracellular signal transduction pathways required in development are located in the genomic regulatory elements that determine expression of genes encoding transcription factors. Second, developmental control systems have the form of gene regulatory networks: Since when they are expressed given transcription factors always affect multiple target genes, and since the control elements of each regulatory gene respond to multiple kinds of incident regulatory factor, the core system has the form of a gene regulatory network. That is, each regulatory gene has both multiple inputs (from other regulatory genes) and multiple outputs (to other regulatory genes), so each can be conceived as a node of the network. Third, the nodes of these gene regulatory networks are unique: Though it is not *a priori* obvious, each network node performs a unique job in contributing to overall regulatory state, in that its inputs are a distinct set, just as the factor it produces has a distinct set of target genes.

Fourth, regulatory genes perform multiple roles in development: The repertoire of regulatory genes is evolutionarily limited, and all animals use more or less the same assemblage of DNA binding domains, which define the classes of transcription factor. However, given factors are frequently required for different processes in different forms of development, and they are often used for multiple unrelated purposes within the life cycle. Thus, both within and among animal species, many regulatory genes must be able to respond to diverse regulatory inputs that are presented in various space/time places in the developing organism.

THE REGULATORY APPARATUS ENCODED IN THE DNA

Genomes, Genes, and Genomic "Space"

Viewing the animal genome as a whole, we may ask how much sequence information is required for the regulatory apparatus, compared to the amount encoding proteins. The question is confounded at the outset by the great variation among animal species in the overall amount of DNA per haploid genome, even within given phylogenetic clades. Examples are the greater than tenfold differences in genome size seen among insects, among fish, and among amphibians. This was already known by the end of the 1960s, from measurements carried out on dozens of species (reviewed by Britten and Davidson, 1971). On the other hand, estimations of the amount of genetic information read out into the mRNA populations of organisms of diverse genome size indicated early on that the large differences in genome size are not reflected quantitatively as differences in expressed mRNA complexity ("complexity" is here total mRNA sequence length in nucleotides if single molecules of each of the different mRNA species represented in a population were laid end-to-end). Two direct sets of measurements led to this conclusion (reviewed in Davidson, 1986). One was a comparison of maternal RNA complexities in eggs of various species of animal, the genomes of which range more than 100-fold in size. The results boiled down to the conclusion that the egg RNAs are all of roughly the same complexity, give or take a small variation. This is of course reasonable, since animal eggs have essentially similar jobs to do with their stored maternal mRNAs. The second set of measurements consisted of cytological and molecular analyses of the number of transcription units active in the extended "lampbrush" chromosomes in the oocytes of two species of amphibian that differ in genome size by a factor of about 10. About the same number of diverse genes is transcribed in the oocytes of these species, though the size of the individual transcription units appears to scale with genome size. The general implication from both data sets was that the complexity of given phases of gene expression is tightly constrained and independent of genome size across species. On the other hand, the amount of transcribed noncoding sequence, i.e., mainly intronic sequence, and of nontranscribed intergenic sequence, seems to have been relatively a "free variable" in animal evolution.

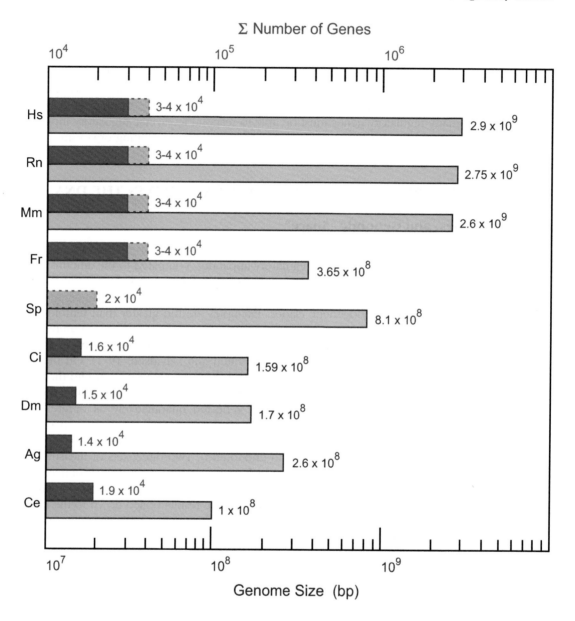

FIGURE 1.1. Representative animal genome sizes and gene numbers. Data are from genome sequencing. Dashed lines indicate larger alternative possible estimates of gene number also consistent with current data, or uncertainties. Genome sizes, indicated by blue bars (bottom scale), are given in

We now have genomic sequence for a number of animal species, and a huge accumulation of gene expression data. Comparison of results suggests that indeed, similarity in complexities of given expressed gene sets, more or less irrespective of genome size, is a basic fact of animal life. A limitation is that the genomic sequences and expression databases available are still clustered in only a few phylogenetic clades, and for obvious practical reasons are somewhat biased toward animals which have the smaller genomes in these clades. In Fig. 1.1 is shown a summary of total gene numbers estimated for different animal species, in comparison to their genome sizes (in this book we are concerned almost exclusively with bilaterally organized animals, the "bilaterians," and occasionally with their cousins the cnidarians, i.e., jellyfish and sea anemones; this excludes protozoans, sponges, and ctenophores). A quick glance at Fig. 1.1 conveys the main import: The basic "package" of genes needed in the genome of a bilaterian is about 15,000, and even the most complex vertebrates, i.e., in Fig. 1.1, rats, mice, and us, do not appear to have more than twice this complement. Compared to this the genome sizes of the animals included in Fig. 1.1 differ enormously (see legend for references). However, a caveat must be noted. This is that the definition of "gene" is in practice not trivial, even considering only protein coding genes. The values in Fig. 1.1 are certainly not overestimates, but particularly in vertebrates they could be to some extent underestimates of the actual gene complement (dashed bars). Estimates of gene number based on expressed cDNA sequences are often higher than those based only on exon sequence homology and computational prediction from the DNA sequence. In amniote vertebrates alternative splicing, and splicing over huge distances, are very common (Johnson *et al.*, 2003), so that determining what exons belong to what gene can be problematical. An important point is that while a large number of diverse exon combinations may be generated by alternative splicing, in regulatory terms given

base pairs in blue numerals; gene number estimates are given in red numerals. Sources: Hs, human: (International Human Genome Sequencing Consortium, 2001; Venter *et al.*, 2001; Johnson *et al.*, 2003; http://maple.lsd.ornl.gov/cgi-bin/GCat/GetOrg.cgi?org = human); Rn, *Rattus norvegicus* (rat): (Rat Genome Sequencing Project Consortium, 2004); Mm, *Mus musculus* (mouse): (Mouse Genome Sequencing Consortium, 2002; The FANTOM Consortium and the RIKEN Genome Exploration Research Group Phase I & II Team, 2002); Fr, *Fugu rubipres* (puffer fish): (Aparicio *et al.*, 2002); Sp, *Strongylocentrotus purpuratus* (sea urchin): (Hinegardner, 1974; Cameron *et al.*, 2000; Sea Urchin Sequencing Consortium, 2006); Ci, *Ciona intestinalis* (ascidian): (Dehal *et al.*, 2002); Dm, *Drosophila melanogaster* (fly): (Ashburner, 1989; Adams *et al.*, 2000); Ag, *Anopheles gambiae* (mosquito): (Holt *et al.*, 2002); Ce, *Caenorhabditis elegans* (nematode): (The *C. elegans* Sequencing Consortium, 1998; Stein *et al.*, 2001).

protein coding genes are tightly defined. Thus there come to mind no examples of genes that have more than one, two, or three alternative transcriptional basal promoters, i.e., locations where productive transcription of mRNA is initiated, and for a given gene, one or another of the exons beginning at these sites are present in all splice variants.

Though the bilaterians have rather similar sized gene toolkits, some of the constituents are specific to each clade, while others are shared with all other bilaterians. As the individual references in the legend to Fig. 1.1 detail, every genome includes sets of genes, often large, paralogous replications of certain gene families, that perform special functions for that kind of organism (common examples are the particular chemosensor and immune function genes that each clade uses). Every one of these genomes also contains a huge set of panbilaterian genes that encode common cytological, enzymatic, and cell type-specific differentiation functions, though these genes are present in diverse numbers in different genomes, and often display strikingly clade specific variation in protein coding sequence. But all these distinctions, as well as the bilaterian gene number constraint shown in Fig. 1.1, are peripheral to an essential fact: If we focus explicitly on the genes encoding transcription factors, and the genes encoding components of signaling systems required for developmental spatial regulation, there is almost no qualitative variation among the genomes of bilaterians. The genetic repertoires of each of these diverse bilaterians include genes encoding every known major family of transcription factor, and components of every known signaling pathway. The point is strengthened by the observation, made anew as each additional genome comes on line, that in each bilaterian clade, though all the regulatory and signaling gene families are represented, these gene families have diversified differently (Ruvkun and Hobert, 1998; Rubin *et al.*, 2000). That is, different bilaterian genomes have different numbers of genes encoding transcription factors belonging, for instance, to the various subfamilies of homeodomain regulators, Ets regulators, T-box regulators, nuclear receptors, or winged helix regulators, or different numbers of genes encoding TGFβ ligands. The replication and diversification of these gene families and subfamilies are always correlated with diversification of their functional roles in development. This major feature can be regarded as direct evidence of the process of reapplication of the same shared gene regulatory toolkit, a process that has occurred endlessly during bilaterian evolution (reviewed by Erwin and Davidson, 2002). Thus we can exclude the proposition that given bilaterian body plans and morphological structures differ from others because each has its own qualitatively specific class of gene regulatory protein and its own set of signaling pathways. Instead, the exact opposite is true.

What of the great majority of the genomic DNA that is not included in genes, here including not only those genes that encode proteins, but also those encoding rRNA and other kinds of RNA that function at posttranscriptional levels? To begin let us think about all forms of known gene regulatory elements in which the genomic sequence is important for function (here and in the following, the term "element" is used broadly, to denote any genomic feature that has a specific regulatory role). The sign of sequence-dependent function is constraint in its rate of change during evolution, relative to the majority basal rate at which selectively neutral DNA sequence diverges (e.g., most intronic DNA, most third base codon sequence, most intergenic DNA). A very approximate estimation can be made based on those few gene regulatory systems that are relatively well known (for cases, turn the pages of this volume). By extrapolation from these cases, this evidence suggests that there is at least as much sequence-dependent gene regulatory information built into the genome as there is sequence included in mature gene products; that is, than in protein coding mRNAs plus all other kinds of functional transcripts that ever appear in the cytoplasm. There could indeed be twice as much gene regulatory as coding information, or more; but not enough is known to recognize most of it *a priori*. It is an amazing comment on the current predilections of molecular biology and genomics that, relatively speaking, only minute attention has so far been devoted to reading the enormous regulatory code carried in the genomes of animals, compared to reading the protein coding capacity. What this has meant is focus mainly on structure/function relations at the outermost layers of animal life systems, whereas it is only at the innermost layers, where the genomic control apparatus operates, that development and evolution can be explained.

Even assuming high-end estimates for the dimensions of the regulatory genome, it would still be true that most of the DNA in the genomes of Fig. 1.1 has no likely sequence-dependent role. Repetitive elements account for some of the sequence-independent DNA. In larger genomes much of the DNA sequence is repetitive, in smaller genomes less. The repeats occur in tens, hundreds, or thousands of copies of more or less related sequence per genome (early studies based mainly on genome-wide DNA renaturation kinetics reviewed by Davidson *et al.*, 1974; current data for sequenced genomes are summarized in references in legend of Fig. 1.1). Repetitive sequences are mainly due to insertions and replications of transposable elements (Moore *et al.*, 1978; Britten, 1984; Deininger and Batzer, 2002). In terms of both the position of these elements in the genome and their frequencies, they change during evolution many times faster than do the underlying syntenic (chromosomal gene linkage) scaffolds of which animal genomes seem basically to be composed (Aparicio *et al.*, 2002; Bourque *et al.*, 2004). Related animal genomes differ more in their repetitive sequence content than in anything else, so by that definition as well as by provenance, repetitive sequences are charter components of the sequence-independent portion of the genome. Even so, they occasionally transpose into a location carrying a gene regulatory element with them, or they mutate to constitute such (reviewed by Britten, 1996; for example, Zhou *et al.*, 2002). To what extent the transposition of repetitive sequence elements has contributed overall to the generation of novel gene regulatory systems during evolution is a question that remains to be resolved.

In most animal genomes the larger part of the sequence-independent, freely evolving genomic DNA is the single or very low copy sequence that constitutes the major extent of intronic and intergenic sequence. Should we think of this major fraction of the inherited genome as having no functional, mechanistic significance? An answer is that instead it has an enormous significance, though of a different kind, one that we are just beginning to imagine. It provides genomic space, space that allows given regions of the genome to be partitioned at given times and places into very large functional domains. Genomic space thus serves as a template for chromatin assembly, and a major component of nuclear three-dimensional structure. Genomic space has a direct significance for the gene regulatory apparatus, our subject here, in that it permits distant, bona fide sequence-specific regulatory elements to loop so as to form intimate contacts with one another, and with the basal transcription apparatus at the head of each gene. In this sense, the space is functionally essential, because it potentiates combinatorial interrelations among sequence-specific regulatory elements. But in its flexibility this is a beautifully free system, in that just how much genomic space is required around a given gene, or just what sequence it is composed of, or just where the space is with respect to the sequence-dependent elements does not much matter, since related genomes differ markedly in these respects. Furthermore, by whichever mechanism the sequence-specific regulatory elements originate in evolution, they form from sequence that did not originally have that regulatory function, and so the space is the raw material for their creation. Huge and sequence-independent genomic space is a general characteristic of every animal genome except for rare and unusually simplified outliers (larvacean urochordates, for example; Seo *et al.*, 2001), and that fact per se requires that the space has functional meaning.

Overview of Regulatory Architecture

The sequence-dependent regulatory genome is so because it is bound by transcription factors that each recognize short DNA sequence motifs. The brevity of these motifs means, however, that they will occur individually at random in the sequence-independent space in enormous number, because the genome is enormous. So right away, we know that for the required exquisite regulatory specificity, the functional distribution of sites must be nonrandom. And so it turns out: Functional regulatory elements that have been physically isolated and experimentally parsed in detail always consist of relatively dense clusters of distinct sites recognized, respectively, by diverse DNA binding proteins. In biochemical terms, the combination of proteins binding in proximity to one another mediates the regulatory output of the site cluster. In informational terms, the cluster of specific target sites specifies exactly what that regulatory activity will be, since the sites define the functions that will be performed. There are various classes of functional regulatory element. Those that perform the greatest amount of developmental information processing are the *cis*-regulatory modules that conditionally control gene expression in time and space. The term "*cis*," (from Latin for "this side of") is commonly used, because these regulatory elements are usually (though not invariably) located on the same DNA molecule as the genes they control. The regulatory state, defined as above to consist of the total set of active diffusible transcription factors in each cell, is often referred to as the *trans* portion of the control system (Latin for "far side of"). The term "module" has a functional significance: *cis*-regulatory modules respond to given regulatory states by producing a unique regulatory output, the consequence of the individual interactions at their clustered internal target sites. Data on genomic *cis*-regulatory organization of a growing number of genes, in diverse bilaterian species (see Chapter 2), suggest that there may be 5–10x as many individual *cis*-regulatory modules as there are genes, or at least one to three hundred thousand in our genomes.

Certain classes of *cis*-regulatory modules are usually called "enhancers." These are *cis*-regulatory modules that, at least some of the time, positively regulate gene expression. They are often located many kilobases distant from the basal transcription apparatus of the gene, to which they must communicate directly or indirectly by looping; therefore their 3′ or 5′ orientation with respect to the distant gene is unimportant. As discussed in what follows, many *cis*-regulatory modules affect gene expression other than by direct activation of the basal transcription apparatus. Some, for example, work indirectly, via interaction with separate, proximal *cis*-regulatory modules located close to the basal apparatus; others function as "insulators," perhaps establishing physical domains within which only the appropriate enhancers can interact with a given gene (e.g., Bell *et al.*, 2001; Yusufzai *et al.*, 2004). Others are dedicated silencers of expression of any gene in their vicinity. But all perform their specific functions when and where they do because they bind proteins that recognize their specific DNA target site sequences.

To understand what *cis*-regulatory elements really do and are, we have to think of them at once in informational and in physical terms. The multiple transcription factors that bind within the *cis*-regulatory module can be regarded as its individual inputs. The regulatory output of the *cis*-regulatory module is the "instruction" given directly or indirectly to the basal transcription apparatus, an instruction that determines whether the gene is to be silent, or active at a specified rate. Both in qualitative and quantitative terms, the output is causally dependent on the degree of occupancy of the individual target sites within the module. If the gene encodes a transcription factor, its output leads to effects on other *cis*-regulatory elements that contain functional target sites for that transcription factor. We speak of these as "downstream" linkages, while the inputs to a regulatory gene are the termini, in its own *cis*-regulatory system, of its "upstream" linkages. Where they can be accurately described in spatial and/or temporal terms, no input into a *cis*-regulatory module is ever found to be the same as its output; downstream never equals upstream. The transformation of the various inputs into the output instruction is the "processing" function that the module executes.

Perhaps this emphasis on input information processing seems unnecessarily baroque: Is it not sufficient just to identify a fragment of DNA, which when introduced into a recipient egg or animal will generate a certain pattern of developmental gene expression? Why are its internal functional architecture and its information processing activities important? There are two direct answers to these questions. The first is that only by experimentally verifying the functional meaning of the specific transcription factor target sites within a *cis*-regulatory element do we understand what the genomic DNA sequence of the element means. The second is that the information processing function of the *cis*-regulatory element constitutes the link between the diverse circumstances presented in each cell, and the response capacities hardwired into the genomic regulatory sequence.

The fundamental requirement for *cis*-regulatory information processing is easy to see *a priori*. Consider the problem faced by a given gene in a given cell at a given moment in development. Somehow, this gene has to "know" what temporal stage it is in the developmental process; what cells are adjacent, and what they are doing or saying; and what is the lineage or developmental status of its own cell. It must "know" whether the cell is in cycle, if that affects the need for its transcripts, and also what regulatory events (mediated by other genes) have occurred earlier which would causally affect its own activity. To bind to the target sites of a *cis*-regulatory module, transcription factors must be presented in the cell nucleus at concentrations that promote occupancy of these target sites a significant fraction of the time, and they must be presented in active forms. Transcription factor concentrations in the cell depend on the immediately prior regulatory state, and their activities constitute the immediately succeeding regulatory state. That is to say that regulatory state depends on circumstances. A factor may be synthesized only in certain spatial domains of the organism, and it may be active, for example, only after a signal transduction pathway has modified either it or a bound cofactor. It is in this direct, mechanistic sense that transcription factors convey circumstantial information to *cis*-regulatory elements. Much of the functional import of transcription factors is executed by effector proteins that form complexes with them once they have assembled on the *cis*-regulatory DNA, and that are essentially recruited or deployed in the regulation of a given gene in consequence of transcription factor binding. But from the point of the hardwired regulatory genome, the ultimate control logic resides not in the biochemical functionalities of either the transcription factors per se, nor in the proteins with which they interact, nor in the widening circles of cellular machinery that thereupon are called into play. Rather it resides in the specification of which effector functions are to be executed; that is, what is encoded in the target site DNA sequence.

Bilaterian organisms are complex, have many parts, and execute many developmental processes. As noted above, many genes of the regulatory repertoire are used at different stages of development to execute diverse and unrelated functions, and the same is true of many other kinds of genes. Such "multiple use" genes must be able to respond accurately to diverse regulatory states. A usual if not invariant feature of *cis*-regulatory architecture in bilaterians is that the diverse phases of expression of given genes are mediated by physically separate *cis*-regulatory modules, each up to several hundred base pairs long. Sometimes the regulatory modules servicing a given gene are strung out over tens of kilobases of sequence-independent DNA, sometimes they are arranged more compactly. A typical organization is indicated in Fig. 1.2A. The various modules controlling a given gene are equally likely to be located 5' or 3' of the gene itself, or in its introns. In the following pages examples appear of genes that are equipped with numbers of characterized, independent *cis*-regulatory modules that range from five to 15 per gene. While this kind of modular regulatory architecture provides the organism with the services of given genes in independent developmental contexts, the "price" is an additional layer of complexity: an absolute requirement for *cis*-"trafficking" controls to ensure that the module relevant in any given spatial and temporal context is the one which communicates its output to the basal transcription apparatus. As indicated in Fig. 1.2D, the way that distant regulatory modules interact with the genes they control is likely to be DNA looping, such that when a given module is loaded with the transcription factors for which it has target sites it is brought into contact with the basal transcription apparatus (or with another module with which it interacts). Several classes of DNA looping protein have been discovered, though it must be said that knowledge of the mechanisms by which these dynamic conformational alternatives are established and maintained remains extremely thin. We revisit this matter in Chapter 2, but note here that intermodule trafficking too is an aspect of sequence-dependent genomic regulatory programming. It is mediated by DNA looping proteins that are targeted to given *cis*-regulatory modules by specific DNA binding sites. Factors that probably perform such "intercommunication" functions appear to operate in both positively and negatively acting *cis*-regulatory modules. Regulatory control is in a general sense measured as the specification of choices among the possible alternative states, and so we see here another level of genomically programmed, sequence-dependent control; choice of alternative intermodule interactions.

Gene Regulatory Networks

The morphological structures of bilaterians are, of course, never the product of single genes or single regulatory systems. The second half of this book deals with examples where we can analyze the causality of specific developmental and evolutionary processes in concrete terms of gene regulatory network architecture. Such networks, as indicated at the outset of this chapter, consist essentially of the genes that encode the transcription factors providing the inputs at each node, and the *cis*-regulatory modules that control the relevant phases of expression of these genes. The term "network architecture" thus refers to the topology of the functional linkages between the genes encoding transcription factors and their target *cis*-regulatory modules within the network. The periphery of a developmental gene regulatory network is defined by the absence of outputs that lead to other genes in the network, that is, by the sets of downstream genes that make products other than transcription or signaling factors. Here are located the sets of protein coding differentiation genes animated in specific domains of the

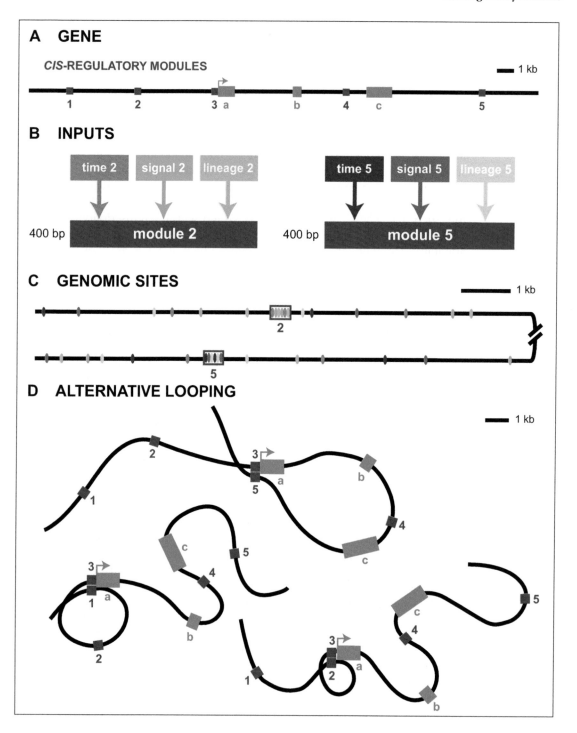

organism by operation of the regulatory network. The subnetworks that operate in each spatial domain of the organism are connected by the signaling inputs that affect spatial expression of key network regulatory genes. We must remind ourselves always that the particular form of a given network architecture is directly specified by the identity of the genomic target sites within the participating *cis*-regulatory elements. Thereby, the architecture is the direct evolutionary heritage of the organism.

The essential point is that the level of regulatory organization that relates directly to processes of development such as differentiation, specification, morphogenesis

FIGURE 1.2 Genomic *cis*-regulatory modules. (A) A typical gene consisting of three exons in blue (a, b, c), serviced by five *cis*-regulatory modules indicated by red blocks. The diagram is approximately to scale such that the modules are 400 bp long, and the gene and its regulatory apparatus are spread out over about 30 kb of genomic DNA. Modules 1 and 2 are located at 10 and 6 kb distances 5′ of the gene; module 3 is what is often termed a "proximal element" located immediately next to the basal transcription apparatus, with which it interacts; module 4 is located in the second intron of the gene; and module 5 is several kb 3′ of exon c. (B) Individual *cis*-regulatory modules receive and process diverse informational inputs, allowing the gene to respond to regulatory states presented in entirely unrelated spatial and temporal developmental contexts. Here modules 2 and 5 each respond to a transcription factor appearing at a particular time, thus serving as a positive temporal gate; to a transcription factor activated in response to an intercellular signaling event; and to a transcription factor presented only in a certain cell lineage of the embryo. As symbolized by the color coding, however, the identities of the sets of factors to which the two modules respond are entirely distinct, depending on the DNA sequences of the specific target sites that each includes. (C) Target site distribution within and without *cis*-regulatory modules. Colored bars represent the DNA target sites for the similarly colored transcription factor inputs of (B); only modules 2 and 5 and their surrounding genomic environments are shown, the remainder of the DNA portrayed in (A) being subsumed within the double broken line. As is typically found (cf. Chapter 2), target sites often occur multiply within modules. The definitive feature of the *cis*-regulatory module that differentiates it from surrounding sequence-independent DNA (see text) is its clustered array of target sites. In this diagram, for comparison, the density of random occurrence of these same sites outside modules 2 and 5 is approximately what would be expected in the indicated length of DNA were each site 6 bp long. To render the bars visible at this scale, their size has been expanded, and so their density inside the modules appears perhaps three times too high. (D) Alternative looping, used to deploy different modules for regulatory function. The three diagrams show, respectively, the conformations when module 1, module 5, or module 2 are in action. The gene is one in which the proximal module (3) is always required, in that the distal modules (i.e., 1, 5, and 2) must interact with elements within the proximal module for function (for real examples of this, see Chapter 2). Alternative DNA looping brings the respective protein complexes assembled on each module into immediate contact with one another.

and the like, is that of the gene regulatory network. The simple reason is that such developmental phenomena depend on the dynamic expression of many genes, which depends at any given moment on the regulatory state. As illustrated in the simplified example of Fig. 1.3, the changing spatial regulatory state is the direct product of the gene regulatory network. In Chapter 4 are examples from

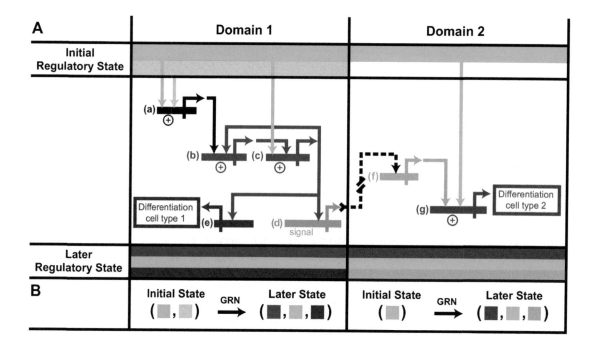

FIGURE 1.3 Gene regulatory network elements generate spatial and temporally changing regulatory states. (A) Cartoon, illustrating initial (I) and later (L) regulatory states, top and bottom of the diagram, in which the constituent transcription factors expressed in each state are qualitatively represented by colored horizontal bars. Two adjacent spatial domains of an embryo are indicated. Within these domains the horizontal thick lines represent genes encoding the color-designated transcription factors (genes a, b, c, e, f, g) and a signal ligand (gene d). The bent arrows emanating from each gene indicate the transcriptional output of the gene; the thin colored arrows show how the outputs of each gene are used as inputs to the *cis*-regulatory elements of the other genes in the network. Inputs ending in arrowheads indicate positive regulatory functions. Only active genes are shown, but it is understood that in the two domains the genomic sequence is, of course, exactly the same, irrespective of the state of gene activity. For simplicity, no more than two inputs are shown into any given

diverse developmental systems in which we can now transit directly from the phenomena of development to the underlying gene regulatory network architecture, and from this architecture to *cis*-regulatory DNA sequence. Essential functions of the developmental system itself arise out of the architectural structure. Such functions include, for instance, the progression from dependence on transient inputs to installation of stable states of gene expression (e.g., as in Fig. 1.3) and setting up exclusionary choices of specification state. These fundamental features of development are generated at the network level, not by any one gene or *cis*-regulatory element.

The bottom line that emerges from this very quick overview of genomic regulatory architecture for development is that input regulatory information is processed at different levels; that is, both at the individual *cis*-regulatory element, and at the system level of networks of *cis*-regulatory modules and the genes they control.

cis-regulatory element. Furthermore, only one kind of information processing function is invoked in this example: This is "AND" logic processing (circled +), in which unless both inputs are present, the element runs at a low level such that the sum of the *cis*-regulatory outputs given each of the inputs alone, is much less than the output when they are all present. The initial regulatory state consists of a positive transcription factor present in both domains (blue-green stripe) plus a transiently presented factor (pink) that appears only in domain 1. These two inputs suffice to activate gene a, also transiently, in domain 1 only. Its product, in turn is a transcription factor that activates a reinforcing feedback system consisting of genes b and c, shown in red; these genes respond to one another's products. The consequence is that once the red system is activated, there is no further dependence on the initial transient input. This sets the condition for the generation of new regulatory states: The targets of the red gene system include the regulatory gene e, symbolized in purple, which serves to activate a cell type-specific differentiation gene battery; and also gene d. The encoded intercellular ligand produced by the mRNA of gene d is received in domain 2 (nested arrowhead and tail at domain boundary), whence it provides a required input to gene f (via an already present transcription factor, not shown, which is activated or mobilized in response to the signal in domain 2; broken black line). The function of the *cis*-regulatory element of gene f is thus to transduce the signal into a new component of the transcriptional state in domain 2. The consequence is to activate gene g, which in domain 2 serves an analogous function to gene e in domain 1, activating another battery of cell type-specific differentiation genes. The outcome of the operation of the gene regulatory network is to produce the later regulatory states in domains 1 and 2, which differ not only from one another but also from the initial states. While this is a highly simplified "toy" model, the reader will encounter all of its elements many times over, in the real developmental systems reviewed in Chapter 4. The basic point is that the spatially and temporally changing regulatory state is the dynamic output of the operation of gene regulatory networks, and that what happens in any given case depends on the particular architecture of the network. (B) The gene regulatory network (GRN) provides the function by which initial regulatory state is transformed into later regulatory state in each domain.

The basic level is that of the individual *cis*-regulatory module, the fundamental unit of developmental control. We might demand to know, even very approximately, how much information processing actually goes on. At the *cis*-regulatory level, one way to approach this issue is by considering the multiplicity of inputs per module. A survey of some well-characterized *cis*-regulatory modules active in developmental processes yielded the conclusion that on the average four to eight different transcription factors service each individual module (Arnone and Davidson, 1997). The operation of gene regulatory networks depends on the conditional status of the multiple *cis*-regulatory modules at its nodes which control expression of transcription factors. It is obvious, therefore, that a tremendous variety of regulatory outcomes could be programmed using networks built from modular control units that function as individual information processing nodes. This is what has made possible the continuous variation in network architecture underlying evolutionary alteration in the body plans of animals.

THE REGULATORY DEMANDS OF DEVELOPMENT

Readout and Generation of Regulatory Information in Developmental Specification

Development is the execution of the genetic program for construction of a given species of organism. Just the dimensions required of such programs are remarkable, for of the many thousands of genes in the animal genome (Fig. 1.1), most are utilized at some time during development (Davidson, 1986; Arbeitman *et al.*, 2002; Baugh *et al.*, 2003) and all must be controlled accurately in space and time. But this is the least of the problem: The essence of development is the progressive increase in complexity which it invariably entails. In informational terms the developmental complexity is measured by the continuing generation of new regulatory states, each produced by readout of a defined genetic subprogram; each arising in a particular spatial domain of the embryo. Meanwhile the populations of cells expressing these regulatory states are being instructed to expand to given extents, by cell growth (and sometimes to contract, by cell death). Thus the spatial components of morphology are laid out.

A general character of genomic programs for development is that they progressively regulate their own readout, in contrast, for example, to the way architects' programs (blueprints) are used in constructing buildings. All of the structural characters of an edifice, from its overall form to local aspects such as placement of wiring and windows, are prespecified in an architectural blueprint. At first glance the blueprints for a complex building might seem to provide a good metaphoric image for the developmental regulatory program that is encoded in the DNA. Just as in considering organismal diversity, it can be said that all the specificity is in the blueprints: A railway station and a cathedral can be built of the same stone, and what makes the difference in form is the architectural plan. Furthermore, in bilaterian development, as in an architectural blueprint, the outcome is hardwired, as each kind of organism generates only its own exactly predictable, species-specific body plan. But the metaphor is basically misleading, in the way the regulatory program is used in development, compared to how the blueprint is used in construction. In development it is as if the wall, once erected, must turn around and talk to the ceiling in order to place the windows in the right positions, and the ceiling must use the joint with the wall to decide where its wires will go, etc. The acts of development cannot all be prespecified at once, because animals are multicellular, and different cells do different things with the same encoded program, that is, the DNA regulatory genome. In development, it is only the potentialities for *cis*-regulatory information processing that are hardwired in the DNA sequence. These are utilized, conditionally, to respond in different ways to the diverse regulatory states encountered (in our metaphor that is actually the role of the human contractor, who uses something outside of the blueprint, his brain, to select the relevant subprogram at each step). The key, very unusual feature of the genomic regulatory program for development is that the inputs it specifies in the *cis*-regulatory sequences of its own regulatory and signaling genes suffice to determine the creation of new regulatory states. Throughout, the process of development is animated by internally generated inputs. "Internal" here means not only nonenvironmental—i.e., from within the animal rather than external to it—but also, that the input must operate in the intranuclear compartments as a component of regulatory state, or else it will be irrelevant to the process of development.

The link between the informational transactions that underlie development and the observed phenomena of development is "specification." Developmental specification is defined phenomenologically as the process by which cells acquire the identities or fates that they and their progeny will adopt. But in terms of mechanism, specification is neither more nor less than that which results in the institution of new transcriptional regulatory states (Fig. 1.3). Thereby specification results in differential expression of genes, the readout of particular genetic subprograms. For specification to occur, genes have to make decisions, depending on the new inputs they receive, and this brings us back to the information processing capacities of the *cis*-regulatory modules of the gene regulatory networks that make regulatory state. The point cannot be overemphasized that were it not for the ability of *cis*-regulatory elements to integrate spatial signaling inputs together with multiple inputs of intracellular origin, then specification, and thus development, could and would never occur.

A definitive feature of specification regulatory states is that they are transient. However, they are irreversible rather than cyclical, in that the states follow one another in progressive succession and never revert to a prior state. The real time dynamics, i.e., the intervals between specification events, how long they last, and quickly they come and go, depends on the system. For example, in *Drosophila* the time constants for change in the populations of mRNAs encoding pair rule gene transcription factors are measured in minutes (Nasiadka *et al.*, 2002;

Jaeger *et al.*, 2004), while in sea urchin embryos the kinetic constants that determine the rates of regulatory state change operate over intervals of several hours (Bolouri and Davidson, 2003). Transience of specification functions means simply that building any part of any bilaterian body plan requires a series of causally linked temporary stages, each mediated by expression of many genes in a given spatial domain, and each directed by a regulatory state that is the output of a specification event. As each step is accomplished, it is used to set up the next transient specification event, often by serving to control the (transient) spatial and temporal transcription of signaling molecules (cf. Chapter 3). Until terminal form is achieved, the specification functions are just way stations along the pathway to the final spatial organization. Their transience implies temporal control, and this in turn requires two things: first, devices for turning regulatory genes off as well as on and second, turnover of the regulatory gene products (mRNA and protein). The first is a major causal aspect of the transience of specification functions. The second is the main kinetic determinant of the dynamics of this transience. The kinetic behavior of gene expression systems, particularly those sorts that generate changing specification functions, is considered briefly in Chapter 4, in respect to the operation of gene regulatory networks.

Spatial Gene Expression

The basic regulatory object in development is control of spatial gene expression, often referred to as "pattern formation." This is a term of many particular uses in phenomenological developmental biology, because spatial patterns appear at every level of biological organization from subcellular to organismal. The early events of pattern formation establish the primary elements of the body plan, e.g., the metameric segmentation which arises early in the embryogenesis of a fly; or the anterior/posterior (A/P) axis and the left-right asymmetry in a mouse. Later pattern formation events define the spatial organization of the main parts of the body plan, and still later ones define more detailed and smaller elements, e.g., the arrangement of sensory bristles in the fly or of limb digits in the mouse. New subdivisions of an existing multicellular domain, or "pattern elements," are created by the expression of new sets of transcription factors within a bounded region, or for short, by "regional specification."

In solving the mechanism of any given pattern formation process the track inevitably leads to *cis*-regulatory elements controlling genes that encode the transcription factors which execute regional specification processes. A clear and classic demonstration that *cis*-regulatory DNA sequence encodes the spatial pattern of regulatory gene expression is reproduced in Fig. 1.4 (Small *et al.*, 1996). This example demonstrates that two different *cis*-regulatory modules, a few hundred base pairs long, each contain the necessary and sufficient sequence information to generate a sharply defined spatial pattern of expression. The pattern produced by each module is a discrete component of the overall pattern of expression of the gene from which they derive. The main points are the following: The gene

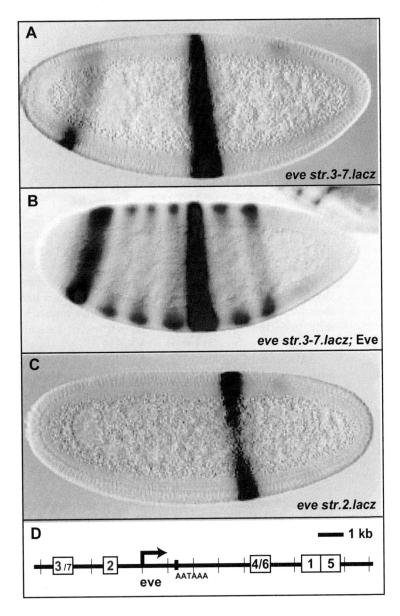

FIGURE 1.4 Accurate expression of *evenskipped* (*eve*) stripe 2, and *eve* stripes 3+7, generated by individual *cis*-regulatory elements. (A) Transgenic embryo

(Continues)

is the *evenskipped* (*eve*) pair rule gene of *Drosophila*, which at the relevant stage is expressed transiently in seven stripes along the axis of the blastula stage embryo (Fig. 1.4B, brown stain). When these expression patterns are set up, the embryo is yet syncytial (i.e., no cell walls have formed to separate the nuclei), so in this case the transcription factor inputs to the *cis*-regulatory modules of the gene can diffuse from cell to cell; intercellular signaling plays no role. In the experiments shown the "Stripe 3+7 enhancer" of Fig. 1.4A, B, and the "Stripe 2 enhancer" of Fig. 1.4C, were isolated from the genomic DNA of the *eve* gene, and linked to a "reporter" gene the transcript of which is visualized by blue staining when it is expressed (see legend). The original dispersed location of the two enhancers in the genomic DNA is indicated in Fig. 1.4D. Each enhancer drives the reporter in the appropriate stripe(s) of embryonic nuclei, in response to local embryonic regulatory states, and exactly how and why it operates thus is taken up in the next Chapter. The point here is just that the experiment excludes any interpretation whatsoever other than that the information required to generate these precise spatial gene expression patterns is entirely encoded in the *cis*-regulatory DNA sequence. When these small fragments of DNA are introduced into the genome (by including them in a transposon that is injected into embryonic germ cells), they are integrated in many different locations, rarely close to where the endogenous gene resides, and there is no particular genomic sequence environment required for the function shown in Fig. 1.4.

Thousands of successful gene transfer experiments of the same basic import have been carried out. The results similarly demonstrate that ectopically incorporated, exogenous *cis*-regulatory modules suffice by themselves to generate correct

FIGURE 1.4 (continued)

carrying a *lacz* expression construct containing a 500 bp stripe 3+7 *cis*-regulatory module (see D for original genomic location); mRNA generated by the *lacz* reporter is visualized by *in situ* hybridization to the *lacz* mRNA (purple stain), followed by chromogenic immunochemical detection of the bound antisense probe. (B) Expression of endogenous *eve* gene, together with pattern of *lacz* reporter expression produced by the same construct as in (A); the brown stain displays the location of the seven endogenous Eve protein stripes, as revealed by reaction and chromogenic detection of anti-Eve antibodies, with *lacz* mRNA detected as in (A). *Lacz* expression can be seen to correspond exactly with the endogenous *eve* stripes 3 and 7. (C) Embryo carrying 480 bp *lacz* expression construct driven by the *eve* stripe 2 *cis*-regulatory module, *lacz* mRNA detected as in (A). By comparison with (B), this construct can be seen to recreate perfectly the endogenous stripe 2 expression domain. (D) Map of *eve* gene displaying genomic locations of *cis*-regulatory modules encoding expression of stripes 1, 5, 3+7, and 4+6 (from Small et al., 1996; with additional material kindly provided by M. Levine).

and predicted developmental patterns of spatial gene expression. Experiments of this kind have been done using diverse gene transfer systems in diverse animal embryos including, in addition to *Drosophila*, sea urchins, nematode worms, fish, mice, chickens, frogs, ascidians, and other creatures (though not many cases are as instantly arresting in visual terms as that in Fig. 1.4). So it is a heavily supported cornerstone of developmental mechanism that spatial patterning of gene expression is determined directly by the heritable *cis*-regulatory DNA sequence code.

To relate regional specification of the morphologically meaningful elements of a developing embryo to *cis*-regulatory function is ultimately to explain how the developmental phenomena we see in the microscope are encoded in the genome. The considerations above tell us this requires resolving the networks of *cis*-regulatory elements and genes which produce the underlying regulatory states. We must keep always in mind that each of the *cis*-regulatory components of these networks functions by executing a spatial control function, just as in the paradigmatic example in Fig. 1.4.

From Regional Specification to Terminal Differentiation

Progressive regional specification processes are seen in all forms of embryonic development. They result in the definition, in terms of regulatory state, of spatial areas and cell populations that are each destined to execute some function required to build the structures of the body plan. In the end, these terminal developmental regulatory states provide the detailed instructions for local deployment of large classes of protein coding genes. Among these are sets of genes that control cell division and reproduce the biochemical machines and the organellar structures of cells; and the classes of genes that cause cells to collaborate in morphogenesis, that is, to form sheets, tubes, ducts, invaginations, tracts, cords, and to carry out migrations. These essential morphogenetic functions are called into play repeatedly in different subparts of the body as they are defined. The identities of the genes needed to accomplish these things, let alone the linkages between these genes and regional specification processes, are only known in a few cases. Thus it is yet difficult to glean a clear image of the genetic control circuitry for three-dimensional form building per se. Deployment of cell replication and morphogenesis gene sets can nonetheless be foreseen to be among the major outputs of developmental gene regulatory networks, controlled by them in time, space, and amplitude.

Much better understood, and directly related to the particular functions of specific body parts, is the control of at least some cell differentiation processes. With the completion of development, populations of specialized cells have been put in place throughout the organism. These express diverse sets of genes encoding the definitive attributes of each cell type. Though there are occasional exceptions, in bilaterians differentiation is usually terminal, in that additional different functions will never be expressed by given cells once they have begun to express cell type-specific genes. If they retain replication capacity they will only produce more of their own sort. Terminal differentiation states can always be defined in terms of a precise set of differentially expressed effector genes, such as the hemoglobins and enzymes of the vertebrate red blood cell, the cell surface receptors and signaling molecules of mature immune cells, the contractile proteins of muscle cells, and the genes encoding the secreted proteins of gland cells. We can say that we understand how a given event of differentiation is controlled in a developmental process if we can trace the *cis*-regulatory interactions in gene regulatory network architecture all the way from the initial specification of a spatial regulatory domain to the deployment and coordinate expression of the effector gene set. Or, turning the matter another way, deployment of differentiation genes is the final output of gene regulatory networks for development. It is the last thing to occur in developmental time in any given domain, and the apparatus for this deployment lies at the peripheral terminus of any given network element, if the network is complete.

A useful concept here is that of the "battery" of differentiation genes, that is, a set of functionally related effector genes expressed in a given cell type. The term "gene battery" was coined by Morgan (1934), and was used by Britten and Davidson (1969) for genes that in their model would be coordinately expressed because the *cis*-regulatory elements of these genes share qualitative target sites for the same transcription factor inputs. Gene batteries that are activated in given differentiation processes due to their shared *cis*-regulatory target sites do exist, and the genes of a given battery do respond to similar transcription factors. Indeed this represents the dominant form of peripheral gene regulatory network organization, that is, the usual way that expression of cell type-specific effector gene sets is controlled. But differentiation gene batteries have a general feature that was not anticipated, and that provides an important constraint in thinking about the evolution of *cis*-regulatory elements (a subject to which we return in Chapter 5). This is that no two *cis*-regulatory modules of the genes of any given battery are ever just the same. The proper description is that they individually utilize different subsets of a given definitive group of factors; they use these in combination with different other factors and the specific target sites they contain never appear in the same order, number, or spacing (Arnone and Davidson, 1997). While the genes of each battery are indeed expressed in the same differentiating cells, there are individual quirks to their expression, in timing, amplitude, and so forth, and at the *cis*-regulatory level we can see why. Figure 1.5 provides examples of the design of *cis*-regulatory elements belonging to three vertebrate gene batteries, encoding respectively muscle contractile proteins, various cardiac proteins, and the crystallin proteins of the eye lens. Here we see variation in design structure equally among the different genes of each battery within a species, and among the control elements of the same gene in different vertebrate species. The genes of each battery are undeniably related in function; that is, the proteins they encode are clearly conjoint members of the respective differentiation gene sets. The main conclusion to be drawn is therefore that the essential property of each set of battery *cis*-regulatory elements has to be defined in combinatorial terms. There is no one type of *cis*-regulatory target site without which a gene cannot be a member of the muscle gene battery, nor is there any one type of target site which all members share. Instead, the structural property that defines this battery is that its constituent *cis*-regulatory elements always use some combination of inputs: for the muscle genes a subset of the MyoD-like, Mef2, and Srf transcriptional inputs (Fig. 1.5A); for the cardiac genes a subset of Gata, Nkx2.5, and MyoD-like inputs (Fig. 1.5B); and for the lens genes, a subset of Maf, Pax6, and Sox1/2 inputs (Fig. 1.5C). This in turn means that upstream, for each differentiation gene battery, expression of the whole set of regulatory genes must be activated in the appropriate spatial domain(s) of the animal by the developmental gene regulatory network.

Sequential and consequential to the DNA sequence-specific *cis*-regulatory interactions defining terminal differentiation states, there are often instituted "lockdown" or stabilization mechanisms that affect the functional status of local regions of the genome over the long term. If *cis*-regulatory interactions have mandated that a given gene or gene battery is to be expressed, such mechanisms may ensure the continuation of that state of activity and enhance it. If a gene or genes are targeted for silencing by *cis*-regulatory interactions with sequence-specific repressors, lockdown mechanisms may later be instituted that preclude any further transcriptional activity in that region of the genome. Furthermore, these same kinds of mechanism are used extensively to transmit states of activity or inactivity to mitotic offspring, across cell generations; that is, to provide a "memory" of the *cis*-regulatory transactions that initially set up the state during development. There are many kinds of such mechanism used in diverse cell types, different organisms, and different kinds of animal. They include methylation of the DNA (for review, Jaenisch and Bird, 2003); methylation and acetylation of nucleosomal histones (for reviews, Jenuwein and Allis, 2001; Felsenfeld and Groudine, 2003); assembly of the enormous Polycomb group repression complexes, which have now been visualized in the electron microscope (Francis *et al.*, 2004); or alternatively, the Trithorax group activation complexes (for reviews, Simon and Tamkun, 2002; Otte and Kwaks, 2003). Of these, DNA methylation is probably the farthest removed from the initial sequence-specific regulatory events (Mutskov and Felsenfeld, 2004).

Particularly in later development, mechanisms further downstream from the primary genomic regulatory interactions are used to modulate the ultimate functional application of transcriptional regulatory states. Such devices include cell type-specific alternative splicing and mRNA-specific interactions with microRNAs. MicroRNAs are utilized on a wide scale, as negative modulators of protein output from pre-existent mRNAs. They typically though not always function at later stages of any given phase of development (Harfe *et al.*, 2005; Leaman *et al.*, 2005; Sokol and Ambros, 2005; Hornstein *et al.*, 2005; Wienholds *et al.*, 2005; see Chapter 4). Their interaction with complementary mRNA target site sequences may result in message destruction or inhibition of translation (for reviews, He and Hannon, 2004; Hobert, 2004; Nakahara and Carthew, 2004; Novina and Sharp, 2004).

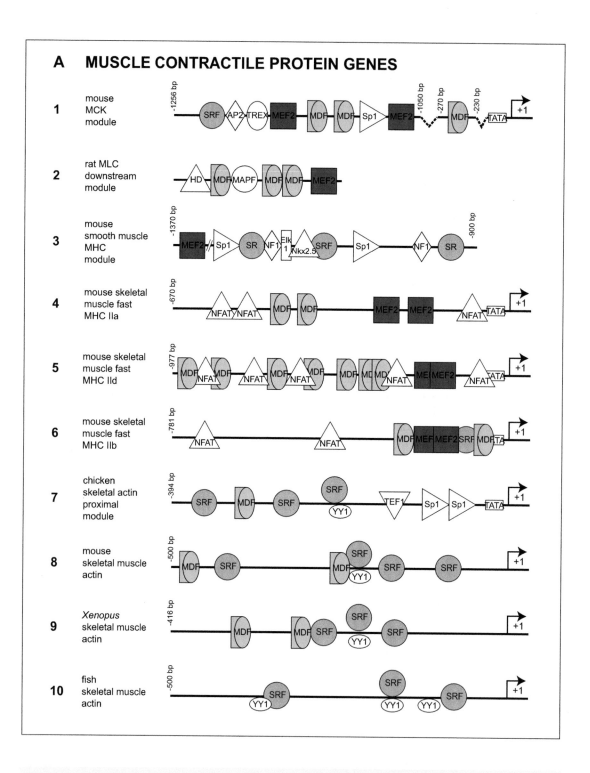

FIGURE 1.5 For legend see page 26.

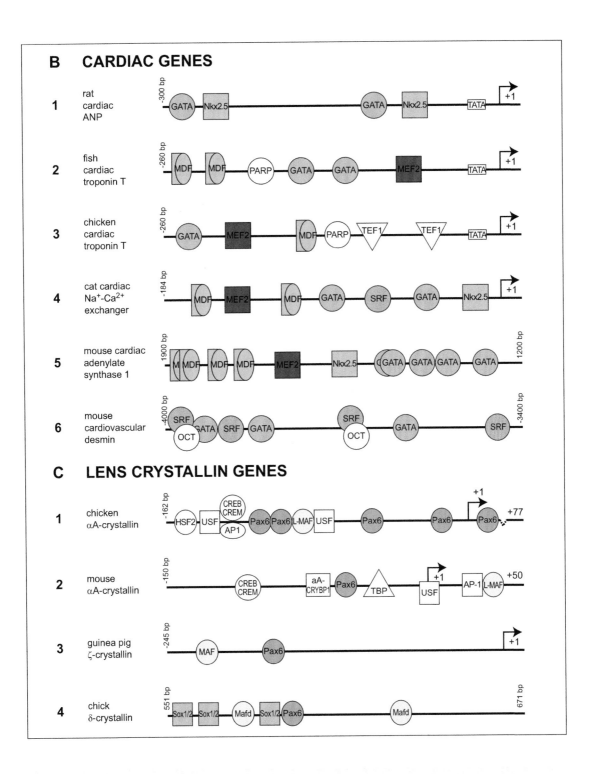

FIGURE 1.5 For legend see page 26.

From a system point of view, genes encoding microRNAs belong to the large class, the function of which is to affect the activity of other gene products in the cytoplasm of the cell. They act to turn down or cancel the activity of specific proteins, when and where the microRNAs are expressed. At the protein level their developmental effects in these cases are somewhat like those of transcriptional repressors. Note that transcription of the genes encoding microRNAs is of course also developmentally controlled by *cis*-regulatory elements, and their expression is part of the output of the nuclear regulatory state (an example is discussed explicitly in Chapter 4). Where the microRNAs are transcribed cell type specifically, they can have cell type-specific effects on the prevalence of target mRNAs, which is thus dependent partially on the *cis*-regulatory elements of the microRNA genes, and partially on the *cis*-regulatory elements of the genes encoding the target mRNAs themselves (e.g., Hobert, 2004; Johnston *et al.*, 2005). Other microRNA species, via proteins that bind to them, may modulate nuclear transcription rate (e.g., Kuwabara *et al.*, 2004). But, unless microRNAs that recognize and bind to DNA target sites by virtue of their complementary sequence are discovered (as once imagined by Britten and Davidson, 1969), note that by definition their functions lie downstream of the primary interaction of transcription factors with DNA target sites, since the RNAs and proteins with which they interact must first be produced.

FIGURE 1.5 *cis*-**Regulatory organization of representative differentiation gene batteries.** (A) Muscle contractile protein gene battery; (B) cardiac gene battery; and (C) eye lens crystallin gene battery. Symbols represent diverse transcription factors, as indicated. Those factors which as a set define the battery in regulatory terms are shown in color; others as open symbols. Positions are given with respect to transcription start site, except for C4, where they indicate position within the third intron of the gene. Sources are as follows. For (A): 1, MCK, mouse muscle creatine kinase, Donoviel *et al.* (1996); Fabre-Suver and Hauschka (1996); and Grayson *et al.* (1998). 2, MLC, rat myosin light chain, Rosenthal *et al.* (1992); Rao *et al.* (1996). 3, MHC, myosin heavy chain from mouse smooth muscle, Zilberman *et al.* (1998); note that MyoD family factors are not expressed in smooth muscle (Olson, 1990) and its target sites are consequently absent from this regulatory element. 4, 5, 6, Mouse fast skeletal muscle MHC IIa, IId, IIb, respectively, from Allen *et al.* (2001). 7, Chicken skeletal muscle α-actin, from MacLellan *et al.* (1994). 8, 9, 10, Respectively, mouse, *Xenopus*, and *Fugu* skeletal muscle α-actin, from Liu *et al.* (2000). For (B): 1, Rat cardiac atrial natriuretic peptide (ANP), Shiojima *et al.* (1999). 2, 3, Respectively, zebrafish and chicken cardiac troponin T, from Tidyman *et al.* (2003). 4, Feline cardiac Na^+-Ca^{2+} exchanger, from Cheng *et al.* (1999). 5, Mouse cardiac adenylate synthase 1, from Lewis *et al.* (1999). 6, Mouse cardiovascular desmin, Mericskay *et al.* (2000). For (C): 1, and 2, Respectively, chicken and mouse αA-crystallin, from Cvekl and Piatigorsky (1996), Ilagan *et al.* (1999), and Yoshida and Yasuda (2002). 3, Guinea pig ζ-crystallin, Sharon-Friling *et al.* (1998). 4, Chicken δ-crystallin, Shimada *et al.* (2003); Yoshida and Yasuda (2002); Kondoh *et al.* (2004).

A most important point, because it defines the causal polarity, is that the initial events that select a given region of the DNA for activity or inactivity are the sequence-specific interactions by which the regulatory genomic code is read and interpreted. Chromatin-level alterations in functionality only follow these interactions. We touch on two examples that provide the paradigm. In *Drosophila*, the complex pattern of *hox* gene expression during development is controlled by specific *cis*-regulatory modules, a large number of which have been identified (discussed in Chapter 2). These respond to diverse transcription factors presented transiently during development. However, the later maintenance and stabilization of many of these patterns of expression depend strongly on the establishment of stable Polycomb and Trithorax protein complexes (Müller and Bienz, 1991; Simon *et al.*, 1993; Gindhart and Kaufman, 1995; Pirrotta, 1997). The intergenic DNA, including some of the relevant *cis*-regulatory elements of the known *hox* target genes, contains sites for Polycomb and Trithorax proteins. So do over 150 other *Drosophila* genes which also utilize Polycomb and Trithorax "lockdown" complexes (Ringrose *et al.*, 2003). But these genes are initially activated by other transcriptional regulators, in various spatial domains of the embryo. The second example concerns mammalian T-cell differentiation. Here the temporal and causal sequence of events leading from transcription factor interaction with cytokine gene *cis*-regulatory elements to changes in histone acetylation, and hence chromatin structure, are known in detail (Yamashita *et al.*, 2002; Omori *et al.*, 2003; Inami *et al.*, 2004; for review, Ansel *et al.*, 2003). For instance, Th2 helper T-cells express the cytokines Il-4, Il-5, and Il-13, and the chromatin of the genes encoding these proteins becomes hyperacetylated when they do so. But this prominent phenomenon depends directly upon the prior interaction of the key transcription factors with the DNA target sites of the respective regulatory modules, viz., Gata3 and Stat6 for the *il-13* gene, and Gata3 and NF-kB for the *il-5* gene. In contrast, in killer T-cells, the same genes are down regulated, and histone acetylation is repressed. This opposing chromatin phenotype is caused by the *cis*-regulatory interactions of a different transcription factor with the same genes, a factor that instead attracts a histone deacetylase.

Lockdown mechanisms are an essential aspect of the later stages of development, and if they are blocked, disasters occur to the organism. But they lie outside the scope of this discussion, because they are indirect, though biochemically essential consequences of regulatory DNA sequence recognition.

EVOLUTION, DEVELOPMENT, AND THE REGULATORY GENOME

The path to scientific understanding of how new body plans arise in evolution rests on a classic syllogism. Since the morphological features of an animal are the product of its developmental process, and since the developmental process in each animal is encoded in its species-specific regulatory genome, then change in animal form during evolution is the consequence of change in genomic regulatory programs for development. This is scarcely a new thought (Britten and Davidson, 1971), but what is new is that the mechanisms of body plan evolution, and therefore causal explanations of animal diversity, have become experimentally accessible. In the end, as the following arguments suggest, from this area of genomic bioscience will arise a more profound understanding of development, as well as solid models for evolutionary process.

Three types of evidence contribute to our knowledge of the phenomena of animal evolution. Paleontological evidence is essential because it provides the only incontrovertible information on what animal forms have actually existed. Taken together with modern geological chronology, this evidence shows us also when they existed, and at what pace given features changed or did not change. Phylogenetic evidence, molecular and morphological, provides weight-bearing surfaces upon which the force of cladistic logic can be applied, so we can tell what forms had common ancestors and what features are ancestral as opposed to derived. At the least, such evidence allows us to pose these questions in a concise way. Paleoecological evidence speaks to the selective constraints that throughout time have, in particular ways, modulated the coming and going of diverse forms, and the degree of diversity allowed. But in evolution, as in development, external phenomenology is not an explanation that reveals internal causes. By the above syllogism, a causal explanation of how novelty in morphological form arises in evolution must be couched in genomic terms; animal diversity is the outcome of changes in development of morphological form. If we knew in functional terms the specific genomic control elements that result in different morphological outcomes in two animals of common ancestry, we would see exactly what essential causal differences distinguish the regulatory DNA sequences of these animals. If we also have available pertinent evidence from a suitably positioned outgroup, then we could know as well what was the ancestral regulatory code in the genome, and thereby the polarity of the evolutionary changes in that code.

Evolutionary diversity occurs at many different levels, and as we shall explore in Chapter 5, these levels depend on the parts of developmental gene regulatory networks that have changed. At lower taxonomic levels, that is, comparing species that are members of the same genus, or genera that belong to the same family, the main differences in general are in properties that are produced by terminal differentiation gene batteries, and in size of given body parts. The last may just depend on the number of rounds of cell division at particular local developmental phases, while many kinds of change occur in differentiation gene batteries. These range from changes in the sequence of individual proteins that alters their properties; to addition or subtraction of genes; to regulatory alterations in time, place, or amplitude of expression of individual genes (cf. Fig. 1.5). For example, the detailed character of the genes and *cis*-regulatory elements that constitute keratin gene batteries affect species-specific properties of the integument, visual pigment gene batteries affect species-specific light response capabilities, digestive enzyme gene batteries affect species-specific dietary requirements, and so forth.

But deeper taxonomic differences, which include differences in the form or presence of given body parts, require that since divergence from a common ancestor genomic regulatory changes have occurred at the level that controls development of the morphological components of the embryo or adult animal. This is the level of the spatial regulatory states generated by developmental gene regulatory networks. Just as development of these components of the body plan cannot be explained except in terms of gene regulatory network architecture, so evolutionary change in animal form cannot be explained except in terms of change in gene regulatory network architecture. A strong conclusion follows: The only way to obtain an explanation of animal diversity in form is by comparative analysis of developmental gene regulatory networks in suitably chosen species. As an experimental enterprise, this is an effort scarcely begun. But the door thus opens on a domain that is equally fruitful for development and evolution, indeed the domain where these subjects merge. For the answers to the fundamental questions of mechanism in both these traditionally separate areas can be extracted by functional analysis of the regulatory genomes of modern animals. In principle, evolutionary and developmental aspects are testable in the same laboratory experiments, in which a portion of a gene regulatory network is "rewired" and its function in development determined.

It is a common subject of research enquiry whether a given regulatory difference between two species is due to "*cis* or *trans*" evolutionary change. Unless the protein sequence of a transcription factor itself has changed, this is not a sensible question. All evolutionary change that causes differences in gene expression within gene regulatory networks is *cis*-regulatory, affecting either input or output, depending arbitrarily on where one happens to be looking. The *cis*-regulatory module is not only the unit of input information processing in development, but also the unit of evolutionary change. The next Chapter is about the internal functional organization of these basic components of the regulatory genome, about what they actually accomplish, and how they do it.

CHAPTER 2

cis-Regulatory Modules, and the Structure/Function Basis of Regulatory Logic

GENERAL OPERATING PRINCIPLES

Though "principle" is a strong word, some broad functional generalities emerge from the wealth of data on *cis*-regulatory elements active in development currently in our hands. These fall within the concept headings of *cis*-regulatory design, *cis*-regulatory repression, *cis*-regulatory logic processing, and external *cis*-regulatory communication.

First, although it may be obvious, one principle is too important not to begin with: The autonomous regulatory functions generated by each *cis*-regulatory module are the direct consequence of its design; that is, of the identity and organization of its transcription factor target sites. In this Chapter we see explicit examples of diverse *cis*-regulatory designs that enable different genes to respond in diverse ways to similar regulatory states. In the end, this provides the most powerful and incontrovertible evidence that it is the genomic regulatory sequence code that directly determines how, where, and when genes are expressed.

Second, in development, *cis*-regulatory modules generate negative (repressive) as well as positive (activating) outputs. With respect to spatial expression, negative regulatory outputs are just as important as positive outputs. This is particularly essential in regional specification processes (Chapter 1), in which the boundaries of regulatory domains are frequently set and maintained by repression.

Third, the output of each module can be considered the combinatorial product of multiple fundamental operations, which the design of the module enables it to carry out. Examples of these are to serve as a regulatory gate at the terminus of a signal transduction pathway; to serve as an "AND" processor such that two different inputs in the same time and place are required for positive function; or to provide a repressive output, conditional on the presence of the repressor. It is most important not to assume that given logic functions like these uniquely indicate the participation of given transcription factors; rather, there are sets of diverse transcription factors that mediate each such function. For example, different signal transduction pathways operate as switches at the *cis*-regulatory level in logically similar ways, even though each pathway ends up affecting the activity of particular dedicated and canonical transcription factors— different factors for different pathways.

Fourth, *cis*-regulatory modules must interact with other regulatory elements, and they must set in train sequential biochemical alterations that do not themselves include sequence-specific DNA recognition, in order for the regulatory instructions they produce to be executed. They are essentially devices for producing conditional outputs for control of gene expression, and they always require other cellular machines to generate effects. For example, enhancers have to interact with the basal transcription apparatus, or with other modules that interact with that apparatus, to actually effect transcriptional activity. So in general, though *cis*-regulatory modules each autonomously process input information, their outputs have to be communicated elsewhere for anything to happen as a result.

Before exploring the consequences of these concepts within the rich vistas of real biological systems, let us glance briefly at what *cis*-regulatory modules are like physically, and what are the properties that we can use to recognize them in the sea of genomic DNA.

MODULARITY, A GENERAL PROPERTY OF GENOMIC *cis*-REGULATORY CONTROL UNITS

When *cis*-regulatory sequences were first proposed to be generally modular in organization (Kirchhamer *et al.*, 1996), there were only a modest number of examples from work in *Drosophila* and sea urchins, and the idea was largely inferential. Now there are literally scores of genes for which detailed experimental analyses have demonstrated sharply modular *cis*-regulatory elements, such that given, nonoverlapping regions of the genomic DNA each control a specific subcomponent of the overall expression pattern. Usually the experimental tests are of sufficiency: The region is isolated and associated with a reporter construct in a gene transfer vector and shown to cause transcription of the reporter in the relevant time and place during development. Such tests indeed do suffice for unequivocal demonstration and dissection of the regulatory code wired into the modular sequence. These tests do not always reveal the uniqueness or necessity of the regulatory module, but in some examples that follow, given *cis*-regulatory modules have been knocked out or mutations in them are known, and the necessity as well as sufficiency of the module can be shown. Sometimes seeming redundancies appear, in that an expression pattern is generated even after knockout of a *cis*-regulatory element that, when isolated and tested in a gene transfer experiment, produces this same pattern. But of course, except where it is of recent evolutionary origin, nothing completely redundant survives in the genome. Even when there appears to be functional overlap between regulatory modules of a given gene, on further examination, their inputs and outputs are discovered not to be identical. For example, it is common to find a module that initiates activity of a gene in a certain domain, and another that is later used to maintain this activity in the same domain.

Although the prediction that genes expressed in diverse places during development will display modular *cis*-regulatory organization has indeed been confirmed many times over, this fact in itself does not begin to convey the elegance and precision of this organization. Some illustrative examples are reproduced in Fig. 2.1, drawn from gene regulatory studies carried out on mouse (Fig. 2.1A), *C. elegans* (Fig. 2.1B), and *Drosophila* (Fig. 2.1C). Many others could have been chosen in their stead (see legend). In each case included in Fig. 2.1, multiple *cis*-regulatory modules have been demonstrated experimentally to direct expression of a gene in diverse spatial domains of the developing animal, at the same or different times. We see that if a gene is expressed in multiple spatial domains during embryonic specification, each domain of expression will be controlled by a distinct *cis*-regulatory module: if in a set of neurons, diverse *cis*-regulatory modules will mandate expression in each neuronal subtype; if in the heart, individual modules will control expression in the different heart regions; if in a given region of the dorsal axis, individual modules will specify expression in the diverse neural and mesodermal domains of the axis. Unless one is thinking about it in the right way, the

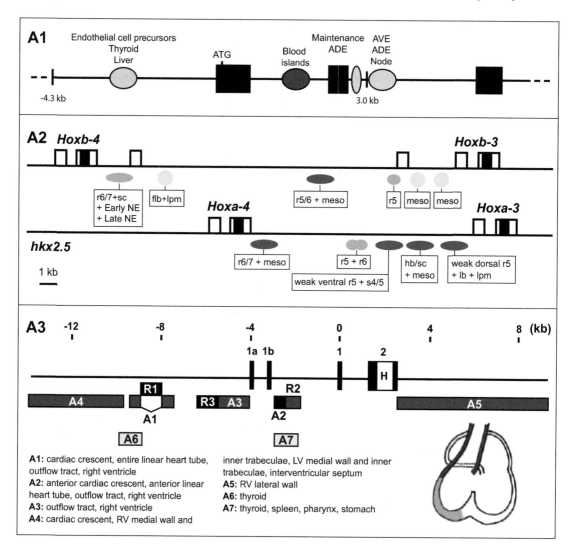

FIGURE 2.1 Examples of modular *cis*-regulatory organization from diverse bilaterians. (A) Mouse: (A1) *hex* gene. Solid rectangles are exons, colored ovals indicate locations of *cis*-regulatory modules. Anterior definitive endoderm (ADE) and anterior visceral endoderm (AVE) of early embryo. Constructs bearing individual modules produce equivalent expression patterns in transgenic mice and *Xenopus* (from Rodriguez *et al.*, 2001). (A2) *hoxb3-hoxb4* and *hoxa3-hoxa4* genes. Rectangles are exons, homeoboxes in black, and regulatory modules, as indicated, are shown by colored ovals: r, rhombomere; s, somite; lpm, lateral plate mesoderm; lb, limb bud; flb, forelimb bud; hb, hindbrain; sc, spinal cord.

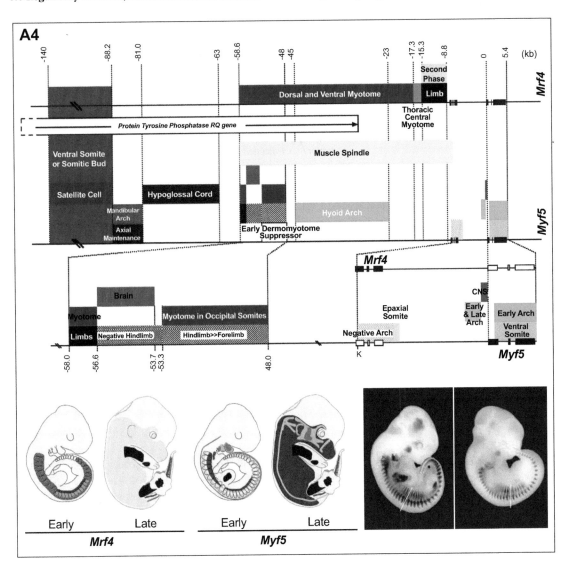

Red denotes a module that acts in both neural and mesodermal tissues; yellow, a mesodermal module; blue, a neural-specific module (from Kwan *et al.*, 2001). *cis*-Regulatory modules controlling specific aspects of spatial expression are also characterized for *hoxa2* and *b2* genes (Tümpel *et al.*, 2002). (A3) *nkx2.5* gene. Solid rectangles are exons, open area (H) denotes homeobox. Modular expression results in transgenic mice are summarized below: A, activating element or enhancer;

(Continues)

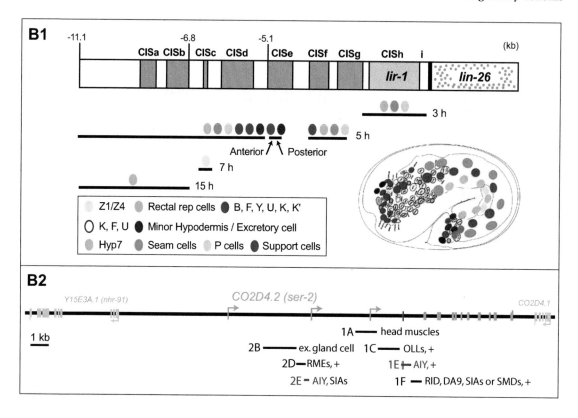

FIGURE 2.1 (continued)

R, repressor element; red indicates a module that works in developing heart, yellow an element that works elsewhere (from Schwartz and Olson, 1999; see also Liberatore *et al.*, 2002; Lien *et al.*, 2002). The modularity of expression is illustrated diagrammatically: green represents expression generated by the element A2 at the stage shown, and red that generated by A5 (from Tanaka *et al.*, 1999). The *gata6* gene is similarly expressed in different regions of the developing heart under control of region specific *cis*-regulatory modules (Davis *et al.*, 2001). (A4) *mrf4/myf5* genes; modular control of the overall expression patterns indicated in the drawings below is color coded with respect to the diagram. Exons are indicated by small solid and open rectangles on lines representing the *myrf4* gene in black on the upper line, and the *myf5* gene in black on the lower line (kindly provided by Peter Rigby). Several of the regions indicated have been further subdivided into discrete *cis*-regulatory modules, e.g., modules that control expression in somites of different axial positions, limb buds and brain lie within the −48 to −58 domain (Buchberger *et al.*, 2003; Hadchouel *et al.*, 2003). In addition some elements operate specifically on *myf5* and others on *mrf4*, as indicated in the diagram by position with respect to the lines representing the genes in black (Carvajal *et al.*, 2001; for a specific case see, e.g., Fomin *et al.*, 2004). Furthermore, the necessity as well as sufficiency of specific modules has been established by their deletion from transgene constructs, e.g., the *myf5* epaxial somite enhancer at −7 kb (Teboul *et al.*, 2002),

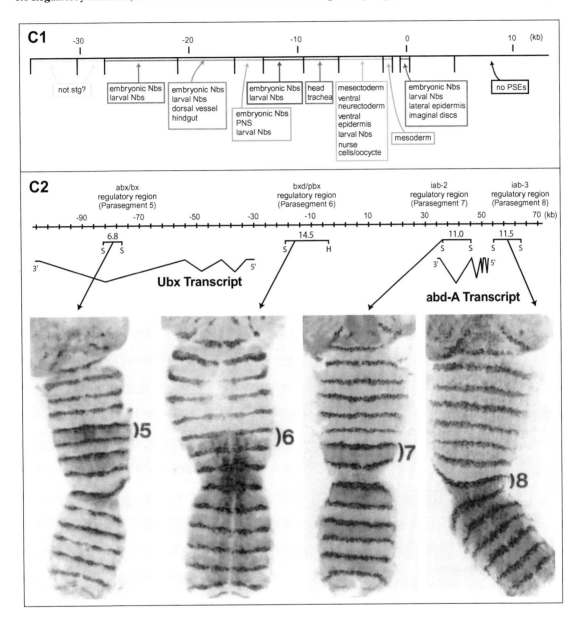

and several of the modules within the −48 to −58 region (Hadchouel *et al.*, 2003). The result of deleting this region is illustrated by the stained embryos shown: that on the left is expressing a BAC containing most of the *myf5/mrf4* locus (blue lacz stain, produced by β-galactosidase reporter), and that on

(Continues)

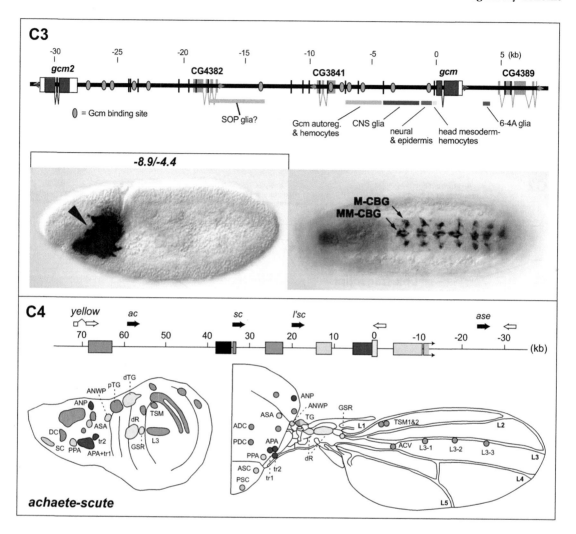

FIGURE 2.1 (continued)

the right by the BAC construct lacking the regulatory sequence from −63 to −48 (from Hadchouel *et al.*, 2003). Many additional examples could be included. Some particularly clear cases among vertebrate genes that have been demonstrated to be controlled by multiple, independent *cis*-regulatory modules responding to distinct regulatory states, and that are of particular interest because these genes encode transcription factors, are the *otx2*, *sox2*, and *pax6* genes (see below); the *neurogenin* gene (Nakada *et al.*, 2004); the *ap-2* gene (Zhang and Williams, 2003); the *pax2* gene (Pfeffer *et al.*, 2002); the *dlx* gene (Park *et al.*, 2004); and the *scl* gene (Sanchez *et al.*, 1999; Sinclair *et al.*, 1999; Göttgens *et al.*, 2000). The same is true of the gene encoding the mouse Fgf4 signaling ligand (Iwahori *et al.*, 2004); the zebrafish *delta* gene (Hans and Campos-Ortega, 2002), and the genes encoding widely expressed cytoskeletal proteins such as the mammalian smooth muscle myosin heavy chain gene (Manabe and Owens, 2001). (A4) From Carvajal *et al.* (2001); photographs from Hadchouel *et al.* (2003). (B) *C. elegans*: (B1), *lin-26* gene, a probable Zn finger transcription factor expressed in all epithelial cells, germ line, and uterus; as color code indicates, expression is controlled by multiple tissue specific *cis*-regulatory modules originally identified by sequence conservation to *C. briggsae* (CISa-g). The CIS elements work together with a proximal element denoted *i* (from Landmann *et al.*, 2004). (B2) *ser-2* gene, which encodes a transmembrane receptor expressed in specific neurons (indicated by three letter capital names, of which the pair of AIY neurons are highlighted in red), and in some other cells. Remarkably, modular control of *ser-2* extends to the level of specifying individual neurons (from Wenick and Hobert, 2004). This study shows that *ser-2* is a member of a large battery of diverse genes, all expressed among other places in the AIY neurons because all include AIY-specific *cis*-regulatory modules that respond to two particular transcription factors expressed in these neurons. Modular *cis*-regulatory elements have been demonstrated in a number of other *C. elegans* genes as well, for instance in the *myod (hlh)* gene (Krause *et al.*, 1994); in *unc-86*, which encodes a POU domain transcription factor (Baumeister *et al.*, 1996); in *lin-11*, a homeodomain regulator required for development of the female reproductive system (Gupta and Sternberg, 2002); in three vulva-specific differentiation genes (Kirouac and Strenberg, 2003); and in the *hox* gene *egl5* (Teng *et al.*, 2004). In the last three cases the regulatory modules were identified by their conservation in the related species *C. briggsae*. (C) *Drosophila*: C1, the *string (cdc25)* gene, encoding a cell cycle phosphatase gene, of which the global expression pattern is composed of cell- and tissue-specific subpatterns generated by the numerous indicated *cis*-regulatory modules; Nbs, neuroblast (from Lehman *et al.*, 1999). (C2) *Ubx* and *abd-A* genes and four *cis*-regulatory modules controlling anterior boundaries of expression in parasegments 5, 6, 7, and 8, respectively. A map of the locus and of the transcription units is shown at the top. The four regulatory DNA fragments indicated were associated with a *β-galactosidase* reporter gene, producing the patterns of expression shown (from Simon *et al.*, 1990). (C3) *gcm* locus; *cis*-regulatory modules are located not only upstream and downstream of the *gcm* gene but also in an adjacent intergenic region; SOP, sensory organ precursor. Expression in transgenic flies of the hemocyte (left) and of the abdominal glial cell (right) enhancers is illustrated; MM-CBG and M-CBG, two specific abdominal glial cells (from Jones *et al.*, 2004). (C4) *achaete-scute* locus (including also *asense* and *lethal of scute* genes); these genes encode bHLH regulators required for peripheral nervous system specification. The colored boxes in the map at the top show the location of *cis*-regulatory modules that cause expression in the respective proneural clusters shown in the same colors of the wing imaginal disc, and the sensory structures to which they give rise in the adult wing; the black box denotes a later-acting module required for expression in the sensory organ precursor cells of all of the sensory structures shown (from Gómez-Skarmeta *et al.*, 1995). Modular *cis*-regulatory organization has been seen in many *Drosophila* genes, particularly pair rule and gap genes (reviewed by Nasiadka *et al.*, 2002). Other examples of note include the gene encoding the *dpp* signaling ligand (Blackman *et al.*, 1991; Jackson and Hoffman, 1994), the *mef2* gene (Nguyen and Xu, 1998), and the *vnd* (Shao *et al.*, 2002), *yan* (Ramos *et al.*, 2003), and *tinman* regulatory genes (Yin *et al.*, 1997). The pair rule stripe modules of the *evenskipped* gene are illustrated in Fig. 1.4, and this gene also has different modules controlling expression in a variety of different neurons, certain heart cells (see below), and other specific tissues (Fujioka *et al.*, 1999). So also is the downstream gene *roughest*, which encodes a cell adhesion protein controlled by multiple *cis*-regulatory modules (Apitz *et al.*, 2004).

results in Fig. 2.1 seem counterintuitive. Why do there have to be so many different modules controlling expression of the *mrf4/myf5* muscle specification genes in similarly functioning muscle cells of the mouse (Fig. 2.1A4); or why does the *cdc25* (*string*) cell cycle control gene of *Drosophila* (Fig. 2.1C1) require a plethora of different regulatory modules to turn on cell division? The patterns of gene expression in both cases are very widespread, in that they include cells in all regions of the body. But these very general patterns turn out to be regulatory mosaics. They are the sums of the outputs of many individual *cis*-regulatory modules dispersed in genomic space, each determining expression in a precise, small subregion of the animal. The meaning of this organization will be obvious from the discussion in Chapter 1. Each region and cell type of the animal is formulated by means of its unique regulatory states. To respond to these different sets of inputs, different assemblages of *cis*-regulatory target sites are required. The requirement is thus modular in logical terms, and the genomic solution to the requirement is modular in physical terms. This is for good functional reasons, as we see below in the discussion of how *cis*-regulatory elements work.

The theory of *cis*-regulatory modularity goes far beyond observation of cases, in that its corollaries have been used in two different ways to generate successful *a priori* predictions of DNA sequence elements that function as *cis*-regulatory modules. By definition, functional *cis*-regulatory modules are nonrandom clusters of certain target sites. Experience has shown that they are usually a few hundred base pairs in length, as illustrated for various examples in Fig. 2.1. Thus where the target sites can be inferred, it should be possible to identify active *cis*-regulatory modules in the genomic sequence by computational methods. This kind of approach indeed successfully recovers a significant fraction of known *cis*-regulatory elements that respond to given regulatory inputs, and successfully identifies some previously unknown ones that on experimental test are seen to operate as bona fide *cis*-regulatory modules. For example, clustering of the known target sites for maternal and gap gene regulators of *Drosophila* was used in two different studies for computational identification in genomic DNA sequence of *cis*-regulatory modules that generate spatial patterns of expression along the anterior-posterior axis (Berman *et al.*, 2002; Rajewsky *et al.*, 2002). In respect to both the fraction of known *cis*-regulatory modules thus recovered, and the activity of previously unknown ones on experimental test, the efficacy of these approaches was about half. This is of course well above random expectation, and whatever the practical usefulness, the main point here is that the simplest structural definition of *cis*-regulatory module, a nonrandomly tight cluster of target sites, has clear predictive value. Additional studies of this kind that were also partially successful have been used to identify a heart-specific enhancer (Halfon *et al.*, 2002), and to find Notch responsive target genes in *Drosophila* (Rebeiz *et al.*, 2002). An elaboration on the basic idea of searching for clusters of known target sites in the DNA has been applied with excellent results to the search for enhancers that specify lateral stripes of expression in response to the ventral to dorsal gradient of the Dorsal (Dl) transcription factor in *Drosophila*, the results of which are detailed below.

Many such *cis*-regulatory modules were found merely by identifying clusters of target sites for this factor (Markstein *et al.*, 2002), and others could be identified by requiring particular combinations between Dl sites and additional sequence motifs observed in the vicinity of Dl sites in known enhancers (Stathopoulos *et al.*, 2002; Markstein *et al.*, 2004). For our present purposes, the success of the computationally guided identification of Dl response enhancers strongly substantiates the predictive value of the *cis*-regulatory module concept. These methods will undoubtedly improve, and as time goes on will be more and more informed by knowledge of particular sites and of the likelihood of combinations of such sites.

The second manner in which the theory of *cis*-regulatory modularity has been applied computationally capitalizes on the implication that these elements should stand out as patches of relative conservation in interspecific sequence comparisons. If *cis*-regulatory elements are indeed discrete and functionally important target site clusters, their evolutionary DNA divergence should be suppressed relative to that of the surrounding, nonsequence-dependent DNA (Chapter 1). Therefore, identification of conserved sequence patches a few hundred base pairs in length should identify *cis*-regulatory elements, providing that the sequence comparison used to detect conservation is at the right distance. The "right distance" would be one in which there has been enough evolutionary divergence so that, unless conserved for functional reasons, the two sequences will be too distinct to be aligned, but not so distant that the functional regulatory elements are no longer similar. Two separate phenomena are responsible for relative conservation of patches of sequence that extend across the *cis*-regulatory module. The target sites themselves are conserved of course, but these usually include a minor fraction of the total conserved sequence length. However, there is also a strong suppression of larger insertions and deletions in the internal sequence between sites relative to the flanking sequence outside the module (Cameron *et al.*, 2005). In the event, this approach has proven very powerful, at least for some organisms, and as more genomic sequence becomes available it will provide an even more useful means of finding *cis*-regulatory modules in the genome. Some remarkable examples from experimental studies in sea urchins and vertebrates are reproduced in Fig. 2.2. In the author's laboratory the recovery of desired *cis*-regulatory modules from within the set of conserved sequence patches in the vicinity of given genes has been successful in more than 10 different cases. Two of these cases are summarized in Figs. 2.2A1 (the *delta* gene), and 2.2A2 (the *otx* gene); see legend for references and details. For both these genes, about two-thirds of the many conserved sequence patches in the noncoding DNA produce expression in the embryo as illustrated (the other patches might also have been found active had the tests extended beyond the first two days of embryogenesis). Note, in Fig. 2.2A1, the exceedingly sharp transition between the conserved sequence of the regulatory module and the surrounding divergent sequence. As also shown in the interspecific sequence comparison plots in the vertebrate examples of Fig. 2.2B, this method has the invaluable practical advantage of indicating the natural extent of

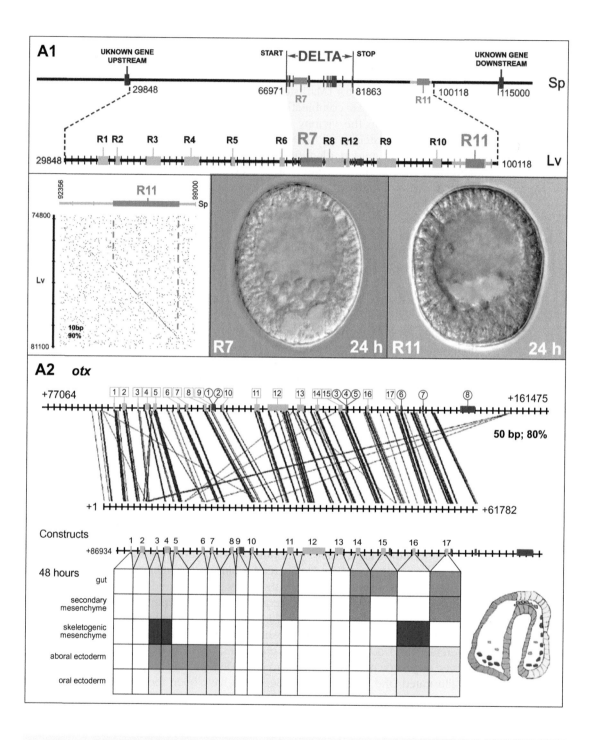

FIGURE 2.2 For legend see pages 43–46.

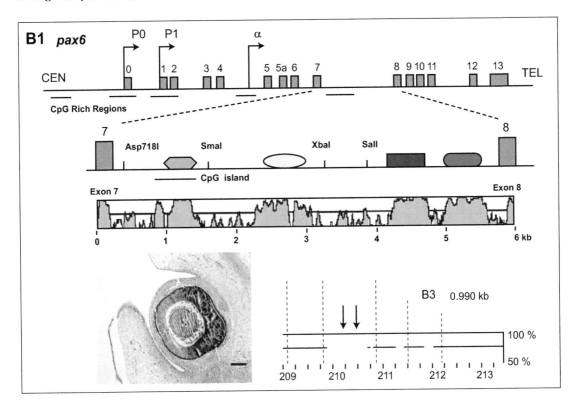

FIGURE 2.2 Identification of multiple *cis*-regulatory modules by interspecific sequence conservation. (A) Sea urchins: (A1) *delta gene*; map of the genomic region, derived from an annotated BAC, is shown at the top, and location of conserved sequence patches below, green boxes. Red boxes denote exons. These were identified in a sequence comparison across 70 kb of orthologous sequence from a *Lytechinus variegatus* BAC (the last common ancestor of these species lived about 50 million years ago). Regulatory modules generating the early- and mid-stage embryo expression patterns of *delta* correspond to conserved sequence patches R7 and R11. A dot matrix plot for R11 using a window of 9 out of 10 bp is shown at lower left, *Strongylocentrotus purpuratus* DNA sequence on top, versus *L. variegatus* DNA sequence on the ordinate. The conserved sequence patch (diagonal line) terminates abruptly on both ends. The expression of GFP driven by the R7 module is shown in a transgenic 24-hr embryo (center) to occur exactly in the precursors of the mesodermal cells of the embryo, the locus of endogenous *delta* expression after 20 hr. In contrast, the R11 module drives GFP expression in skeletogenic cells (right), the locus of earlier *delta* gene expression. These cells are seen here in another 24-hr embryo within the blastocoel, following their ingression (from Revilla-i-Domingo *et al.*, 2004; and unpublished data of R. Revilla-i-Domingo and Davidson). (A2) *otx* gene. A comparison across about 60 kb of genomic sequence surrounding the gene using BACs from *S. purpuratus* and *L. variegatus* is shown.

(Continues)

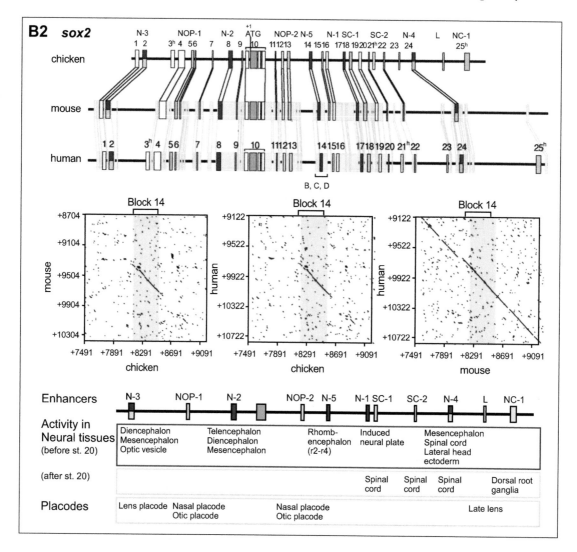

FIGURE 2.2 (continued)

Solid diagonal bars give locations of consecutive windows within which the sequences match at 40 or more out of a 50 bp moving window. The analysis identifies 17 patches of conserved sequence shown as green boxes; exons are indicated in red. The conserved patches were isolated from the *S. purpuratus* BAC, associated with reporter genes in an expression construct, and tested by gene transfer into *S. purpuratus* eggs, with the results indicated by the color code (for late gastrula-stage embryos): 11 out of the 17 patches were active in the various domains of the embryo, together accounting for most of the overall pattern of endogenous *otx* expression (from Yuh *et al.*, 2002). (B) Vertebrates. (B1) *pax6* gene (from Kleinjan *et al.*, 2004). The 13 exons of the gene are shown as gray boxes in the locus map at the top: tel, telomere; cen, centromere. Below the map are indicated intron enhancers for late expression in the eye (green), hindbrain (red), forebrain, and heart (yellow). These correspond to

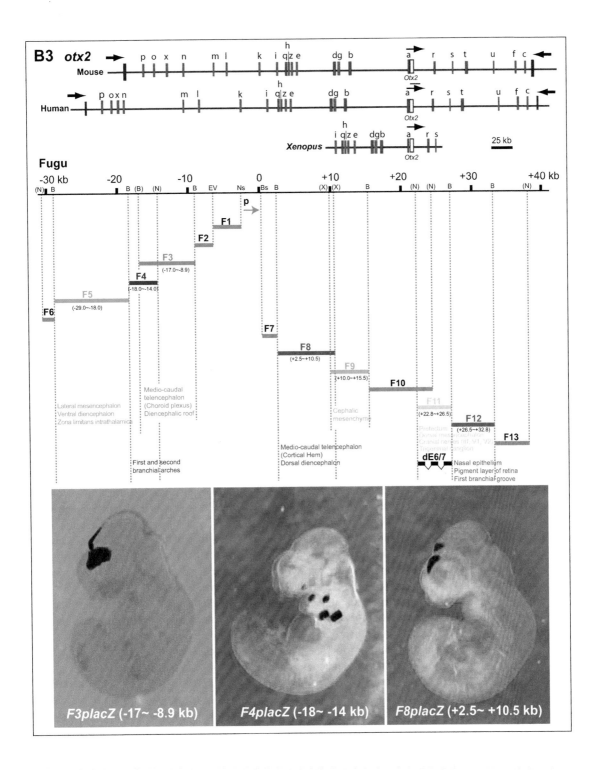

B3 *otx2*

Mouse

Human

Xenopus

25 kb

Fugu

-30 kb   -20   -10   0   +10   +20   +30   +40 kb

p

F1

F2

F3
(-17.0~-8.9)

F4
(-18.0~-14.0)

F5
(-29.0~-18.0)

F6

F7

F8
(+2.5~+10.5)

F9
(+10.0~+15.5)

F10

F11
(+22.8~+26.5)

F12
(+26.5~+32.8)

F13

dE6/7

Lateral mesencephalon
Ventral diencephalon
Zona limitans intrathalamica

Medio-caudal
telencephalon
(Choroid plexus)
Diencephalic roof

First and second
branchial arches

Cephalic
mesenchyme

Prefectum
(Dorsal mesencephalon)
Cranial nerves (III, V1, V2)
Trigeminal ganglion

Medio-caudal telencephalon
(Cortical Hem)
Dorsal diencephalon

Nasal epithelium
Pigment layer of retina
First branchial groove

F3placZ (-17~ -8.9 kb)   *F4placZ* (-18~ -14 kb)   *F8placZ* (+2.5~ +10.5 kb)

FIGURE 2.2 (continued)

(Continues)

the *cis*-regulatory modules it identifies; that is, the termini of the regions of conservation. But even more important, the results confirm an argument implicit in the conceptual idea of a module: It has specific boundaries.

Certain modular conservation patterns can be perceived all across the vertebrates, as indicated in Fig. 2.2B, for the *pax6*, *sox2*, and *otx2* genes. However,

FIGURE 2.2 (continued)

sequence patches conserved between human and mouse genomes. Conservation is shown in plots (pink areas) that display sequence identity from 50 to 100% (ordinate) as a 50 bp window is moved along the sequence. The conserved sequence patches were tested by incorporating them in a β-galactosidase expression vector, and their regulatory function was determined in transgenic mice. No function was identified for the patch labeled in blue. Lower right, additional conserved sequence patches conserved between human and mouse genomes, located about 80 kb downstream from the last exon of the *pax6* gene, and active in other regions of the brain. The "B3" patch confers expression specifically in the retina in transgenic mice, as indicated by the blue lacz stain in the sectioned embryo shown (from Griffin *et al.*, 2002). The *pax6* gene is controlled by many modular enhancers, some of which are conserved even in fish/mammal genomic comparisons (*op. cit.*; Kammandel *et al.*, 1999). (B2) *sox2* gene; three way human-mouse-chicken genomic comparisons are shown. The gene is in green and the locus maps indicate that all conserved patches are present in all three amniote genomes except for the chicken-specific enhancer L. The dot matrix plots (6 out of 10 bp identity required) show that in this case mouse and human genomes are too close to discern the conserved patch at this criterion, but it is easily observed in comparison of either to chicken. At the bottom are indicated the spatial functions of the modules identified by sequence conservation, as determined in transgenic chicken embryos, early neural enhancers in red (from Uchikawa *et al.*, 2003). (B3) *otx2* gene (from Kurokawa *et al.*, 2004). A map of conserved sequence patches surrounding the gene in mouse, human, and *Xenopus* is shown. The six patches in red were demonstrated to have regulatory activity in transgenic mice in various regions of the forebrain and midbrain, as well as in certain early embryonic domains. Deletion was used to demonstrate essentiality of some of these enhancer regions. Conservation of *otx2* regulatory modules extends to fish at both functional and sequence levels. The map of the *Fugu otx2* gene shown below indicates it to be densely surrounded by enhancers, active in the domains given in color when introduced as reporter constructs in transgenic mice (from Kimura-Yoshida *et al.*, 2004). Three examples of these fish regulatory modules operating accurately in transgenic mice, stained for the lacz reporter product, are illustrated: m, mesencephalon; d, diencephalon; t, telencephalon; ma, mandibular arch; hy, hyoid arch; v, trigeminal ganglion. Both mouse and fish enhancers can be isolated by identification of sequence patches conserved between them. Many remarkable cases of *cis*-regulatory modules that are conserved across mammals and other vertebrates have been reported in which conservation provides an immediate practical guide to their experimental isolation: striking examples include the *scl* gene, required in hematopoietic development (Göttgens *et al.*, 2000, 2001; Chapman *et al.*, 2003); and the globin gene family (Hardison, 1997, 2000). Conservation between two congeneric species of ascidian also successfully identified multiple *cis*-regulatory modules responsible for different spatial domains of expression of the *pitx* regulatory gene (Christiaen *et al.*, 2005). For reviews of this area, see Margulies *et al.* (2003) and Pennacchio *et al.* (2003).

different regions of the genome diverge at different rates; for example, while human-mouse genomic comparison suffices for identification of *cis*-regulatory modules in the *pax6* and *otx2* genes, the modular boundaries cannot be distinguished except by a comparison out to chicken sequence in the case of a *sox2* module (Fig. 2.2B). There are many other examples noted in the legends of Figs. 2.1 and 2.2 in which *cis*-regulatory modules have been found by interspecific sequence comparison in both vertebrates and nematodes. Because of its generality and the lack of need for any prior knowledge of binding sites, recognition of conserved sequence patches is likely to remain for some time a very useful method for primary identification of putative *cis*-regulatory modules in the genome.

INSIDE THE *cis*-REGULATORY MODULE: LOGIC PROCESSING AND INPUT/OUTPUT RELATIONS

Now that we see the generality and the reality of *cis*-regulatory modules, as they have been revealed experimentally in diverse bilaterian genomes, it is time to face the detailed issues of how they work and what they do. As stressed in the discussions of Chapter 1, their role boils down to the essential function of information processing, where the "information" consists of their regulatory inputs. For the remainder of this Chapter we deal with how *cis*-regulatory information processing functions are encoded in the DNA sequence, and how these functions underlie the spatial control of developmental gene expression.

At the outset, there is an important distinction to be made between two different classes of input. Those inputs, which determine specifically when and where a target gene is to be expressed (by definition), must do so by their variation in time and space. But there are other kinds of input to *cis*-regulatory modules as well, mediated by DNA-binding proteins which function as necessary parts of the information processing machinery, that do not necessarily vary in a determinate way in time and space. The time- and space-varying inputs can be considered the functional "drivers" of developmental *cis*-regulatory modules. In physical terms, the drivers are the active transcription factors which are components of specific regulatory states, and hence they appear only at given developmental situations in given cells. It is these drivers, their origins and their destinations, which are figured in the linkages of a gene regulatory network (e.g., see Fig. 1.3). In many analyses only the driver inputs are revealed, or are required to be revealed. The remaining functions are usually lost in the black box obscurities of qualitative *cis*-regulatory observations. But the total information processing capacity encoded in the DNA sequence of a *cis*-regulatory module is in fact much greater than that mediated exclusively by driver target sites. We can see this experimentally only by a comprehensive approach to *cis*-regulatory analysis. Rather than by focusing only on driver effects, suppose we begin by finding all the target sites within a *cis*-regulatory module where transcription factors bind sequence-specifically, and then determine the effects on regulatory output of site-specifically modifying every target site. The first thing to emerge from this kind of approach is that there are many more tightly and specifically bound target sites than are occupied by driver factors. The next thing to emerge is that each species of site has some influence on module output: As revealed by the consequences of mutation, they unmistakably all "do" something (Kirchhamer and Davidson, 1996; Davidson, 2001; Yuh *et al.*, 1998, 2001). Mutations of target sites and combinations of such mutations can be considered perturbations of *cis*-regulatory structure, and as the following examples illustrate, the functional effects of these perturbations are best seen by including quantitative kinetic measurements in assays of module output.

cis-Regulatory Logic Processors: Examples

The most thoroughly examined examples we have thus far are *cis*-regulatory modules from the sea urchin. These are examples in which the functional significance of every detectable DNA-protein interaction has been examined in context in gene transfer experiments. The primary case remains the *endo16* gene of *Strongylocentrotus purpuratus* (Yuh and Davidson, 1996; Yuh *et al.*, 1994, 1996, 1998, 2001, 2005). What emerges from the years of study to which this regulatory system has been subjected is the demonstration of a network of conditional logic interactions programmed into the DNA sequence, which amounts essentially to a hardwired biological computational device.

The *endo16* gene is expressed in the endoderm of the embryo. It encodes a secreted protein which in the late embryo is found on the inner wall of the midgut (Soltysik-Espanola *et al.*, 1994). At first glance its regulation appears a rather typical tissue-specific process, and indeed, it probably is rather typical. The spatial pattern of *endo16* expression is shown in Fig. 2.3A (a diagrammatic review of endoderm formation in this embryo is to be found in Chapter 3). The *endo16* gene is activated long before there is a midgut, or any gut at all, in mid-to-late cleavage, during or soon after the initial processes of endoderm specification (Godin *et al.*, 1996; its product could have a gastrulation as well as a midgut function). Figure 2.3A1, 2 shows the initial expression of *endo16* in the cells that will give rise to the endodermal and mesodermal cell types of blastula stage embryos. The skeletogenic progenitors of the embryo (plus a few cells that will contribute to the future coelomic sacs) appear in Fig. 2.3A2 as a polar patch of nonexpressing cells lying within the ring of expressing endomesodermal cells. Thus two boundaries must be established: one within, between the endomesodermal cells and the skeletogenic cells; and one without, between the endomesodermal cells and the overlying ectoderm. It will not surprise that these boundaries depend on spatial repression (Yuh and Davidson, 1996). The subsequent panels of Fig. 2.3A show that *endo16* is later expressed throughout the archenteron, remaining silent in the skeletogenic mesenchyme and the surrounding ectoderm. Expression is then extinguished in the foregut, and later in the hindgut, while the gene remains active in the midgut. This is its terminal locus of activity (Nocente-McGrath *et al.*, 1989; Ransick *et al.*, 1993).

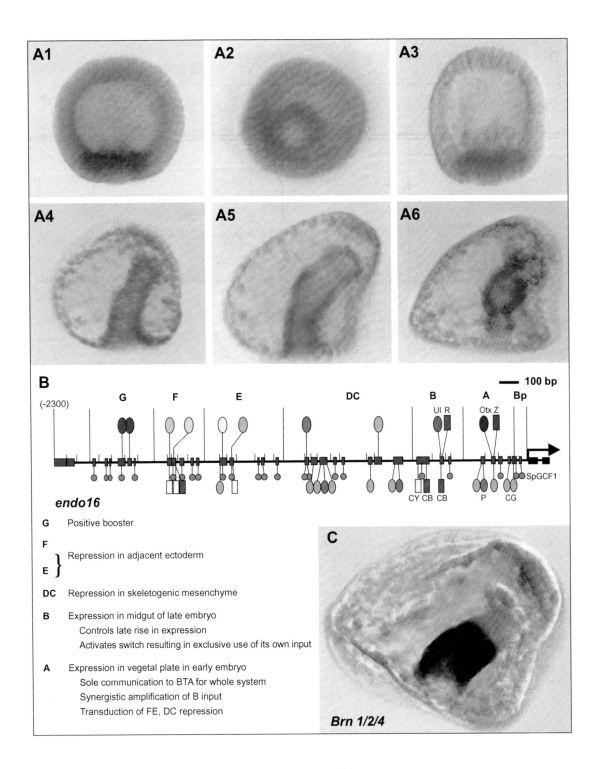

FIGURE 2.3 For legend see pages 51–52.

(Continues)

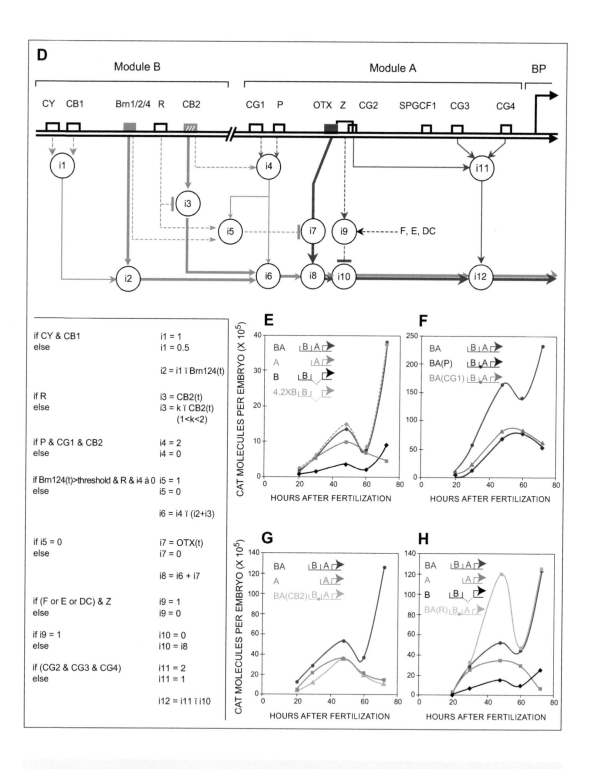

FIGURE 2.3 (continued)

FIGURE 2.3 *cis*-Regulatory logic in the *endo16* gene. (A) Whole mount *in situ* hybridization (WMISH) display of *endo16* expression pattern in *Strongylocentrotus* embryos (from Ransick *et al.*, 1993). The gene is expressed in the vegetal plate (A1), but not the skeletogenic cells and small micromeres located at the vegetal pole in the early blastula (A2), nor is it active in the skeletogenic cells as they ingress (A3). After gastrulation it is expressed throughout the archenteron (embryonic gut; A4). Expression is later shut off in the foregut (A5), and the hindgut (A6) so that the terminal pattern of expression is confined to midgut. (B) Map of protein interactions in the *endo16 cis*-regulatory system (from Yuh *et al.*, 1994). The 2300 bp DNA sequence indicated as the horizontal line is necessary and sufficient to provide accurate expression of a reporter construct after injection into the egg. Proteins that bind at unique locations are shown above the line, and proteins that bind at several locations are indicated below. Different colors indicate distinct proteins. "G–A" indicate the functional regions or modules of the overall system as defined in gene transfer experiments (Yuh and Davidson, 1996). At lower left the functions of each region are briefly indicated. The inner boundary beyond which expression is precluded in the skeletogenic cells (A2) requires repressive interactions in the DC region; for the boundary with the overlying ectoderm interactions in both regions E and F are required. The protein responsible for repression in the F region is a factor of a class which is usually associated with signaling systems, *viz.* a Creb factor, so the boundary with the ectoderm is at least in part likely to be established by a signaling interaction (as implied by other evidence as well; Yuh *et al.*, 1998). (C) Locus of expression of the transcription factor Brn1/2/4, the driver of Module B, visualized by WMISH (from Yuh *et al.*, 2005). The expression domain is the same as that of *endo16* (A5). (D) Logic model for Modules B and A (from Yuh *et al.*, 2001). The regulatory DNA of *endo16* is shown as a horizontal strip at the top of the diagram; individual binding sites are indicated by labeled boxes. Module B and its effects are shown in blue; Module A and its effects are shown in red. Logic functions (i) are indicated by numbered circles. Each follows from a specific regulatory interaction, the consequences of which are modeled as a logic operation. Note the two types of regulatory input: time-varying driver interactions (colored boxes) which determine the temporal and also spatial pattern of *endo16* expression, and time- and probably space-invariant interactions (open boxes) which affect the level of expression and control internal regulatory output traffic. In the diagram interactions that can be modeled as Boolean are shown as dashed lines, those which are scalar as thin solid lines, and those which are time-varying quantitative inputs as heavy solid lines. The individual logic interactions are defined in the set of statements below the diagram in the box at left. Here, for statements of the form "If X," where X is the name of a target site, and the value of the corresponding "i" function is not zero, the meaning is that this site is present and occupied by the respective factor. If the site has been mutated (or if the factors were inactivated or eliminated) this is denoted by zero or as the alternative ("else") to the site being present and occupied. The statements afford testable predictions of the output for any given mutation or alteration of the system, as shown by Yuh *et al.* (2001). Briefly, the system works as follows (from left): CB1 and CY1 interactions together (i1) synergistically increase the output of the positive spatial and temporal regulator of Module B which binds at the Brn1/2/4 site. The output of the Brn1/2/4 subsystem is at (i2). An additional smaller time-varying positive input, which peaks at about 40 hr, is generated by the interaction at CB2 (i3). An interaction at R is required for the B-A intermodule input switch, which shuts off Otx input (i5, i7), but this switch operates only if there is input from the Brn1/2/4 site, and if the CB2 site is present and occupied (i5). Furthermore, the proteins binding at the adjacent R and

(Continues)

Figure 2.3B shows a protein-binding map of the *endo16 cis*-regulatory system, which is included in a 2.3 kb fragment of DNA that suffices to reproduce the normal pattern of expression when linked into an expression vector (Ransick *et al.*, 1993). Thirteen different proteins bind with high specificity within this DNA sequence (Yuh *et al.*, 1994), which has been divided experimentally into six functional regions (G–A). The role played by each region is indicated briefly below the map, as established by kinetic measurements of the output of reporter gene product after injection of appropriate expression constructs into sea urchin eggs (Yuh and Davidson, 1996; Yuh *et al.*, 1996). The repressive interactions, which set

FIGURE 2.3 (continued)

CB2 sites apparently interact, in that if the R site is mutated, CB2 input (i3) is somewhat enhanced. In Module A the CG1 and P sites together with CB2 in Module B are all required for linkage of Module B to Module A (i4), and for synergistic amplification (by a factor of about 2) of the Module B input (i6). If the switch mediated by R does not function (i.e., in an R mutation) the summed input of Modules A and B at i8 is observed. If CB2, CG1, or P are mutated, Module B is unlinked from Module A. That is, i4 = 0, so i6 = 0 and in this case i8 is just the output of the Otx interaction (at i7). If the gene is in a cell where the repressive interactions in any of the upstream F, E, or DC modules are operating, the system is shut off via an interaction at the Z site of Module A (i9, i10). Finally, the CG2, CG3, and CG4 sites combine (at i11) to boost whatever output is present (i12) by an additional factor of 2. The result at i12 is transmitted to the basal transcription apparatus. (E–H) Kinetic experiments in which expression of a reporter gene encoded in the injected constructs (the chloramphenicol acetyl transferase, or CAT gene) is measured as a function of time after fertilization, in batches of 100 transgenic embryos per time point. The eggs were injected with expression constructs consisting of Modules B and A linked to one another and to the basal promoter as in the normal gene (BA; red curves); with a construct driven by Module B alone (B; black curves); with a construct driven by Module A alone (A; blue curves); or with constructs in which specific target sites in Modules B or A were mutated, as indicated by parentheses. In the color-coded cartoons in each figure the dots indicate the location of the mutated target sites given in the parentheses. (E) Wild-type constructs. The dashed line shows the time function generated if the measured B curve is multiplied by the factor 4.2 at each point. (F) CAT activity profile in batches of eggs injected with BA, with BA in which the P site had been mutated [BA(P); violet]; or with BA in which the CG1 site had been mutated [BA(GC1); pink]. The latter two time curves are almost identical to that generated by wild-type A module in isolation from any other modules, as in E, blue curve (E and F are from Yuh *et al.*, 1998). (G) CAT activity profiles generated by BA control; and by BA in which the CB2 site was mutated [BA(CB2); orange]. The profile of expression generated by this construct can be seen to be almost identical to that produced by Module A alone, just as predicted in the logic table in (D): if CB2 = 0, i4 and i6 are 0, and the only input is from Module A, i.e., now i8 = i12. (H) Derepression of Module A when the intermodule switch function is canceled by mutation of the R site of Module B [BA(R); green]. The output is the sum (i8) of the enhanced CB2 input (i3), Otx input (i7), plus the amplified input from Module B, including (i3) the enhanced CB2 input: i.e., [A + i4 · (i2 + i3)]; see logic table of (D); (G and H are from Yuh *et al.*, 2001).

the boundaries during the earlier phase of *endo16* activity are mediated by target sites in the upstream domains called DC, E, and F (see legend). There are three positively acting regulatory regions as well. These are, from distal to proximal, the relatively unimportant Module G, which acts throughout as a booster; Module B, which is largely responsible for the late accumulation in the product of the reporter gene; and Module A. The positive regulatory function of Module A is mediated by an Otx factor, which acts as its driver (βOtx; Yuh *et al.*, 1998; Li *et al.*, 1999; Yuh *et al.*, 2005). Module B is activated by an endoderm-specific driver, the Brn1/2/4 transcription factor (Fig. 2.3C; Yuh *et al.*, 2005). Of the target sites in Module B, the site where Brn1/2/4 binds conveys the main late spatial and temporal activation input. Module A not only activates the gene earlier, but throughout acts as the central processing unit for the whole upstream system, G–B. We now focus on structure/function relations within Modules A and B, and on B to A interactions.

The basal promoter (Bp) of the *endo16* gene consists of the sites where the transcription apparatus assembles, plus a few proximal target sites (Fig. 2.3B). This apparatus is entirely promiscuous with respect to the inputs it will service, and it is almost inactive without upstream inputs (Yuh *et al.*, 1996, 1998; Cameron *et al.*, 2004). The only inputs the Bp receives from the whole *endo16* *cis*-regulatory system are normally channeled through Module A. That is, the input to the Bp is the output of the whole *cis*-regulatory system, after processing in Module A. The value of this output determines whether transcription of the gene will occur, and if so at what rate, in every cell at all times (Yuh *et al.*, 1996, 1998). Module A also serves as a terminus for the upstream modules that mediate spatial repression, in the specific sense that in its absence, or mutation of a certain Module A target site, the repressive interactions have no effect. The outputs of these repressor subsystems (i.e., in the DC, E, and F regions) can be considered negative inputs to Module A. In biological terms, the repressive controls on spatial expression are required because the Otx driver of Module A is present and active in many regions of the early embryo (Mao *et al.*, 1996; Li *et al.*, 1997; Yuh *et al.*, 2002). However, later in development, the DC, E, and F regions can all be discarded, as can Module G, and a construct consisting only of B and A runs accurately in the midgut at almost the same rate as does the whole system. By this stage the gene mainly utilizes the gut-specific Brn1/2/4 input of Module B as a driver. But this requires two additional interactions of Modules A and B. The first is a switch that is sensitive to the level of input of this positive regulator of Module B. The Brn1/2/4 gene itself is activated only after gastrulation, that is, only after the gut has formed, but once this input into Module B becomes significant the switch turns off the input of the Otx regulator. This ensures that only the driver input of Module B will count. The second interaction is an amplification function by which Module A steps up the amplitude of the Module B input. These various functions of Modules A and B could never have been enumerated, except by kinetic assessment of the significance of each of their target sites (Yuh *et al.*, 1998, 2001), and they require a variety of sites within these modules, in addition to those at which bind the Otx and Brn1/2/4 drivers.

The sites of Modules A and B are indicated individually in the model reproduced in Fig. 2.3D (Yuh *et al.*, 2001). This is a logic model that shows all the inputs into Modules A and B, and that indicates explicitly their functional effects; i.e., each DNA–protein interaction can be treated as producing an input of some kind into the regulatory system if and when this interaction occurs. The way these inputs affect the operation of the regulatory system can be symbolized, as in a computer program, by a series of conditional logic statements, as shown in the Table in Fig. 2.3D (see legend). Each function was established by measurement of the output kinetics for the system after mutating one or more target sites, or substituting synthetic target sites for the natural sequence (Yuh *et al.*, 1998, 2001).

The intimate linkages between Modules B and A are also shown individually in the model (Fig. 2.3D). Here we see how the output of Module B runs through Module A to the Bp, and therein is multiplied severalfold. Interactions with three proteins that bind adjacently, CB2 of Module B, and CG1 and P of Module A, are required for these linkage and amplification functions. Another interaction between Modules A and B requires the site R of Module B, the site mediating the switch which cuts off driver input into Module A when the driver input into Module B attains a certain level (i.e., functions i5 and i7, of Fig. 2.3D).

Examples of experimental kinetic output data leading to these conclusions are shown in Fig. 2.3E–H. Figure 2.3E demonstrates the augmentation by Module A of the input from Module B: We see that (after very early in the profile, not visible on this scale) the output of a BA construct is equal to about a fourfold linear amplification of the output generated by Module B alone. The output of Module A alone is also comparatively low, but has a different time course; in contrast to that of Module B its activity dies away late in development. Figure 2.3F demonstrates that the functional linkage of Module B to Module A, and the late step-up in total activity, depend specifically on interactions at both the CG1 and P sites of Module A. Thus the experiment shows that if either site is mutated, the output of BA is about equal to that of A alone and has the same kinetic form; although the physical linkage between B and A is undisturbed in these mutants, the functional linkage is gone. Figure 2.3G similarly shows that if the CB2 site is mutated, the B-A linkage is again cut. Therefore, exactly as predicted, the output is just that of Module A alone, even though the remainder of Module B is present, including the site where the Brn1/2/4 factor binds. In Fig. 2.3H an experiment is shown demonstrating the effect of mutation of the R site of Module B: Now the switch which normally cuts off Otx input is inactive and the output is higher than normal, i.e., it includes the normal output of BA plus that of A (see legend).

Modules A and B of *endo16* regulate a garden variety, downstream gene, and they are unlikely on the face of it to be exceptional in functional complexity. The apparatus represented in Fig. 2.3D serves notice of the nature of the logic processing functions generally carried out by *cis*-regulatory systems and of the significance of their internal organization. There is a lesson to be drawn in respect to the way these systems have to be considered. Only two of the nine different factors that bind sequence-specifically in these regions of the control system function as drivers (a third may have a minor time-varying input as well; see legend). All of the others, to summarize their various roles in a single phrase, are used for processing the spatial and temporal data carried to the gene by its drivers. But we cannot forget that the deployment of these processing functions in the *endo16* regulatory system is encoded in the genomic target site sequences where these factors bind, on the average, just as tightly as do the driver proteins (Yuh *et al.*, 1994).

As a further example, a similar analysis of an entirely different gene is encapsulated in Fig. 2.4. This is the *cyIIIa* gene, which encodes a cytoplasmic actin, and which is expressed very specifically in the aboral ectoderm of the sea urchin embryo. The *cis*-regulatory system of this gene, again defined as the genomic DNA fragment necessary and sufficient to confer accurate expression on injected expression constructs, is about 2500 bp long. It contains target sites for at least 10 specifically binding proteins, all but one of which has been identified (for reviews and evidence, see Kirchhamer *et al.*, 1996; Kirchhamer and Davidson, 1996; Coffman *et al.*, 1997; Coffman and Davidson, 2001; Davidson, 2001; C. T. Brown and Davidson, unpublished data). There are two major points to be made about the *cyIIIa cis*-regulatory system. The first concerns the nature of spatial control. Just as for the early phase of *endo16* expression, the activators of the *cyIIIa* gene are widely expressed, and the boundaries are set by spatial repression (see diagrams at the top of Fig. 2.4). Module A of this system is a proximal unit of spatial control, dependent for accurate expression on a Zn finger repressor (P3A2; Fig. 2.4). The more distal Module B is used to step up expression, and it has its own activators and repressors (Runt, Myb, and Z12-1; Fig. 2.4). Module B, however, requires the target site for a factor (Oct1) binding in Module A for function; that is, essentially for communication of its regulatory import to the basal promoter. Second, and directly apropos of present concerns, we see that there are again two classes of input, both of which contribute to the logic processing functions of the overall system. These are on the one hand, the time- and space-varying drivers, including the spatial repressors indicated by solid symbols in Fig. 2.4; and on the other, the inputs mediated by target sites required for amplification functions (Tef1 and CCAAT sites) for linking the distal and proximal modules (Oct1 site), and for assisting internal communication by causing DNA looping (Gcf1 sites, of which more below). All of these latter proteins are likely to be ubiquitous or very widespread in the embryo. Therefore, as above, their deployment in this regulatory system as functional processors of the driver inputs just depends on the presence of the target sites at which they bind.

In summary, to relate quantitatively the driver inputs of a *cis*-regulatory module to its output, requires that we understand its encoded input processing capacities. Of course if the objective is entirely qualitative, as it sometimes legitimately is, these capacities can be more or less ignored or considered as black box functions. But exclusively qualitative analysis fails to encompass what evolution has produced in animal *cis*-regulatory modules. Nor is it possible to obtain a closed explanation of what these systems actually do with their

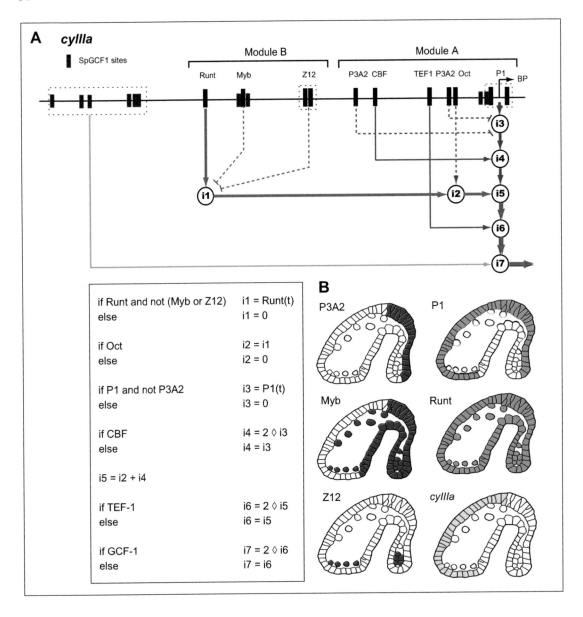

FIGURE 2.4 *cis*-Regulatory logic model of the *cyIIIa* gene. The *cyIIIa* gene encodes a cytoskeletal actin expressed exclusively in the aboral ectoderm of *S. purpuratus*. Expression constructs containing the 2.3 kb regulatory region 5′ of the gene recapitulate normal expression. This region contains over 20 transcription factor target sites bound, respectively, by nine distinct proteins, eight of which have been

regulatory driver inputs, without taking into account their hardwired conditional logic operations.

Target Site Occupancy and Transcriptional Output

The concept of target site "occupancy" relates the nuclear concentrations and intrinsic properties of driver transcription factors to what happens on the *cis*-regulatory DNA. These relations are most clearly expressed in the quantitative terms of macromolecular physical chemistry. However, a quick and qualitative sketch of the mechanisms will suffice for the arguments that follow. In the simplest situation, where the ratio of target site DNA to any other DNA sequence is high enough so that only the target site DNA need be considered, the occupancy of *cis*-regulatory target sites by specific transcription factors depends essentially on three parameters: the concentrations of the factors that bind to these sites, the energetics of their interactions with the specific DNA sequence (expressed by the equilibrium constants for these interactions), and the energetics of any interactions with adjacent specifically bound factors (given by the cooperativity constants). The stability of specific DNA-protein complexes is directly measured by the rate at which they decay (off rate), and this varies over orders of magnitude for different DNA target sites and proteins. In contrast, the rate at which DNA-protein complexes assemble (on rate), varies relatively little from one protein to another at given protein concentrations, since diverse proteins diffuse into range

identified (Kirchhamer *et al.*, 1996; Coffman *et al.*, 1997). (A) Interactions and logic processing functions (from C. T. Brown and Davidson, unpublished data). The horizontal line at the top represents the regulatory region, the basal promoter (BP) at the right. Below is a *cis*-regulatory logic diagram, as in Fig. 2.3, summarizing the results of binding site deletion and mutation experiments (Coffman *et al.*, 1996, 1997; Kirchhamer and Davidson 1996; C. T. Brown and Davidson, unpublished data). Inputs from Module A are in red, from Module B in blue. Time-varying driver inputs are indicated by solid lines extending to the interaction nodes of the diagram, and inputs that behave in a Boolean fashion (including the dominant repressors) by dashed lines. The logic encoded by the diagram is represented in programmatic form on the lower left, with the output of each node (i1–i7) assigned a value according to the presence or absence of the respective factors on the DNA. (B) Diagrams indicating spatial expression in the late gastrula embryo of the five spatial regulators of *cyIIIa* expression, *viz.* P1 (unidentified) and Runt activators in green; and P3A2, Z12, and Myb repressors in red (the widely expressed CCAAT-binding factor and the Tef1 factor act as amplifiers, and the ubiquitous Oct factor is a required linker between Modules A and B; see logic chart). These input spatial patterns can be compared to the expression pattern of the *cyIIIa* gene (yellow), the output of the whole regulatory system. Spatial expression is restricted to the aboral ectoderm, with temporal modulation controlled by the prevalence of Runt, P1, CCAAT, and Tef-1 factors (from C. T. Brown and Davidson, unpublished data).

more or less similarly. Therefore the equilibrium constants, which can be computed as the ratio of the "on" to the "off" rate constants, differ for different DNA-protein interactions mainly according to differences in complex stability, i.e., to the off rates.

A very useful computational partition of the probabilities of individual and multiple site occupancies (i.e., the expected statistical distribution of occupied sites), based on these same parameters, was worked out for λ-phage operators by Ackers *et al.* (1982). But in animal cells the situation presents additional complexities: There is a relative ocean of nonspecific DNA (each base pair in the enormous genome can be considered as the beginning of another nonspecific site), compared to a few to perhaps a few hundred specific sites of each kind; typically there are thousands of transcription factor molecules of each active species per nucleus, sometimes tens of thousands. This is many more than there are specific target sites (for example, see factor prevalence data for sea urchin embryos compiled in Bolouri and Davidson, 2002). Most of the factor molecules are bound ephemerally at any given moment to the nonspecific sites (i.e., primarily to the negatively charged phosphate groups of the DNA backbone), and only a minority are bound more stably to their specific sites, where they form various kinds of chemical bonds with the nucleotides of the target site DNA sequence. As pointed out by Emerson *et al.* (1985), the essential parameter in determining site occupancy for animal cells is the relative specific to nonspecific affinity of the factor for its target sites, i.e., the ratio of the equilibrium constants for specific to nonspecific interaction, rather than just the specific equilibrium constant for target site interactions per se. Occupancy probabilities for the multiple target sites in *cis*-regulatory modules of animal cells can be calculated for given driver concentrations, by using such relative equilibrium constants (Bolouri and Davidson, 2002).

The transcriptional activity of the basal transcription apparatus of an active gene depends on occupancy of the relevant *cis*-regulatory target sites. This must of course be true in principle, since the *cis*-regulatory modules control expression. It has been shown directly by kinetic measurements of the output of expression constructs bearing many different mutated sites, in both the *cyIIIa* and *endo16* analyses reviewed above (e.g., Fig. 2.3F). Because these sites affect nothing unless they are occupied, this is in effect to say that for positively acting factors transcription initiation rate is related to the level of site occupancy (Bolouri and Davidson, 2002; that is, once initiation has occurred, the polymerase translocation rate is relatively invariant; reviewed by Davidson, 1986). Standard kinetic treatments for macromolecule synthesis directly relate the transcription initiation rate, in turn, to the levels of primary transcript (nuclear RNA). Since intranuclear processing is usually not rate-limiting (e.g., Cabrera *et al.*, 1984), the transcription initiation rate is also the exit rate of the mRNA to the cytoplasm. Thus the amounts of mRNA and protein product at any given time can be obtained, for given transcript species or populations of transcripts, providing the respective turnover rates are known (absolute measurements, calculations, and earlier applications in several developmental systems are to be found in Davidson, 1986;

for a recent treatment, and simulations, see Bolouri and Davidson, 2002; Ramsey *et al.*, 2005).

What all this means is that explicit and quantitative causal relations link driver input concentration functions to *cis*-regulatory target site occupancy, to transcriptional activity, and to accumulation of the cytoplasmic effector products, at least in principle and on average. We revisit these kinetic linkages in considering the operation of gene regulatory networks (Chapter 4), but for now our focus is on the implications of these ideas for concepts of *cis*-regulatory performance.

Generalization: The Combinatorial *cis*-Regulatory Logic Code

The overall functions of *cis*-regulatory modules can be considered the combinatorial outcome of "unit operations" that take place upon occupancy of their individual target sites. This is an interesting way to think about *cis*-regulatory modules, because it speaks directly to the concept of the genomic regulatory code. As with any code, the significance could be read if we had an appropriate key. In this case, the key that would enable a translation of the genomic *cis*-regulatory code on (computational) inspection of the DNA sequence would provide: first, identification of target sites, that is, of the factors binding to them; second, interpretation of the functions these factors will mediate, individually and/or in combination with their neighbors; and third, rules for combining these functions so as to infer the overall *cis*-regulatory output. For many reasons, the biochemical knowledge requirements of such a code are yet in the gloaming. For example, it has not yet been possible to formulate sufficiently useful, discriminatory, and general target site databases. Furthermore, multiple factors of diverse function that belong to given transcription factor families often bind under different conditions to given similar target sites, as often seen in vertebrate systems. Indeed, given factors sometimes act positively and other times negatively, depending on their combinatorial partners. To cite only a few examples among many possible, the usually positive Dorsal transcription factor collaborates in repression in a *cis*-regulatory module of the *zen* gene of *Drosophila* (Jiang *et al.*, 1993; Valentine *et al.*, 1998); and the mammalian Gata-1 transcription factor functions in either a positive or a repressive fashion depending on its *cis*-regulatory context and its cofactors in diverse erythroid and platelet cell genes (Wang *et al.*, 2002; Letting *et al.*, 2004). The greatest deficiency may be that so far, the decisive importance of particular site and factor combinations has only been sporadically recorded. But the problem of the genomic regulatory code can be approached from the other end, so to speak. We can utilize *cis*-regulatory functional data to ask what sorts of operations are combined within the *cis*-regulatory module, and to sort out the repertoire of the operations that the code contains.

Some of these operations are obvious, such as transcriptional activation and repression, others less so; for instance some of those discussed in the context of *endo16*, such as intermodule linkage and amplification of regulatory inputs (Fig. 2.3). Operations by which different levels of positively acting driver occupancies are communicated to the transcriptional apparatus must in some measure be quantitative and continuous (Bolouri and Davidson, 2002). So are operations such as scalar amplification of driver activation functions (Fig. 2.3D, E). But others of the operations mandated in the *cis*-regulatory code are conveniently treated as Boolean in nature (Istrail and Davidson, 2005). For purposes of representing experimental data, a general class of these is mediated by proteins that may always be present in excess of the levels needed for function. Such factors will execute their function in a given *cis*-regulatory module if it contains sites for them, while if the site is experimentally mutated that function will be absent (Yuh *et al.*, 2001; for examples, see Figs. 2.3, 2.4).

There are also certain unit regulatory operations that are combinatorial in that they are mediated by multiple sites, which normally behave in an intrinsically Boolean manner (Yuh *et al.*, 2001; Istrail and Davidson, 2005). Among these, two are of particular importance for development. The first is the "AND" *cis*-regulatory logic operation. As noted in Chapter 1, this device is often used in spatial *cis*-regulatory integration functions to ensure a positive output only in domains where two different regulatory factors are coincidently expressed. Many examples are touched on in the following section. If both factors are not present at effective levels the output is zero, and the level of either factor is irrelevant if the other is absent; if both are present significantly, then the output is given by the driver levels. A second very important class of developmental operation that is usually combinatorial with respect to target sites, and that may act in a Boolean, switch-like manner, is signal transduction in response to intercellular ligands. Several different such pathways operating through diverse biochemical pathways that terminate in different transcription factors act similarly at the DNA level as "toggle switches" (reviewed by Barolo and Posakony, 2002): If the signal ligand is absent, and the mediating transcription factor is present, then this factor acts as a dominant repressor; otherwise, if the ligand is present, it functions (usually together with a co-activator) to promote transcription. Other kinds of *cis*-regulatory function can be considered as Boolean logic operations as well (Istrail and Davidson, 2005). For example, sequence-specific transcriptional repressors generally behave in a dominant fashion, that is, if they are present they shut the *cis*-regulatory module down, transforming its output to zero regardless of what other inputs are present (examples and references abound in the next section of this Chapter). Or consider DNA looping, mediated by combinations of module target sites, which might bring into play a given *cis*-regulatory module under one developmental circumstance and another in a different developmental circumstance (as in Fig. 1.2). From the standpoint of any one of these modules, this is a Boolean choice of states.

The bottom line is that the *cis*-regulatory code specifies Boolean or discrete as well as continuous operations, all of which are directly implied in the *cis*-regulatory DNA sequence. It is probably true that *cis*-regulatory modules always execute a mix of Boolean logic operations and processing of continuous driver inputs, and their total information processing capacities can be considered the product of the unit functions mediated by the interactions at their individual target sites (Yuh *et al.*, 2001; Istrail and Davidson, 2005). Most of the individual operations the regulatory code specifies will probably turn out to be mediated by diverse transcription factors, and there will clearly be no simple one-to-one correspondence between a given functional operation and a given target site recognizing a given species of factor. In other words, because it is specified in a combinatorial manner, and because in differentiated cells it is also strongly influenced by prior events (cf. Chapter 1), the jungle of biochemical effector functions is richer and more tangled, and harder to generalize, than is the source genomic regulatory code. Thus there is likely to be a modest number of different unit regulatory logic operations, but a far greater number of factors and combinations of factors that execute these operations. Ultimately, functional combinations of target site combinations will become recognized, and it will become possible to formulate canonical *cis*-regulatory "grammars." The code will have become transparent when we know enough to associate each type of operation with the variety of different sites that mediate it, and the diverse biochemical effectors that interact at these sites. However, in the mean time, the "operations repertoire" of the genomic regulatory code is already accessible to us, as shown by the *cis*-regulatory analyses in Figs. 2.3 and 2.4, and others in the next section of this Chapter.

cis-REGULATORY DESIGN

Integration of Spatial Inputs at the *cis*-Regulatory Level

A concept highlighted in Chapter 1 is that a major developmental function of *cis*-regulatory information processing is integration of spatially diverse driver inputs, so that the output pattern of expression is always different from that of any of the individual inputs to the module. This is a main byway for the progressive generation of new spatial patterns in development. Like many other information processing functions of *cis*-regulatory modules, this one is an AND logic operation. In the following we review examples of spatial integration, in life rather than as abstractions, and we see that its always similar logic may be executed by entirely different combinations of spatially expressed factors operating in different biochemical ways. Various mechanisms can account for the requirement for two (or more) factors to be present (at significant levels) at once: among these is a need for cooperative interaction in order to ensure complex stability; or a need for both to provide a platform for interaction with a certain effector cofactor; or a need for synergistic stimulation of different components of the basal transcription apparatus (BTA). Independently of which mechanism is used, novel spatial pattern formation by AND logic provides us with canonical examples of *cis*-regulatory design for developmental function, in that these key developmental functions are encoded directly in the genomic target sites that specify the drivers of the module. As we now shall see, this particular design feature is utilized in all sorts of developmental contexts.

In Fig. 2.5 are our examples, a few among many possible, but particularly clear in their import. Figure 2.5A concerns rhombomere specification in the amniote

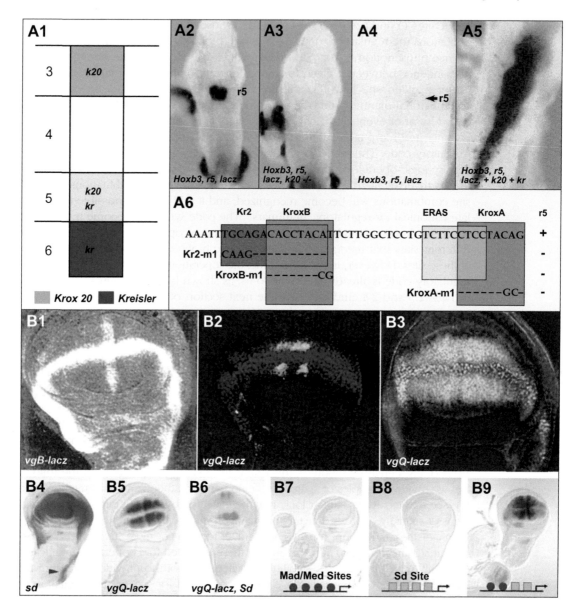

FIGURE 2.5 *cis*-Regulatory integration of diverse spatial inputs. (A) *cis*-Regulatory module of the mouse *hoxb3* gene, integrating Krox20 (K20) and Kreisler (Kr) inputs to produce expression where they overlap in rhombomere 5 (r5). (A1) Diagram showing domains of input expression, K20 in r3 and r5, and Kr in r5 and r6 (overlap in yellow; modified from Barrow *et al.*, 2000). (A2) Expression in r5 of

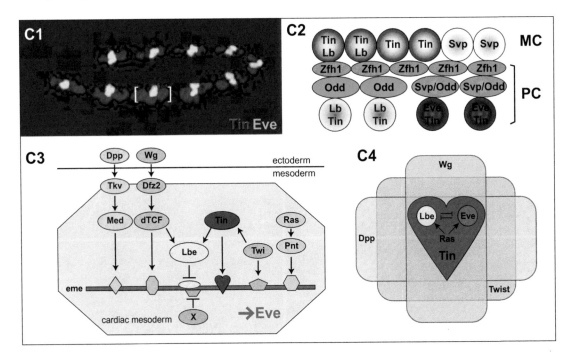

hoxb3 cis-regulatory constructs driving lacz reporter in transgenic mice. (A3) Same, except K20 sites mutated. (A4, A5) Demonstration that K20 and Kr inputs are necessary and sufficient for expression of this *hoxb3* module: The construct was electroporated into a chick embryo neural tube alone (A4), or together with K20 and Kr expression vectors, thus producing widespread ectopic expression (A5); neither of these vectors by itself has any such effect. (A6) Summary of effects of indicated mutations of K20 (KroxA and B sites are both required) and of Kr sites, indicated as expression in r5 (+) or silence of construct (−). ERAS is a partly overlapping site at which binds an Ets-related factor (A2–6, from Manzanares *et al.*, 2002). (B) Integration of spatial inputs by the quadrant enhancer (*VgQ*), a *cis*-regulatory module of the *Drosophila vestigial* (*vg*) gene expressed in the developing wing disc. The relevant inputs consist of the wing disc transcription factor Scalloped (Sd), and of the Mad/Medea factors activated along the anterior/posterior (A/P) boundary of the disc by Dpp signaling. (B1) Expression of *vg* boundary enhancer (*vgB*), visualized by lacz immunofluorescence (from Kim *et al.*, 1996); this enhancer responds to inputs from Dpp signaling along the A/P boundary and from Notch signaling along the dorsal/ventral (D/V) boundary. (B2, B3) Expression of *vgQ* enhancer, lacz in blue, together with endogenous vg gene product in red. (B2) Early third instar, endogenous *vg* pattern mainly due to *vgB* activity; *vgQ* enhancer is initiating activity in center of wing blade (B2 from Kim *et al.*, 1996). (B3) Later pattern; *vgQ* expression now largely fills wing blade, overlapping *vg* expression domain at this stage (B3 from Kim *et al.*, 1997). (B4) Expression domain of *sd*, visualized by β-Gal staining generated by lacz insertion in the endogenous gene (from Halder *et al.*, 1998). (B5) Expression of normal *vgQ-lacz* and (B6) of *vgQ-lacz* lacking all four Sd sites. (B7–B9) Expression of synthetic *cis*-regulatory

(Continues)

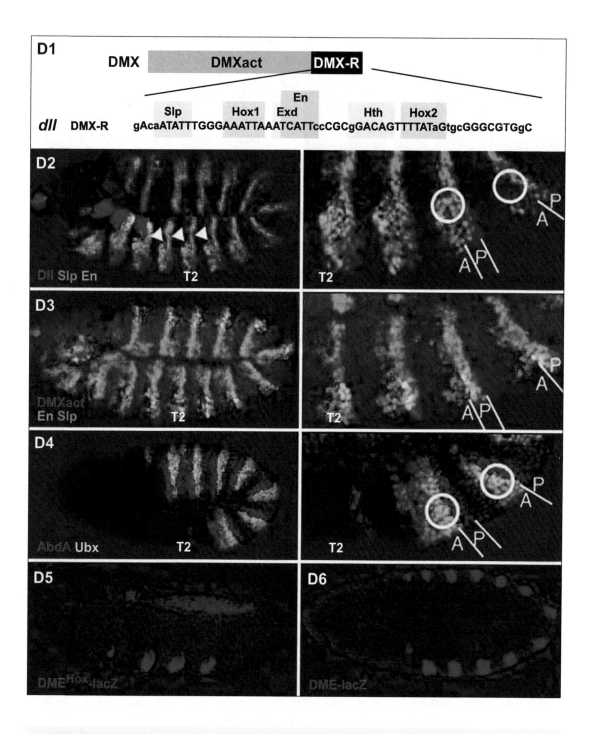

FIGURE 2.5 (continued)

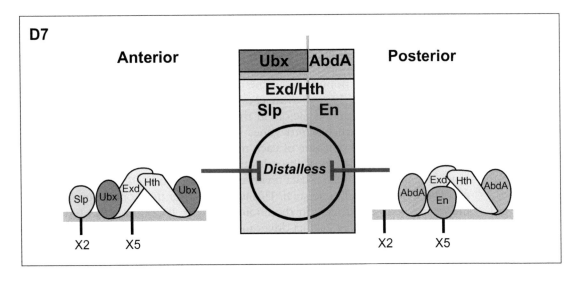

constructs driving lacz reporter: (B7) including four Mad/Medea sites (the transcription factors mediating Dpp signal transduction); (B8) four Sd sites; (B9) two Mad/Medea sites and two Sd sites (B5–9 from Guss *et al.*, 2001). (C) *eve* pericardial cell enhancer of *Drosophila*. This regulatory module integrates Dpp and Wg signaling inputs (expressed, respectively, in an A/P dorsal stripe and in metameric D/V stripes) and other inputs consisting of the heart mesoderm factor Tinman (Tin), the ventral factor Twist (Twi), and the anterior pericardial cell repressor Ladybird (Lb). (C1) Immunofluorescence display of Tin (red) and Eve (green) gene products in cardiac cells: the Eve cells (yellow) are a subset of the Tin cells. (C2) Diagram of eight cardiac cell types (stage 16) distinguished by expression of different sets of transcription factors: MC, myocardial cells; PC, pericardial cells. Other factors are Odd-skipped, Odd; Seven-up, Svp; Zfh-1. (C3) Diagram of inputs into *eve* enhancer. The Ras pathway input is an early signal response required to relieve a prior state of repression. X is an additional unknown repressor. (C4) Schematic indication of spatially noncoincident domains of expression of these individual inputs (C1–4 from Han *et al.*, 2002, incorporating information discovered by Halfon *et al.*, 2000; Knirr and Frasch, 2001, and prior studies referred to therein). (D) An integrating *cis*-regulatory spatial repression device from the *distal-less* (*dll*) gene of *Drosophila*. (D1) The *dll cis*-regulatory system (DMX) responsible for activation at metameric boundaries (DMXact) and for repression of expression in abdominal segments (DMX-R), with relevant target sites indicated in the repression domain: Sloppy-paired, Slp; Ubx and AbdA sites denoted Hox1 and Hox2, respectively; Extradenticle, Exd; Homothorax, Hth; Engrailed, En. (D2–4) Expression domains of relevant factors, as indicated, shown in whole embryos in left panels, and in thoracic (T) segments (T2 and T3) and abdominal segments (A1 and A2) at higher magnification in right panels. (D2) Posterior En and anterior Slp expression, plus Dll expression at the A/P boundaries, only in T segments. (D3) Expression of DMXact, not confined to T segments, plus En and Slp expression. (D4) Ubx and AbdA expression. At this stage Ubx levels are higher in anterior compartments and AbdA levels are higher in posterior compartments. (D5, 6) Abdominal

(Continues)

(here bird and mammal) hindbrain. Rhombomeres are transient metameric units set up along the anterior/posterior axis of the developing hindbrain, each of which is specified distinctly from its neighbors by establishment of a unique, transient regulatory state. From each rhombomere derives particular facial and lower head ganglia, and specific populations of neural crest (Lumsden and Krumlauf, 1996; Rossel and Capecchi, 1999; Barrow *et al.*, 2000; Marin and Charnay, 2000). *Hox* genes are key transcription factors in rhombomere specification, since the correct expression of particular *hox* genes is required for installation of the subsequent developmental outcome of each rhombomere (*op. cit.*). The *cis*-regulatory modules controlling some of the relevant *hox* gene expression domains in the hindbrain are known (earlier data reviewed by Davidson, 2001; Kwan *et al.*, 2001), and Fig. 2.5A deals with one such. This is an enhancer that controls expression of the *hoxb3* gene exclusively in rhombomere 5 (r5). It performs this function by integrating two spatially distinct driver inputs, provided by the Krox20 (K20) and the Kreisler (Kr) transcription factors (Manzanares *et al.*, 2002). The *k20* gene is expressed in r3 and r5, and the *kr* gene in r5 and r6 (Fig. 2.5A1). These patterns overlap in r5, and this information is "read" by means of the integrating function executed by the *hoxb3* enhancer, to produce its r5 specific expression (Fig. 2.5A2). Thus mutation of either K20 or Kr sites in the enhancer destroys its activity, and ectopic expression of both sites produces corresponding ectopic expression of a reporter driven by this enhancer in the hindbrain (Fig. 2.5A3–6).

Figure 2.5B concerns a completely different developmental system, the *Drosophila* wing disc, but exactly the same regulatory logic. Again an enhancer "reads" and responds to the overlap geometry of two disparate inputs, to generate expression only where both are represented. Here the gene is *vestigial* (*vg*), which encodes a nuclear transactivating protein that functions on interaction with the DNA-binding transcription factor Scalloped (Sd; Halder *et al.*, 1998; Vaudin *et al.*, 1999). The *vg* gene is expressed in the region of the wing imaginal disc from which arises the wing blade itself (Williams *et al.*, 1994). Expression of *vg*, required for growth and subsequent development of the wing, is mediated by two complementary *cis*-regulatory modules (Kim *et al.*, 1996, 1997). The "boundary" enhancer responds to signaling cues that mark the orthogonal anterior/posterior and dorsal/ventral boundaries of the future wing (Fig. 2.5B1); and the

FIGURE 2.5 (continued)
repression of *dll* by Ubx: (D5) a portion of DMX driving lacz in T segments only; (D6) same, with site where Ubx binds deleted. Expression now spreads to all abdominal segments. (D7) The DMX-R regulatory device. In the abdomen, where the Hox proteins are expressed, Exd and Hth form a tetrameric complex with the respective Hox proteins, and in anterior compartments this enables binding of the Slp repressor; in the posterior compartments binding of the En repressor. (D1–D4) and (D7) are from Gebelein *et al.*, 2004, where evidence supporting the model in D7 is reported; (D5, 6) are from Gebelein *et al.*, 2002.

"quadrant" enhancer is responsible for *vg* expression throughout the remainder of the blade (Fig. 2.5B2, 3). The quadrant enhancer responds to its own gene product, i.e., to the Vg-Sd complex, as well as to transcription factors (Mad, Medea) mediating response to the Dpp signal ligand. The signal is emitted along the anterior/posterior axis and thereupon diffuses outward across the epithelium. We can see in Fig. 2.5B4 that the domain of expression of the *sd* gene is broader than is the domain of expression of the *vg* quadrant enhancer per se (Fig. 2.5B5), and that the enhancer becomes almost dead if its Sd target sites are destroyed (Fig. 2.5B6). The key demonstration of spatial input integration that is executed by this regulatory module is reproduced in Fig. 2.5B7–9. Here it is shown that, neither synthetic target sites for the factors transducing the Dpp signal, nor for the Sd factor, alone suffice to produce expression, but both together do, and in a pattern similar to that of the natural quadrant enhancer (Guss *et al.*, 2001).

The third and fourth examples of spatial input integration in Fig. 2.5 differ from the first two in that both positive (i.e., activating) and negative (i.e., repressive) inputs are combined within the same *cis*-regulatory module. The *cis*-regulatory module of Fig. 2.5C controls expression of the *eve* gene in a specific set of pericardial cells, and as can be seen in the diagrams of Fig. 2.5C3, 4, this enhancer integrates a number of diverse driver inputs, each of a certain developmental significance (Halfon *et al.*, 2000; Knirr and Frasch, 2001; Han *et al.*, 2002). The mechanism underlying the spatial activation of the respective genes encoding all these different factors is a problem in gene network analysis (see Chapter 4), while our focus here is on the integration of all these inputs within the *eve* pericardial *cis*-regulatory module. There are five primary sources of transcriptional spatial information impinging on this one regulatory system, four acting positively: the Twist factor, expressed in all mesodermal and some other ventrally located cells (see below); the Tinman factor, expressed in dorsal mesodermal cells enabled thereby to become heart progenitor cells; the Wg signaling ligand, expressed in overlying ectoderm in a series of stripes along the anterior/posterior axis (which causes activation of the Tcf transcription factor in the heart cells); and the Dpp signaling ligand, which is expressed in the dorsal ectoderm. As the diagram in Fig. 2.5C4 shows, these inputs are spatially noncoincident, and together they define an overlap domain where the enhancer may be active. But the enhancer also includes sites for a repressor, Ladybird, which is expressed in anterior pericardial cells, as shown in the diagram of Fig. 2.5C2. Therefore the *eve* gene is not expressed in those cells (Han *et al.*, 2002; Jagla *et al.*, 2002). Like those discussed canonically in Chapter 1, this is a conditional processor. Since as a general rule repression is dominant, it is "off" in cells where the Ladybird repressor is present, but "on" where its several activators overlap in expression.

A spatial repression system from another positive–negative integrating control system is portrayed in Fig. 2.5D, which represents a piece of a *Drosophila cis*-regulatory module, the function of which is to direct expression of the *distalless* (*dll*) regulatory gene to the leg imaginal disc primordia. The *dll* gene is activated at the metameric boundaries of the thorax in response to Wg signaling and other information (Cohen *et al.*, 1993; Panganiban *et al.*, 1997), but is repressed in all the abdominal segments (in adult insects, the legs form only in the thorax). Abdominal repression is mediated by *hox* genes expressed there (but not in the thorax at this stage), *Ubx* in each anterior compartment, and *abdA* in each posterior compartment (Fig. 2.5D4; Gebelein *et al.*, 2004, and prior references therein). The diagram (Fig. 2.5D7) shows in both structural and logical terms how this very precise spatial control system may work. The respective Hox proteins form a tetramer with two other DNA binding proteins, with which they are often found associated (Exd, Hth). A dual repression mechanism operates such that in the posterior compartments this tetrameric complex is required to facilitate binding of the Engrailed (Eng) repressor to an adjacent target site; while in the anterior compartments the same role is played by the Slp repressor. Slp and Eng function as repressors by interaction with cofactors that have dedicated transcriptional silencing activity, and as shown in Fig. 2.5D2–4, these factors are expressed, respectively, in the anterior and posterior compartments.

Intertwined in the examples in Fig. 2.5 are the themes of the sequence based genomic regulatory code, and the variety of developmental control jobs executed by integrative *cis*-regulatory logic processors. But is it fair to think of these as fundamentally common examples, or are they rather to be regarded as extreme and uncommonly baroque cases of spatial gene regulation? The answer has become obvious: Integrative *cis*-regulatory design has been shown to be the cause underlying the generation of new spatial patterns of gene expression in the most diverse developmental contexts. Other examples include the *end1* gene in gut lineage specification in *C. elegans* (Calvo *et al.*, 2001; Maduro *et al.*, 2005); the growth hormone gene in mammalian pituitary (Scully and Rosenfeld, 2002); an early acting chicken brain specification gene (*anf*; Spieler *et al.*, 2004); the mammalian *pax5* gene in formation of the midbrain-hindbrain boundary (Pfeffer *et al.*, 2000); and among many additional cases in *Drosophila* (in addition to those considered below), another which clearly illustrates this same spatial control mechanism is the *cis*-regulatory element of the *knirps* gene responsible for its expression specifically in vein 2 of the developing wing disc (Lunde *et al.*, 2003). Most of these examples include both positive and negative functions, such that, as in the cases above, the detailed expression boundaries are established by binding of repressors within the *cis*-regulatory module, which thereupon turn off the gene outside these boundaries.

Repression and the Diversity of *cis*-Regulatory Design

We now consider on an expanded scale the use of different repression functions in a set of *cis*-regulatory modules controlling the same gene, each of distinct design. Our case is again the *evenskipped* gene of *Drosophila*. We saw in Fig. 1.4 the precise pattern generated by the *cis*-regulatory module that is responsible for expression of *eve* stripe 2, a thin circumferential stripe of gene expression a few nuclei wide, located at an exact position along the A/P axis. The minimum required regulatory interactions encoded in the sequence of this module are given in the diagram of Fig. 2.6A1 (Small *et al.*, 1991, 1992; Arnosti *et al.*, 1996). The stripe 2 module activates transcription in response to the positive Bicoid (Bcd) and Hunchback (Hb) regulators (the *bb* gene is activated by Bcd as well). The Bcd protein is distributed in a broad concentration cline that decreases from anterior

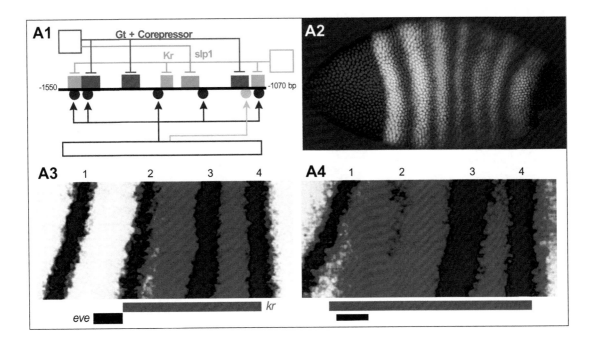

FIGURE 2.6 *cis*-Regulatory repression in the definition of spatial boundaries of pair rule gene expression in *Drosophila*. (A) *cis*-Regulatory control of *evenskipped* stripe 2. (A1) Diagram of "minimal stripe element" (−1070 to −550 of the upstream sequence; see Fig. 1.4). Target sites for key transcription factors are indicated. Repressors are Krüppel (Kr), Giant (Gt), and Sloppy-paired1 (Slp1), indicated as solid rectangles above the line representing the DNA. Activators, Bicoid (Bcd) and Hunchback (Hb), are shown as solid circles below the line. A, anterior; P, posterior (data from references cited in text) (from Small *et al.*, 1992). (A2) Immunofluorescence display of the location of the Hb activator (dark green) and the anterior Bcd domain (blue fading to magenta), with respect to the stripes of Eve protein (orange-red), 14th cleavage. The overlap of Eve and Hb appears yellow; thus the anterior domain of Hb includes Eve stripes 1, 2, and 3 (image kindly provided by J. Reinitz).

(Continues)

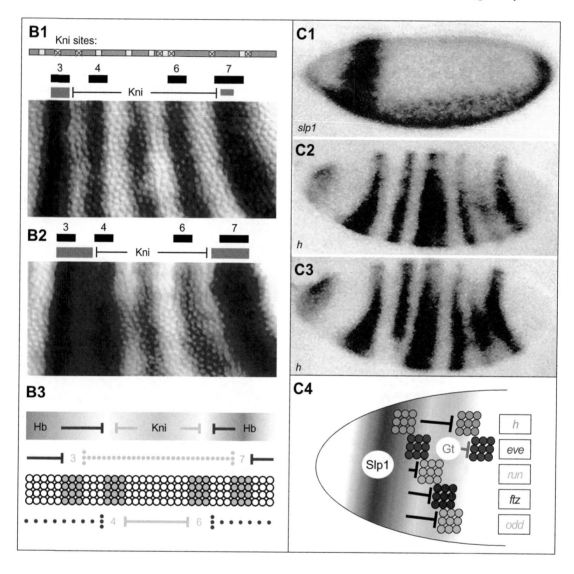

FIGURE 2.6 (continued)

(A3, 4) Establishment of the posterior boundary of stripe 2 by Kr repression. Here *eve* mRNA (black) and *kr* mRNA (red) are displayed by *in situ* hybridization; numerals indicate *eve* stripes. (A3) The Kr domain directly abuts the posterior boundary of *eve* stripe 2 in a normal embryo. (A4) Forced expression of *kr* in the region overlapping *eve* stripe 2 (by placing the *kr* gene under control of the stripe 2 *cis*-regulatory element) almost completely represses transcriptional expression of *eve* stripe 2 (from Wu et al., 1998). (B) Establishment of the posterior boundary of *eve* stripes 3 and 7 by the Kni repressor.

to posterior, and the Hb protein is present in about the anterior 40% of the early embryo (Fig. 2.6A2). As Fig. 2.6A1 shows, there are multiple Bcd target sites in the *eve* stripe 2 module, some of which bind the factor more tightly than others. The Bcd activators binding at these sites work synergistically with one another, and with the Hb factor; thus these sites all contribute, but some are more important than others. Their synergism is required to obtain a strong activation response to relatively low endogenous activator concentrations (Arnosti *et al.*, 1996).

The activating functions of the *eve* stripe 2 enhancer would by themselves suffice for expression extending from far anterior of *eve* stripe 2 to far posterior of it, i.e., throughout almost the whole anterior half of the embryo. The sharply confined boundaries of *eve* stripe 2 therefore depend entirely on repression, mediated by the Krüppel (Kr) and Giant (Gt) target sites shown in Fig. 2.6A1. The posterior boundary is produced as a response of the *cis*-regulatory element to the Kr repressor (Small *et al.*, 1992; Gray and Levine, 1996), and if the Kr expression domain is caused to expand in an anterior direction, so that it overlaps rather

Black bars indicate normal positions of *eve* stripes 3, 4, 6, and 7; tan bars the locations of the domains of expression of stripe 3+7 construct lacking Kni sites. At the top is a diagram of the stripe 3+7 *cis*-regulatory module, showing only the Kni sites. Those marked by red x's were mutated in the experiment in (B2). Orange-brown stain indicates lacz reporter mRNA produced by stripe 3+7 construct; black stain indicates *eve* mRNA. (B1) Normal stripe 3+7 construct; (B2) construct with 6 of 12 Kni sites destroyed. Stripe 3 expands posteriorly, and stripe 7 expands anteriorly. (B3) Diagram summarizing roles of Hb and Kni repression in pair rule gene expression: the Kni sites in the *cis*-regulatory module for stripes 3+7 suffice to determine both the posterior stripe 3 and the anterior stripe 7 boundaries; similarly the Kni sites in the stripe 4+6 enhancer suffice to determine the posterior stripe 4 and the anterior stripe 6 boundaries, while Hb determines the opposing boundaries of these stripes, given the domains of expression of these regulators shown at the top of the diagram (B1–3 from Clyde *et al.*, 2003). (C) Slp1 repressor contributes to anterior boundaries of expression of four different pair rule genes. Slp1 is expressed at declining anterior-to-posterior concentrations, and the *cis*-regulatory modules programming expression of the anterior stripes of these pair rule genes are differentially responsive to Slp1 concentration. The boundaries of no other pair rule gene stripes are directly affected by Slp1 (Andrioli *et al.*, 2004). (C1) Embryo-bearing transgene in which *slp1* coding sequence is driven by the snail *cis*-regulatory module generating expression in ventral mesodermal cells (cf. Fig. 2.7); *slp1* mRNA is displayed by *in situ* hybridization, visualizing both normal *slp1* head stripe and ectopic ventral stripe of *slp1* expression. (C2) Normal stripes generated by endogenous *hairy* (*h*) pair rule gene, visualized by *in situ* hybridization. (C3) Same, except in embryo in which Slp1 is expressed ventrally as in (C1); expression is wiped out ventrally in *h* stripe 2, but not in other anterior *h* stripes (i.e., only in stripes generated by enhancers that can respond to Slp1). (C4) Diagram summarizing Slp1 effects on boundaries of the anterior stripes of *h*, *eve*, *runt* (*run*), *ftz*, and *odd-skipped* (*odd*) genes; immediate boundaries of *eve* stripes 1 and 2 and *h* stripe 1 are not affected (from Andrioli *et al.*, 2004).

than abuts the usual position of *eve* stripe 2, it wipes out all expression of the *eve* stripe 2 element (Fig. 2.6A3, 4). The anterior boundary is positioned by the adjacent location of a band of expression of the Gt repressor, working together with another as yet unknown corepressor which requires Gt binding for function (Small *et al.*, 1992; Arnosti *et al.*, 1996; Wu *et al.*, 1998). The quiescence of the *eve* stripe 2 module further anterior of the stripe boundary itself is produced by interactions with a different repressor, Sloppy-paired 1 (Slp1, the same as we just encountered in repressive control of the *dll* gene; Fig. 2.5D), and a third repression mechanism prevents activity at the anterior tip (Andrioli *et al.*, 2002). Note that the expression of both *gt* and *slp1* genes in turn depend upon activation by Bcd, just as does the *eve* stripe 2 module (Andrioli *et al.*, 2002). Response specifically to these spatially expressed anterior and posterior repressors is the hardwired feature of the *cis*-regulatory design that produces the exquisitely precise stripe 2 pattern, creating a novel spatial regulatory domain, which did not previously exist.

A similar strategy is used to control the borders of the other *eve* stripes, though none have yet been analyzed at the level of *eve* stripe 2, and the activators are as yet in general not identified. A single enhancer controls expression of *eve* stripes 3+7 (Fig. 1.4), and another of stripes 4+6 (Small *et al.*, 1996; Fujioka *et al.*, 1999). The reason this can work is the distribution of the respective repressors in the *Drosophila* embryo, here Knirps (Kni) and Hb. Hb is expressed in broad anterior and posterior peaks on the body axis, and Kni in a single more central band. At high concentrations Kni generates the interior boundaries of stripes 4+6, and at more distal lower concentrations the interior boundaries of stripes 3+7; while at high concentrations Hb forms the exterior boundaries of stripes 3+7 and at more central lower concentrations the exterior boundaries of stripes 4+6 (Fig. 2.6B1–3; Clyde *et al.*, 2003). Details aside, the point here is that the general design strategy of *eve* stripe 2 is applied over again, to different inputs, in the *cis*-regulatory modules that control other *eve* stripes: The design logic is the general feature. The definitive feature of this design logic is the incorporation of target sites for particular repressors, thereby endowing each enhancer with an autonomous capacity to set its own specific boundaries of expression.

Response to the Slp1 repressor also positions the anterior borders of several different pair rule gene expression domains in the future head region of the animal (Andrioli *et al.*, 2004; Fig. 2.6C1–4). These are all genes expressed metamerically in seven stripe patterns as is *eve*, but offset from one another throughout the body plan by one or two cell diameters at the stage considered (early cellular blastoderm). Thus, the Slp1 repressor is presented in a graded anterior-to-posterior concentration cline, and the spatial gene regulatory code of the modules controlling these particular anterior stripes specifies response to different levels of this repressor.

These are conceptually satisfying examples because they take down to the regulatory DNA sequence level mechanisms by which spatial pattern elements are specified. The domains of expression produced by the enhancers of Fig. 2.6 have the properties we see whenever we look carefully at spatial expression domains of endogenous genes during development: they are novel with respect to their inputs, or to any previous divisions of cell fate and spatial gene expression; they are exactly positioned; and they have precise boundaries. It is experimentally clear that *cis*-regulatory information processing is the level at which these properties are generated, and no other. In summing over the outputs of the several different patterns produced by different *cis*-regulatory modules involved in the same process, we also see that they diversely use a common set of relatively imprecise inputs to create a larger set of particular subpatterns: Put another way, they directly encode the creation of increasing biological complexity. The two main design strategies in these examples are their choice of (target sites for) specific activators and repressors, and their use of internal devices that enable response to different concentrations of activators and repressors. The diversity of designs enables many qualitatively different *cis*-regulatory responses to a few graded inputs, a property of particular importance in early development that we now further explore.

cis-Regulatory Design and the Creation of Spatial Complexity in Development

Development always progresses by institution of new regulatory states, the overall complexity of which increases throughout (Chapter 1). But to fully appreciate the one-to-one relationship between the generation of developmental complexity and the genomic *cis*-regulatory design code, we need to look at a whole system of genomic control modules that differentially interpret a prior regulatory state. An excellent case to this end is the set of *cis*-regulatory modules which respond to the gradient of nuclearized Dorsal (Dl) transcription factor in the early *Drosophila* embryo. Their target site design features predict their different domains of expression in the embryo, and the diversity of their expression patterns provides the basis for division of the embryo into what will become the distinct body parts that form at specific locations along the dorso-ventral axis.

In the syncytial *Drosophila* embryo the declining ventral-to-dorsal gradient of nuclearized Dl factor can be regarded as a simple input, in that to a first approximation it has a unique, localized source, and a diffusion-limited form. This gradient results from the activation of the Toll receptor along the ventral surface of the egg, initially triggered by the regional proteolytic activation of an extracellular ligand (Stein and Nüsslein-Volhard, 1992). Diffusion of Toll pathway signaling components ultimately results in the phosphorylation and then degradation of a protein to which the maternal Dl factor is bound in the cytoplasm. It is thereby released for transit into the nuclei, and a graded ventral-to-dorsal concentration cline of nuclear Dl protein is produced (Anderson *et al.*, 1985; Steward, 1989; Steward and Govind, 1993; Rusch and Levine, 1996; Fig. 2.7A1, 2). The most ventral nuclei in the embryo, where the Dl concentration is highest, are specified as mesodermal precursors. In these and adjacent nuclei, *cis*-regulatory modules able to respond strongly only to relatively high Dl levels are activated, among them is one controlling embryonic expression of the gene encoding the Twist activator

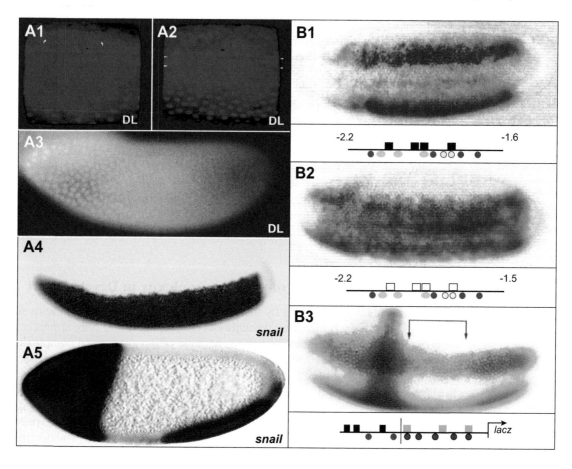

FIGURE 2.7 *cis*-Regulatory design in specification of ventral-to-dorsal developmental domains in *Drosophila*. (A) Gradients of nuclear Dorsal (Dl) transcription factor, displayed by fluorescence immunocytology. (A1, 2) Formation of normal endogenous ventral-to-dorsal gradient of nuclear Dl (the factor is of maternal origin and is initially present throughout the egg cytoplasm): (A1) 12th cycle, Dl nuclearization just beginning; (A2) 13th cycle. Higher nuclear Dl concentrations are ventral. Nuclei lacking Dl appear dark, and solid red stain indicates persisting cytoplasmic Dl (adapted from Rushlow *et al.*, 1989). (A3) Experimentally generated anterior-to-posterior gradient of nuclearized Dl, displayed in a late syncytial-stage embryo as yellow-green immunofluorescence, dorsolateral view. Highest nuclear Dl concentrations are anterior in consequence of forcing ectopic Toll receptor activity at the anterior end (Huang *et al.*, 1997; the endogenous Dl gradient is invisible). (A4, A5) Expression of endogenous *snail* gene, displayed by *in situ* hybridization. (A4) Normal *snail* expression, in ventral (mesodermal) domain. (A5) Expression of *snail* gene both in an ectopic anterior cap and in the normal ventral domain, in an egg in which an anterior-to-posterior gradient of nuclearized Dl has been superimposed on the endogenous

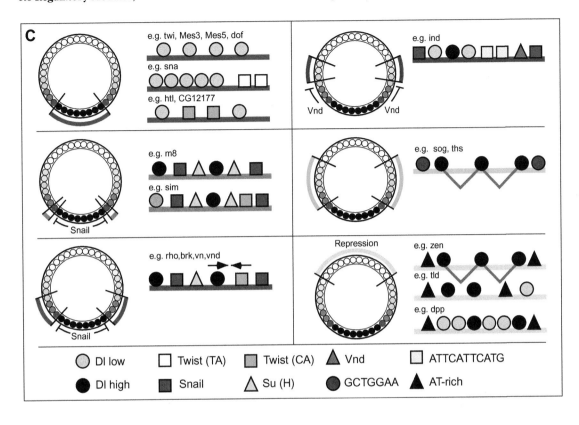

gradient, as in (A3). (A3–5) are from Huang *et al.*, (1997). (B) Expression of *cis*-regulatory constructs from the *rhomboid* (*rho*) gene, *lacz* reporter mRNA visualized by *in situ* hybridization. (B1) Normal expression pattern, similar to that of endogenous *rho* gene, ventral view of an early cellularization-stage embryo. The construct is expressed in two broad lateral stripes directly abutting the ventral mesodermal domain, in which it is silent. Positively-acting factors (shown below the line representing the DNA) are Dorsal (red); bHLH factor (green); Twist (yellow); the Snail repressor (black boxes) is indicated above the line. (B2) Expression of mutated *rho* construct in which all Snail sites are destroyed (open boxes): Ectopic activity now fills in the ventral mesodermal domain. Other experiments show that if Snail sites are replaced at positions no more than 50 bp outside the terminal Dl sites then mesodermal repression is restored (Gray *et al.*, 1994; B1, 2 are from this work). (B3) Autonomous function of *rho* and *eve* stripe 2 modules combined in a single lacz expression construct (map; *eve* factors indicated are Kr repressor, blue; and Bcd activator, black: cf. Fig. 2.6A). The *eve* and *rho* modules operate independently and additively to produce a crossed pattern. Note that the Snail repressor of the *rho* module does not prevent the *eve* stripe 2 module from expressing in the mesoderm, and conversely that the Kr repressor of the *eve* stripe 2 module does not prevent the *rho* module from expressing in the neuroectoderm, although it has a mild weakening effect (bracket):

(Continues)

(Jiang and Levine, 1993). A second key regulatory gene turned on exclusively in what will become the mesodermal domain encodes the Snail repressor, as illustrated in Fig. 2.7A4. The *cis*-regulatory module responsible for the *snail* spatial pattern synergistically uses both Dl and Twist as drivers (Rusch and Levine, 1996); that is, inputs that are themselves, respectively, directly and indirectly dependent on the graded Dl nuclearization process.

We might ask at the outset whether this is really the sole causal pathway by which the mesoderm-specific *snail* expression pattern is to be understood. The answer is shown dramatically in Fig. 2.7A3, 5: Here is reproduced a clever experiment in which the gradient of nuclearized Dl was axially redirected so as to produce an ectopic anterior-to-posterior gradient of nuclear Dl, by expressing Toll receptor RNA at the anterior pole (Huang *et al.*, 1997; Fig. 2.7A3). In these embryos a prominent anterior cap of endogenous *snail* gene expression now appears (Fig. 2.7A5), and indeed the expressing cells are respecified as mesoderm (Huang *et al.*, 1997). The normal ventral *snail* mRNA stripe persists, but we see that there is a gap between the ectopic and the original *snail* expression domains. This betrays

FIGURE 2.7 (continued)

These repressors execute functions confined to their own modules. (B3) is adapted from Gray and Levine (1996). (C) *cis*-Regulatory design features for classes of gene expressed at given ventral-to-dorsal positions in the embryo, putative identities of factors as indicated (diagram provided by A. Stathopoulos and M. Levine, from experimental *cis*-regulatory and computational data of Stathopoulos *et al.*, 2002, 2004; Markstein *et al.*, 2004; and references therein). Six spatial expression outputs are illustrated in order of response to decreasing levels of nuclearized Dl factor. The first pattern, confined to ventral-most regions, is generated in response to highest nuclear Dl levels by use of low-affinity Dl sites and synergism between Twist and Dl (including *snail*, which is mesodermal, and *twist*, expressed in mesoderm and adjacent ectoderm). The second, third, and fourth patterns are generated in various regions of the prospective neuroectoderm, utilizing (among other factors) the Snail repressor for ventral repression and higher affinity Dl sites sensitive to the lower levels of that factor. Among these genes are *singleminded* (*sim*), encoding another transcription factor expressed at the midline of the CNS following mesodermal invagination (second pattern); *rho* (part B of this Figure), representative of the third pattern; *vnd*, encoding another transcriptional regulator expressed in what will be the more lateral regions of the CNS, as the *cis*-regulatory module portrayed is responsive to lower levels of Dl still (fourth pattern). The fifth (e.g., the gene *thisbe*, *ths*, encoding an FGF ligand), and sixth patterns (e.g., the *zen* regulatory gene), are dependent on presence of multiple high-affinity Dl sites that endow responsiveness to the lowest Dl levels, but they differ in that modules controlling the response to the lowest Dl levels include AT-rich sites that bind factors working together with Dl to produce repression in the ventral regions of the egg, rather than activation (see text). Note that dorsal boundaries of patterns 1–5 probably all require repressive interactions, many of which are not yet known: For example, for Snail, see (A5) and text; and the differences between patterns 2, 3, and 4, are probably due to dorsal repressors; a likely case is the factor indicated for *ind* by the yellow box, which acts as a repressor binding at this site.

the existence of a repressor which is apparently also under Dl control, and which must normally delimit the distal boundary of *snail* expression (i.e., what would be the dorsal boundary of the mesoderm in an unperturbed embryo). The main point made by the redirection experiment here is that *cis*-regulatory interpretation of nuclear Dl levels does indeed constitute the necessary and sufficient explanation of the mesodermal spatial expression domain, though this happens in two steps: in the Snail *cis*-regulatory module, and in that controlling expression of the (still unknown) repressor on Snail.

The *snail* expression pattern is used in turn to provide a *cis*-regulatory input to the control modules of other genes that also use Dl as a driver. These are genes that are expressed in lateral stripes just above the ventral mesodermal domain, but not within this domain. In general, their Dl target sites bind the factor more tightly, so they are occupied in locations farther up on the egg where nuclear concentrations of Dl are not as high as those to which the lower affinity sites of the Snail gene respond. An example is the *rhomboid* (*rho*) gene (Fig. 2.7B), which encodes a cell surface component of a signaling pathway, the expression of which is a mark of potential neuroectodermal specification (about one-fourth of this stripe of cells give rise to neuroblasts, the remainder to ectoderm). The expression of *rho*, and (Fig. 2.7B1), of a *rho-lacZ cis*-regulatory expression construct in the lateral neuroectodermal domain, is due not only to higher affinity Dl sites, but also to the inclusion of sites for additional factors that synergize with Dl. The neuroectodermal domain must be exclusive of the mesodermal domain; that is, it must express a different regulatory state, and this is here visualized by the ventral gap in *rho* gene expression. The mechanism by which neuroectodermal gene expression is excluded from the mesodermal domain is encoded at the genomic level by inclusion of multiple sites for the Snail mesodermal repressor. Thus, as shown in Fig. 2.7B2, if the Snail sites of the *rho cis*-regulatory construct are mutated, expression fills in the mesodermal domain. If they are replaced anywhere within 50 bp of the activator sites of the module, correct spatial repression is restored (Gray *et al.*, 1994). This is a dedicated repression design such that the Snail repressor works specifically on the activation functions of the enhancer, as shown by its autonomous performance when combined in the same construct with another enhancer (Fig. 2.7B3; see legend). The Snail repressor works via an effector cofactor (CtBP), which it binds by means of a short motif (PXDLSXK). Several otherwise unrelated transcription factors that also function negatively contain similar motifs and bind the same CtBP corepressor (Nibu *et al.*, 1998; Courey and Jia, 2001).

There are at least 30 Dl-responsive enhancers active in the *Drosophila* embryo. A majority has now been looked at experimentally. They were identified as genes that respond to manipulated levels of ectopic Toll (hence Dl) expression, and by computational methods, i.e., searching for clustered Dl and other sites (Stathopoulos *et al.*, 2002, 2004; Markstein *et al.*, 2004; Papatsenko and Levine, 2005). Correlation of the various *cis*-regulatory designs with the spatial patterns of expression that they direct reveals the genomic regulatory codes by which the set of initial developmental regulatory states is set up from ventral to dorsal in the cellular blastoderm of the embryo. Six different classes of spatial pattern, and of *cis*-regulatory design, are summarized in Fig. 2.7C. Note that they occur in sets, so that given codes underlie given expression domains: the most ventral rely mainly on low affinity Dl sites, and the neuroectodermal modules all utilize the Snail repressor and higher affinity Dl sites plus other factors, probably including repressors that set their different dorsal boundaries. While they all include devices for response to given levels of nuclear Dl, almost all require other activators as well. Some genes expressed in the dorsal-most regions, and not ventrally, such as *zen* and *dpp*, utilize the Dl factor as a part of a ventrally active silencing complex in which adjacently binding proteins are also essential (Jiang *et al.*, 1993; Cai *et al.*, 1996; Valentine *et al.*, 1998; Courey and Jia, 2001). As do many other transcriptional silencing apparatuses, this one operates by means of the Groucho (Gro) cofactor, which is bound through a specific motif (WRPW or a related sequence) found in many transcription factors that function (indirectly) as dominant repressors, or silencers. Silencers may have long-range effects, repressing transcription from a promoter many kilobases away in the genome so as to prevent it from responding to any enhancer whatsoever; hence the term "silencer" (for mode of action of Gro, see Chen and Courey, 2000; Courey and Jia, 2001).

Let us step back for a moment and consider the large and general significance of a genomic *cis*-regulatory code that suffices to generate diverse spatial domains out of simple input patterns. The inputs are here the ventral-to-dorsal distribution of nuclear Dl and the expression domains of the small number of other factors indicated in the diagrams of Fig. 2.7C. Experimental demonstration proves that the target site sequences of the enhancers, in which the code is hardwired, are necessary and sufficient to produce their specific patterns of expression. Out of the primary function of these *cis*-regulatory modules devolves the definition of spatial regions from which the body parts of the embryo respectively develop at different dorso-ventral levels. In respect to information flow, it can be concluded that all other aspects of the mechanisms required to effect the developmental transcriptional states along this axis of the embryo lie downstream of these *cis*-regulatory interpretation functions. While this is the most complete set of functionally related *cis*-regulatory modules yet recovered and compared, the principle that emerges is of course general, and (as the foregoing illustrates over and over) many examples of the exquisite developmental precision afforded by specific *cis*-regulatory designs are available. For instance, striking and detailed evidence is appearing from across the Bilateria of *cis*-regulatory modules that target expression to specific neurons, e.g., in *Drosophila* (McDonald *et al.*, 2003), *C. elegans* (Wenick and Hobert, 2004), and fish (Uemura *et al.*, 2005).

CONTROL BY SEQUENCE-SPECIFIC INTERACTIONS AMONG DISTANT *cis*-REGULATORY ELEMENTS

The control functions built into the modular regulatory code are conditional, as considered briefly above (Chapter 1), not only on the inputs for which the module itself contains target sites, but sometimes also on inputs impinging on external elements with which the module must interact. Here we consider two classes of external interaction which in specific experimental situations have been demonstrated to be functional. The first is interaction of *cis*-regulatory modules with the BTA, and the second is their interaction with other, distant regulatory modules. These external interactions occur via genomic looping, shown to be mediated by specific DNA sequence elements.

Development and BTA Specificity

A large number of diverse proteins (>40 polypeptides) assemble on an active BTA (for brief reviews, Kaiser and Meisterernst, 1996; Malik and Roeder, 2000; Hochheimer and Tjian, 2003; Müller and Tora, 2004). Most general transcription complex components are not variable cell to cell and are encoded by single copy genes. However, there is some diversity in both the DNA and protein constitution of BTA complexes, including different classes of promoter DNA sequence and organization and differences in certain specific protein components that are correlated with particular developmental situations. But at the outset it must be stressed that the diversity in BTA complexes is minor compared to the enormous diversity of the repertoire of developmental *cis*-regulatory functions.

Certain genes are equipped with given kinds of DNA promoter sequence, and their *cis*-regulatory modules are found to work more or less specifically with that rather than with other promoters in the same genomic vicinity ("promoter choice"; reviewed by Smale, 2001). However, this is essentially an aspect of regulatory genomic organization, rather than of conditional developmental information processing. An interesting experiment on this subject was done by Butler and Kadanoga (2001). Alternate promoter forms equipped with GFP reporters were inserted at a number of positions in the *Drosophila* genome, and their expression was monitored. These transgenes expressed in response to whatever *cis*-regulatory modules active during embryogenesis were local to the random sites of insertion. In the large majority of cases the different types of promoter inserted at any given location behaved just the same, but a few instances were observed in which one or the other promoter type was clearly preferred by the nearby enhancers. Much evidence suggests that this result is indicative: Promoter choice is occasionally encountered, but is not the rule (see reviews *op. cit.*). In any case, most of the proteins that constitute the pre-initiation complexes formed in known cases of promoter choice are common, and their presence is certainly not a developmentally important variable.

The most interesting variations in the specific protein components of transcription initiation complexes are those occurring specifically in certain cell types. Here there is occasional causal evidence for developmental significance, though most observations are correlative. At the outset two arguments overshadow this whole area of discussion, independently indicating that such variations can account for only a minor aspect of regulatory diversity. First, in all developing animal systems in which gene transfer experiments are done, it has been common to utilize expression vectors that include a given promoter taken originally from some convenient extraneous source. When associated with diverse *cis*-regulatory modules, these promoters are for the most part found to work promiscuously and reasonably accurately. This in itself proves that, as far as the level of that kind of evidence permits, specificity in the BTA generally contributes little to the variety and specificity of developmental *cis*-regulatory performance. Second, despite extensive genomic and biochemical searches, there have turned up only a few additional forms of promoter-recognizing factors (TRFs, "TATA box-binding protein [TBP]-related factors") and a relatively small variety in the 15 or so known TAFs (TBP associated proteins, Albright and Tjian, 2000; Müller and Tora, 2004). Here the word "relatively" is used with reference to the almost infinite variety in *cis*-regulatory design and function.

That having been said, there are indeed some examples of contributions to developmental specificity in the BTA, and more specific cases are sure to be discovered. Most appear to identify whole classes of developmental function. For example, there is the interesting case of TAF105 of mice, the gene encoding which is expressed strongly in ovaries and testes. Female mice that are homozygous for its absence are sterile, because many follicle cell genes are not expressed (Freiman *et al.*, 2001). Similarly there are five testis-specific TAFs in *Drosophila,* of which the first to be discovered is encoded by the *cannonball* gene. Null mutants for this gene fail to carry out spermatogenesis and fail to express some male differentiation genes (Hiller *et al.*, 2001; Hochheimer and Tjian, 2003). The same defects in spermatocyte transcription occur when the genes encoded by the other four variant TAFs are mutated, and all five may form a spermatocyte transcription complex (Hiller *et al.*, 2004). Curiously, in mice (but not other vertebrates), a special form of TBP, called TRF2, is also required for spermatogenesis (Zhang *et al.*, 2001). Furthermore, in *Drosophila* and some other organisms TRF2 seems to be required for expression of certain large sets of early embryonic genes (for references, see reviews, *op. cit.*). In general it may be said that most evidence to date is accommodated by the generalization that these functional variations in the transcription complexes assembled on the BTA act as class gates, which select for expression sets of gene batteries utilized in gametogenesis and perhaps very early embryogenesis. As such they can be considered external controllers of *cis*-regulatory module functions (Istrail and Davidson, 2005).

Control Functions of Sequence-Specific, Intra- and Intergenic Genomic Looping

The common existence of multiple *cis*-regulatory modules per gene (Figs. 1.2, 2.1, 2.2), located distantly from the basal promoters with which they must alternatively interact, implies that genomic looping is a major aspect of developmental gene regulation. This essential concept was discussed at the outset in Chapter 1. There is concrete evidence for genomic looping, and here we focus on cases that concern specific sites in known genes, and where the regulatory significance of the loops is experimentally clear. Not considered here are the larger scales on which loop-like structures probably also exist, defining giant chromatin domains that include multiple genes, within which conditions diminish or enhance throughout the domain the possibilities of their expression (e.g., see discussion of West *et al.*, 2002).

First, there is clear evidence for loop formation due to specific DNA sequence-recognizing proteins. This is a kind of watershed in the discussion, for it shows that looping architecture and its consequences are among functions programmed in the genomic regulatory code. Looping can occur by binding of a protein to target sites located apart, contact between which is required for function (for instance a *cis*-regulatory module and the basal promoter), if once bound the protein then multimerizes homotypically. Or, less easy to analyze, loops could also be formed by heterotypic interaction between different DNA-binding proteins or their cofactors. Direct visual and experimental evidence for homotypic loop formation at meaningful *cis*-regulatory sites is shown in Fig. 2.8A. This example concerns the same control system as is the subject of Fig. 2.4, that of the sea urchin *cyIIIa* gene. Within the 2.3 kb *cis*-regulatory system of this gene are multiple occurrences of sites for a factor called Gcf1 (Fig. 2.8A1). Its potential looping function was indicated by its ability to form higher order multimers when added to target site oligonucleotides *in vitro* (Zeller *et al.*, 1995). Figures 2.8A2–4 illustrate alternative loops formed as Gcf1 is brought into the presence of *cyIIIa cis*-regulatory DNA; those shown bring different components of the system into the immediate proximity of one another. The common mammalian transcription factor Sp1 behaves very similarly (Su *et al.*, 1991; Mastrangelo *et al.*, 1991). Sequence-specific looping functions are at least in some cases thus encoded in *cis*-regulatory DNA.

A sequence-specific genomic looping function is illustrated in Fig. 2.8B (Calhoun *et al.*, 2002; Calhoun and Levine, 2003). A map of the *Sex combs reduced* (*Scr*)–*Antennapedia* (*Antp*) region of the *Drosophila* genome is shown in Fig. 2.8B1. The key points for our purposes are that expression of the *Scr* gene in the head (Fig. 2.8B2) is driven by the "T1" *cis*-regulatory module, about 20 kb away from the start point of transcription; while the *ftz* gene, which lies between T1 and the *Scr* gene, and has a completely different pattern of expression, is controlled by its own *cis*-regulatory module, "AE1." The 7-stripe pattern of *ftz* expression generated by the AE1 enhancer is shown in Fig. 2.8B3. The action of T1 on the *Scr* BTA rather than on the closer *ftz* BTA is not due to promoter choice, but rather to a cluster of specific 6 bp sequences occurring just upstream of the *Scr* gene. This acts to "tether" the T1 enhancer to the *Scr* proximal region; that is, to cause a loop to form that brings the enhancer into proximity to *Scr*. This interpretation is demonstrated by experiments in which synthetic multimers of the 6 bp sequence were placed just upstream of a *ftz* BTA-*lacz* reporter: When introduced into flies a distantly located T1 enhancer now activates the *ftz* BTA in the same head stripe in which *Scr* is normally expressed (Fig. 2.8B4, 5). Multimers of the 6 bp element are necessary, and it has no enhancer activity on its own; nor is it needed if the T1 enhancer is placed right next to the *ftz* BTA. These and additional observations separate out the looping communication function, from other *cis*-regulatory

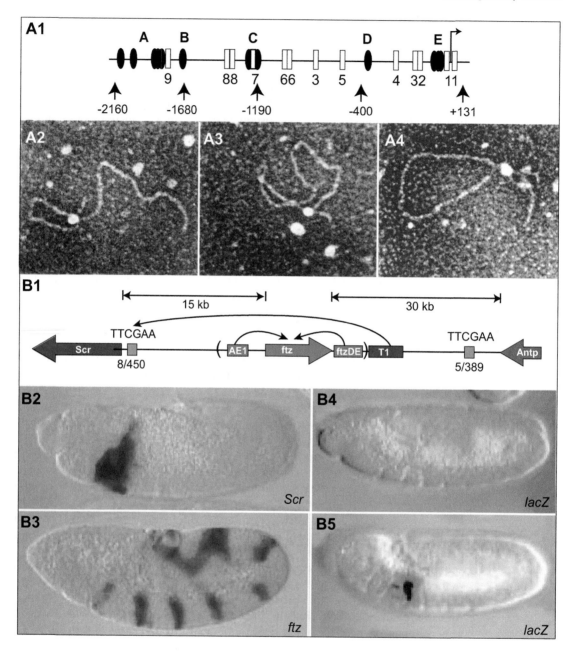

FIGURE 2.8 External interactions of *cis*-regulatory modules by genomic looping. (A) Loop formation by a purified transcription factor interacting with *cis*-regulatory DNA that contains multiple sites for

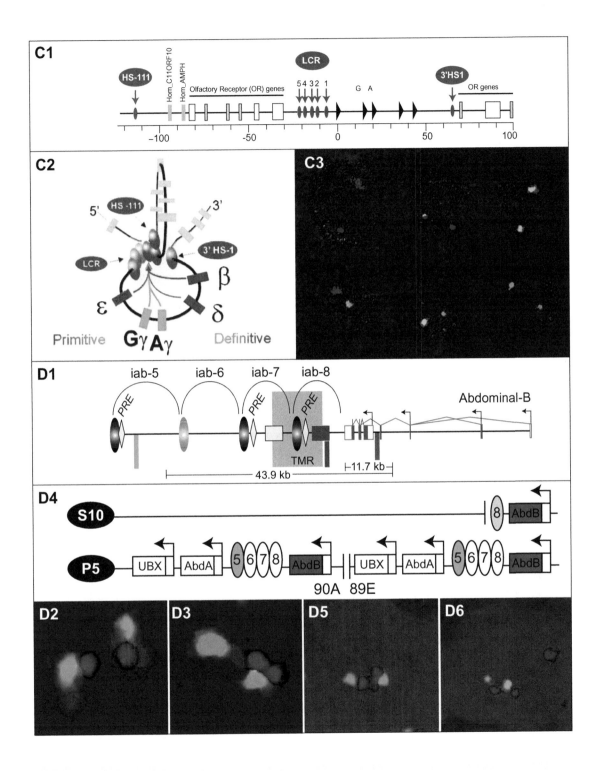

FIGURE 2.8 (continued)

(Continues)

FIGURE 2.8 (continued)

this factor (from Zeller *et al.*, 1995). The factor, Gcf1 of *Strongylocentrotus*, binds to itself and multi-merizes. (A1) Map of *cyIIIa cis*-regulatory domain (cf. Fig. 2.4), indicating in black the Gcf1 sites. (A2–A4) Alternative loops formed when protein and DNA are mixed *in vitro*, visualized by electron microscopy: (A2) site A-site C loop; (A3) site A-site E loop; and (A4) site B-site E loop. (B) Specific "tethering" sequence required for distant enhancer-promoter interaction in the *Sex combs reduced* (*Scr*) gene of *Drosophila*. (B1) Map of locus, displaying relevant parts of *Scr hox* gene in red, and of *ftz* gene in blue: TTCGAA, tethering sequence elements (occurrences/bp); AE1, 7-stripe *ftz* enhancer; ftzDE, another *ftz* enhancer; T1, enhancer of *Scr* responsible for embryonic head expression pattern; *Antp, Antennapedia hox* gene (from Calhoun and Levine, 2003). (B2, 3) Visualization of endogenous gene expression by *in situ* hybridization (from Calhoun *et al.*, 2002): (B2) *Scr* expression under T1 control; (B3) *ftz* expression, under AE1 control. (B4, B5) Expression of synthetic lacz constructs in head region under T1 control, visualized by *in situ* hybridization. Here the promoter is from the *ftz* gene, separated from T1 by a 1.6 kb spacer. (B4) No expression in the absence of, but (B5), head stripe expression in the presence of a 6x multimer of the tethering sequence (from Calhoun and Levine, 2003). Multimers are necessary for this expression, and the 6 bp tethering sequence has no enhancer activity on its own; nor is it needed if the T1 enhancer is placed right next to the *ftz* BTA. (C) Genomic looping in the mammalian globin locus. (C1) Map of 190 kb human globin locus; LCR (locus control region) and HS111 and 3′ HS1 (DNAse1 hypersensitive sites) are distant modules that themselves bind transcription factors, and that are required to be in the active state for developmental activation of individual globin genes. These are the ε, G-γ, A-γ, δ, and β genes. (C2) "chromatin hub" (CH) required for function, formed by looping of genome so as to generate a complex of proteins bound to LCR, HS111, and 3′ HS1; the individual genes are activated as their own *cis*-regulatory modules and are brought into required contact with the CH by gene-specific interactions (arrows). (C1, 2 from Patrinos *et al.*, 2004). (C3) Demonstration of dynamic interaction (i.e., looping) between CH and regulatory components of genes, resulting in the transcriptional activation of individual globin genes. Gene expression is visualized by fluorescence *in situ* hybridization with gene-specific intron probes: red, γ genes; green, β gene; yellow, both. Individual mouse cell nuclei are seen, expressing either or both genes from the two allelic globin clusters. Temporal analysis shows that the CH-gene associations are formed anew every 15–30 min (C3 is from Wijgerde *et al.*, 1995). (D) Intra- and inter-chromosome looping required for activation of *abdB hox* gene in *Drosophila* (from Ronshaugen and Levine, 2005). (D1) Map of region 3′ of *abdB* gene (red), indicating regulatory domains that contain specific enhancers (iab-5,-6,-7,-8). These enhancer regions are themselves transcribed (Bae *et al.*, 2002). Polycomb response elements required for stability of expression patterns (PRE) and insulator elements (dark ovals) are shown. TMR is a 10 kb region required for interchromosomal interaction (Ronshaugen and Levine, 2005). (D2, 3) Fluorescent *in situ* hybridizations identifying lab-5 transcript (green), lab-8 transcript (blue), and *abdB* promoter region (red) in individual abdominal cell nuclei of late embryos. (D2) Loop formation is indicated in the left chromosome by apposition of promoter (red) and distant iab-5 region (green), while the proximal iab-8 region is off to one side. (D3) Chromosome on top is in extended unaltered conformation (green, blue, red), while the other is folded or condensed. (D4) Map of heterozygous genotype in which one chromosome (S10) contains a deletion of all downstream sequences in the *abdB* locus, retaining only the promoter proximal region, while the other (P5) contains two normal copies of the *Ubx-abdB* region (i.e., the Bithorax Complex).

functions of T1 and the Scr proximal region, by demonstrating that the specific tethering sequence element is necessary and sufficient to mediate it. This is likely a heterotypic looping function, since the 6 bp element is not present within the T1 enhancer.

A prominent feature of *cis*-regulatory systems in vertebrate genomes is control of whole clusters of related genes that are linked in the genome, often at distances tens of kilobases from one another. Each gene in such clusters has its own *cis*-regulatory apparatus as well as basal promoter, but in addition there is a distant "locus control region" (often termed the LCR), which must be in an active state for any of the individual genes to be expressed. LCRs are *cis*-regulatory modules of a specific kind. They have the following properties: they are distantly located from at least some of the genes of the cluster; their state of activity depends initially on binding of the sequence-specific transcription factors for which they have target sites; and once activated in this way, they alter chromatin structure within the cluster so as to facilitate activity of the genes therein. Operationally, they are required to be in an active state in order for the individual genes to be activated, the mechanism of which almost certainly requires looping so as to form proximal contacts between the LCR and proximal elements of these genes. That is to say, the function of LCRs essentially depends on intermodule communication, most likely via genomic looping. The canonical and most intensively studied case is the globin gene cluster (Fig. 2.8C). A summary of the 200 kb globin locus in humans, and its developmentally significant structural organization, is shown in Fig. 2.8C1, 2, as confirmed by the consequences of genomic deletion of all the major working parts of the structure (Patrinos *et al.*, 2004, and references therein). The LCR is upstream of the five-gene cluster (ε-β), each member of which has its own expression profile during development (for review, Stamotoyannopoulos and Grosveld, 2001), but in addition there are two other distant elements required to form the potentially active structure. This is termed the "chromatin hub," represented in Fig. 2.8C2. The formation of the active structure requires erythroid-specific transcription factors (e.g., EKLF1; Drissen *et al.*, 2004), and precedes actual globin gene expression. When this does occur, a dynamic process of loop formation between each gene and the chromatin hub takes place, as is visualized dramatically in Fig. 2.8C3. Here we see evidence that alternative globin genes are activated sometimes in the same erythroid cells, as detected by gene-specific intron probes.

(D5, 6) Visualization of productive interchromosomal looping. S10/P5 nuclei were hybridized with lab-5 (green) and *AbdB* promoter (red) probes. (D5) Unpaired chromosomes: the P5 chromosome has the expected two promoter and two lab-5 signals, and the S10 chromosome only the one promoter signal. (D6) The transcriptionally active promoter region of the S10 chromosome is physically associated with one of the two lab-5 enhancers of the P5 chromosome.

There are many other mammalian gene clusters that are likely controlled by similar kinds of LCR-like mechanisms, among which might be mentioned the genes at the "posterior" end of the *hoxD* cluster (Spitz *et al.*, 2003), the six genes of the *distal-less* cluster (Sumiyama *et al.*, 2002), and a cluster of genes required for limb morphogenesis (Zuniga *et al.*, 2004). The main consequence for this discussion is the strong conclusion that long-range regulatory function requires regionally-specific, and therefore sequence-specific, interaction between various kinds of *cis*-regulatory modules. The functions of LCR-like modules are determined by the factors they bind, of course, and essential to these functions is specification of functional architecture: This is likely to occur by mobilization of looping proteins.

Until tethering sequence elements, looping factors, and extensive mutational evidence are in hand, as in the above examples, it is difficult to exclude absolutely the alternative that there is some form of "action at a distance" that travels along the chromatin and that could account for intermodule interactions. But such arguments are null when the functional interaction occurs between an enhancer on one DNA strand and the BTA of a gene on another. This can only be understood as the result of an interaction that would usually function to generate an intrachromosomal loop, but that can also cause formation of a loop between the enhancer and the promoter of the corresponding allele on another chromosome. Such is the case illustrated in Fig. 2.8D (Ronshaugen and Levine, 2005). As described in the legend, the key observation is that a defective chromosome containing only a promoter of the *abdominalB hox* gene is able to associate with a distant enhancer on a different chromosome; the evidence is provided by direct visualization using *in situ* hybridization. In principle, there is no reason that distant genomic targets of looping proteins should have to be on the same rather than different chromosomes so long as they can find each other, and interchromosome regulatory interaction events may be less uncommon than usually assumed.

These examples of sequence-specific intermodule communication process are all functional in import. The code for their execution is part and parcel of the genomic regulatory apparatus. From the standpoint of the individual *cis*-regulatory modules that they affect, they are essentially permissive, or nonpermissive, events mediated in part by structural elements external to these modules (Istrail and Davidson, 2005). While *cis*-regulatory modules are indeed the autonomous source of unique patterns of developmental gene expression, it is only through external communication functions that they are linked into the overall apparatus that controls life.

The "unit of the developmental regulatory code," the *cis*-regulatory module, is considered in this Chapter in structural and functional terms, in diverse contexts of various aspects of animal development. This has been a DNA sequence-level view, focused on the genomic regulatory logic where it lives, so to speak. We now remove to a more system scale view of developmental processes, a necessary prelude to analysis of the genomic regulatory code at the gene regulatory network level.

CHAPTER 3

Development as a Process of Regulatory State Specification

DEVELOPMENTAL PROCESSES OF THE PREGASTRULAR EMBRYO

Regulatory state is the lowest organizational level of the genomic control system that provides overall causal explanation of what we perceive as the process of development. We could go to more complex levels, and for example, consider an embryo in terms of what all the proteins that it is utilizing are doing in each of its spatial domains. This would indeed be a far higher level of organization, because the number of interrelationships that would need to be considered is exponentially related to protein diversity, due to the multiple interactions in which proteins engage, particularly those building three-dimensional structures, or serving as components of cellular "machines." However, to a first approximation, once the proteins appear in a given domain of the animal, they are going to interact due to their intrinsic properties, so the basic issue is getting them there. This depends directly on differential expression of the genes encoding all these proteins, and that in turn depends on the regulatory state in each nucleus, at each point in developmental time. "Regulatory state," as we recall, is the sum, total set of expressed and active transcription factors presented in any given nucleus. But we cannot go to a lower level of the control apparatus than regulatory state and still learn how development works. We might, for example, figure out how a *cis*-regulatory module controlling a particular gene operates, and thus generate an illuminating new understanding of some aspect of the genomic regulatory code. Many instances were reviewed in the last Chapter. What we thus learn of the developmental process overall is only by inference, however, based on a tiny example: No single gene, however necessary to a developmental process, produces an embryo or a territory, or a state of differentiation, or a regulatory state.

Furthermore, regulatory state is the direct output of the gene regulatory networks encoded in the *cis*-regulatory DNA controlling transcription factor gene expression. To understand the *cis*-regulatory inputs that are required for the conditional function of these networks in developmental time and space is to understand why the animal develops as it does, in terms that can be related directly to the genomic regulatory code. Our object in this Chapter is to reduce canonical, widely occurring, developmental phenomena to inputs into the mechanisms of regulatory state specification.

Following are some developmental processes almost always observed in early to mid-stage embryos and the manner in which they impact the first regulatory state specifications. In the terms of this discussion, the initial staging point in animal development is the institution of diverse zygotic regulatory states in different cell nuclei. The term "zygotic" denotes the requirement that these states be generated by new transcription of regulatory genes, occurring in spatial domains of the egg, which will ultimately give rise to different parts of the embryo.

Vectorial Embryogenesis

The embryonic process is entirely unidirectional and internally controlled. It does not change direction, oscillate, or alter course depending on environmental factors; as for example, do many biochemical processes in prokaryotes and yeast, or many physiological responses in animals. Among the characters shared by all Bilateria are some very general aspects of embryogenesis. As was already noted in the 19th century for many bilaterian groups, this process traverses a series of more or less similar morphological stages that can be recognized externally.

As do non-bilaterian metazoan groups (sponges, ctenophores, and cnidarians), the bilaterians begin life with the union of an extremely unequal pair of gametes. The sperm contains little more than the packaged, quiescent male genome, plus apparatus to effect its approach to and fusion with the egg, the injection of the genome into the egg, and local metabolic activation. The egg, on the other hand, is an enormous cell, typically equal in cytoplasmic volume to between a few hundred and several thousand cells of the later embryo. It contains vast stores of proteins and preassembled organelles and components of membranes and nuclei, which are soon divided up among the embryonic cells to form their working constituents, so that these cells can hit the ground running. Most important for what follows is that eggs always also contain proportionately large amounts of mRNA transcribed during the process of oogenesis (maternal mRNA), as well as the maternal genome itself, and a complex store of maternally translated transcription factors plus the intracellular components of signal transduction systems. In other words, the egg contains a logistically complete regulatory apparatus that is awaiting differential spatial instructions. Only recently, as taken up in the following, has it become clear that some of the gene regulatory components of eggs are distributed in particular spatial domains of these huge cells. Once the triggers have been pulled by processes set in train during fertilization, these regulatory components serve as crucial initial cues that cause the nuclei of the new embryo to function differently from one another.

By about 1910 it was generally recognized that the first and most important thing that happens following fertilization is a continuing series of divisions of the now diploid egg nucleus, so that replicates of it containing complete copies of the genome are distributed throughout the egg cytoplasm (reviewed by Wilson, 1896, 1925). In bilaterian embryos this process (cleavage) typically proceeds until the nucleus to cytoplasm ratio approaches that of the more ordinary cells of later tissues, when the rate of nuclear replication slows down. The ultimate significance of the cleavage process is that it distributes active or soon to be active transcriptional expression factories throughout the egg cytoplasm. There follows a morphologically heterogeneous stage (blastula), which in some kinds of embryo is a hollow sphere, the wall of which is composed of closely apposed cuboidal cells (e.g., among the model embryological examples considered below, the sea urchin blastula); while in others it is made of layers of cells either lying upon (as in birds or cephalopods), or encompassing (as in fish or *Drosophila*), a mass of yolk. In others the blastula is a solid, closely packed assemblage of cells (as in nematode or ascidian embryos). But these differences in morphology are basically irrelevant: The blastula is the stage that ends, before any morphological aspects of the phylum-specific body plan have formed, with many or most of the cells engaged in spatially specified, differential regulatory gene expression. This fundamental and general fact became obvious only when molecular probes and methods of *in situ* hybridization became available.

There follows the morphological reorganization of the embryo, a process which always involves the expression of motility genes, the products of which permit cells to migrate with respect to one another. The outcome is the physical separation of cell populations that will form endodermal, mesodermal, and neural and ectodermal derivatives. Because of the prominence in this process of gut formation, it was early on named the "gastrula" stage. In many bilaterians other immediate results of gastrulation include the positioning of head forming cell populations, of structures composed of mesodermal cell layers, and of cellular precursors of the axial central nervous system. An alternative mode, as discussed below, is an indirect process of development in which the embryo forms only a simplified larva and puts off the business of generating complex organs such as its definitive nervous system until later. Either immediately following gastrulation, or in indirect development following establishment of the larva, the next stage is the generation of adult body parts. Some body parts are definitive for the adult forms of all Bilateria (e.g., the anterior-posterior organization and the through gut with mouth and anus), other body parts are characteristic of the bilaterian phylum to which the animal belongs (e.g., the dorsal central nervous system in chordates). It is body part formation that is by far the most demanding and complex of developmental phenomena. We shall take the view that there is an underlying, common kind of regulatory mechanism used in postgastrular development by all bilaterians to form their adult body parts.

So embryonic development is indeed a vectorial series of canonical morphological phenomena, but these are just the visible outcome of an underlying vectorial series of canonical regulatory state transformations. The deconstruction of embryonic phenomenology requires that we think about the means by which the regulatory state transformations of the embryo are programmed and executed, and this is the object of the following treatment.

Maternal Anisotropy and External Cues for Axial Polarity

In almost all animal groups there are two kinds of mechanism used to start the process by which given parts of the egg become given parts of the embryo. First, there are internal anisotropies built into the egg during oogenesis that result in asymmetric distribution of various macromolecule(s) attached to the egg cytoarchitecture, as in the diagram of Fig. 3.1A. Second, there are cues that result from the advent of the sperm in fertilization, or other localized stimuli that originate externally to the egg surface. Considered in detail, the early spatial cues are often qualitatively peculiar to each animal clade, at least partly because these details depend on the adaptive lifestyle of the species, which presents a variable set of conditions and opportunities that can be used to mark given regions of the egg as distinct from the remainder.

For example, in *Drosophila*, the egg forms within a highly structured, differentially functioning complex of accessory cells (nurse cells and follicle cells) distinct from anything found in the other species mentioned in the following. The anterior/posterior (A/P) axis of the *Drosophila* embryo depends mechanistically on localization of maternal mRNAs, most importantly one encoding the transcription factor Bicoid (Bcd) at the anterior end (Fig. 3.1B). This maternal mRNA is pumped into the growing oocyte after being synthesized in the adjacent nurse cells, and results in an anterior concentration of the Bcd transcription factor as development begins (illustrated in Fig. 2.6A). Other maternally encoded proteins affect the posttranscriptional survival of maternal mRNAs at the posterior end,

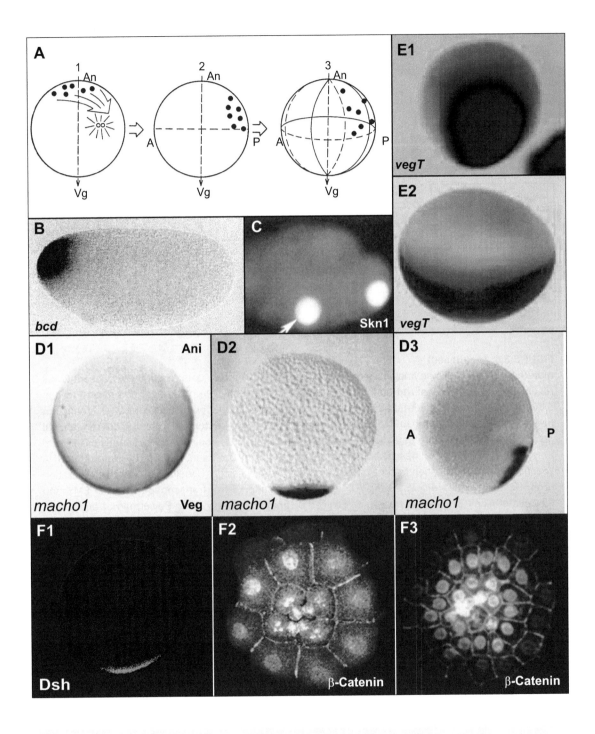

FIGURE 3.1 Anisotropies in animal eggs and the differential establishment of early regulatory states. (A) Autonomous specification of cell lineage by inheritance of an anisotropically distributed

(Continues)

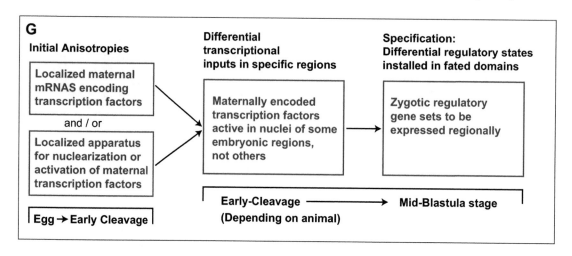

FIGURE 3.1 (continued)

maternal mRNA. This is an imaginary case in which the mRNA (black dots; encoding a transcriptional regulatory protein) is localized during oogenesis at the vegetal pole of an egg, and moved to one side as a result of cytoskeletal asymmetries occasioned by formation of the aster after pronuclear fusion. At third cleavage the mRNA is inherited by a vegetal (Vg) posterior (P) blastomere (An, animal pole). The regulatory state established in the descendants of this cell will consequently differ from that of other blastomere lineages (diagram reproduced from Davidson, 1990). (B) Maternal mRNA encoding Bicoid transcription factor at anterior end of *Drosophila* egg, visualized by *in situ* hybridization (from Ochoa-Espinosa *et al.*, 2005). (C) Maternally encoded Skn1 transcription factor in EMS and P2 cells of 4-cell *C. elegans* embryo, visualized by immunocytology (from Bowerman *et al.,* 1993). (D) Anisotropic localization of maternal mRNA encoding Macho1 transcription factor in oocyte (D1) and in early (D2) and later (D3) first cleavage cycle zygote of the ascidian *Halocynthia*, visualized by *in situ* hybridization (compare A; from Nishida, 2002a). (E) Vegetal localization of maternal mRNA encoding VegT transcription factor in mid-stage oocyte (E1; vegetal view) and fertilized egg (E2; lateral view) of *Xenopus* (from Zhang and King, 1996). (F) β-catenin nuclearization in sea urchin embryos (F1) lateral view, sharply vegetal localization of Dishevelled protein in vegetal cortex of sea urchin egg, visualized by immunocytology. This protein is required for release of β-catenin from its docking station inside the cell membrane (image kindly provided by C. Ettensohn). (F2) and (F3) Vegetal views, nuclear β-catenin in 32- and 60-cell stage *Lytechinus variegatus* embryos, respectively, visualized by immunocytology (from Logan *et al.,* 1999). β-catenin nuclearization occurs in every cell, the descendants of which participate in formation of mesodermal (including skeletogenic) and endodermal territories. (G) Diagrammatic summary of process by which initial anisotropies lead to expression of spatially differential zygotic regulatory states in early embryos.

and many additional asymmetries along this axis are established during oogenesis (Gavis and Lehman, 1992; St. Johnston and Nüsslein-Volhard, 1992; Roth and Schüpbach, 1994). The dorsal/ventral (D/V) axis of the *Drosophila* egg is set up by entirely different means. It depends on a train of extra-embryonic enzymatic reactions triggered by ventrally located follicle cells that eventually produce a localized external ligand (Stein and Nüsslein-Volhard, 1992). Interaction of surface receptors with this ligand ultimately causes the ventral-to-dorsal cline of maternal Dorsal transcription factor within the egg that was discussed in Chapter 2 (see Fig. 2.7). Both the A/P and D/V polarity generating mechanisms are in detail unique stratagems not known in other animal clades. Yet in the abstract sense of their regulatory outcome, all bilaterian eggs have to face and solve the same basic problem that these mechanisms solve: installation of asymmetric transcriptional regulatory states in the egg that define the future morphological coordinates of the embryo.

In *C. elegans*, the A/P axis is specified by asymmetries that very early in cleavage result in the presence in posterior blastomeres of maternally encoded transcription factors that are absent from anterior ones, e.g., the Skn1 (Fig. 3.1C) and Pal1 transcription factors. The D/V axis is established in consequence of cytoskeletal reorganization following sperm entry (Schnabel and Preiss, 1997; Bowerman, 1998; Newman-Smith and Rothman, 1998). In ascidian eggs the animal-vegetal axis is set up during oogenesis, and at fertilization at least 17 different maternal mRNAs are localized at the vegetal pole (Nakamura *et al.,* 2003; Yamada *et al.,* 2005). The one illustrated in Fig. 3.1D encodes a transcription factor (Macho1) that is required for muscle cell specification. Certain of these localized maternal mRNAs are associated with elements of the cytoskeleton. After fertilization these are translocated to the future posterior pole, due to cytoplasmic reorganization in the process of pronuclear fusion and to movements of the cleavage asters (Nishida and Sawada, 2001; Sardet *et al.,* 2003; for review, Nishida, 2005), just as in the imaginary diagram of Fig. 3.1A. In *Xenopus*, animal-vegetal polarity depends on storage at the vegetal pole of a maternal mRNA encoding the transcription factor VegT, which is localized to that pole during growth of the oocyte (Fig. 3.1E; Stennard *et al.,* 1996; Zhang and King, 1996; Heasman *et al.,* 2001; Xanthos *et al.,* 2001), and on many additional vegetal-specific cytoarchitectural features. These anisotropies are later used to install transcriptional activities in certain blastomeres that are not expressed in other blastomeres. The D/V axis of the *Xenopus* egg depends on transport of some originally vegetal components to the future dorsal side along oriented microtubules, following a cytoskeletal reorganization that originates in the sperm pronucleus, and that is facilitated by a cortical rotation toward the future dorsal side (Gerhart *et al.,* 1983; Weaver and Kimelman, 2004). In sea urchin eggs, the endoderm and mesoderm form from the vegetal cells of the early embryo, the nuclei of which uniquely receive a localized β-catenin/Tcf transcriptional input (Wikramanayake *et al.,* 1998; Logan *et al.,* 1999). β-catenin localization is a widespread mechanism for installing polarity in eggs, reported, for example, in ascidians (Imai *et al.,* 2000), *Xenopus* (Heasman *et al.,* 1994; Miller *et al.,* 1999), and even cnidarians (Wikramanayake *et al.,* 2003), though it is used for diverse kinds of tissue specification in these forms. In sea urchin eggs, the β-catenin/Tcf input reflects the primordial animal-vegetal polarity of the egg, a polarity indicated by the striking vegetal localization of Dishevelled illustrated in Fig. 3F; this protein is required for the vegetal concentration of nuclear β-catenin (Weitzel *et al.*, 2004). The orthogonal oral/aboral polarity of the sea urchin egg is set up quite differently. It is probably due to localized accumulation of mitochondria on one side, producing a redox gradient, which later differentially affects the activity of maternal transcription factors (Coffman and Davidson, 2001; Coffman *et al.*, 2004).

All these examples illustrate two features. The less important is the variable specifics of the initial cytoplasmic bases of spatial anisotropy. The other feature is of ultimate importance: This is the common functional endpoint of these very diverse initial stratagems for the spatial indication of future developmental domains. The principle is that whatever the basis of the anisotropies, however they come into being, whatever the cell fates that derive from what they set in train, they end up causing certain maternal transcription factors to be present and active in some spatially defined embryo nuclei, but not in others. These factors then interact with the *cis*-regulatory modules of zygotic regulatory genes in those nuclei. In this way, as summarized in the diagram of Fig. 3.1G, the initial spatial regulatory states of the animal embryo are differentially imposed.

A note should be added here about mammalian embryos, which are absent from the foregoing discussion. Mammalian embryos do not proceed at once to the specification of the axial coordinates and major internal parts, such as the CNS, gut, and the mesodermal structures of the body plan, as do the directly developing embryos of all other vertebrates, or of insects and nematodes. The indirectly developing embryos of many marine invertebrate clades also go right to the business of specifying the parts of their microscopic larval body plans, i.e., gut, mouth, anus, ciliated ectodermal specializations, neurogenic territories and so forth. In contrast mammalian embryos, resident in their protected internal environment, at a relatively leisurely pace generate a specialized extra-embryonic epithelium that will be used for nutrient transfer and, some days later, for implantation in the uterine wall. But the cells that will build the body plan of the embryonic animal after implantation (for review, Rossant and Tam, 2004) are maintained for these pre-implantation days in an entirely pluripotent and unspecified state. Therefore, it is not relevant to mammalian embryos to consider the means by which maternal anisotropies lead directly to the embryonic territories of the forming body plan. This is a mammalian peculiarity, for though there is yet little molecular evidence, it is clear that in other amniotes such as the chicken anisotropically localized maternal determinants are crucial in the initial phases of embryonic development (Callebaut, 2005).

Spatial Regulatory Complexity and Interblastomere Signaling

The spatial domains deriving from anisotropies present in the very early embryo, say by second or third cleavage, are always crudely defined and of low diversity relative to the outcomes of even the simplest forms of bilaterian embryogenesis. And so another very general process of animal embryogenesis is the rapid institution of signaling among cells of the cleavage stage embryo (blastomeres). In essence this is used to increase spatial complexity, by adding new conditions to those arising from the first anisotropies. Thus as the egg cytoplasm is divided up during cleavage, it is usual that the descendants of blastomeres, which early on owned special cytoplasmic domains, emit ligands that when received by their neighbors become specified in new and distinct ways. This is not a single step mechanism, but only the beginning of a major mode of developmental regional specification that continues to be elaborated throughout embryogenesis, postembryonic development, and body part formation. Once specified, each spatial domain of a developing organism typically engages in new signaling interactions of several kinds. It may undergo additional spatial subdivision by integrating inputs (cf. Chapter 2) from its own state with diverse signals arriving from across its different borders; it may issue new signal ligands to its neighbors, which may be either positive or negative in import (i.e., either preventing these neighbors from assuming the same state as itself or inducing in them a new state); and it will also likely signal from cell to cell within its domain to keep its constituent cells on the same page, so to speak (this has been termed the "community effect"; Gurdon, 1988). Signaling interactions, like the anisotropies of the egg, are diverse. In comparing species, we see that any given signal transduction system is always used for multiple unrelated developmental purposes in any given animal, and even among animals belonging to the same clade there are frequently observed striking differences in signal system utilization. Of course, the most important thing to keep in mind is that in development, signals *per se* are meaningful only because they affect transcriptional regulatory state.

The early role of interblastomere signaling as a mechanism for increasing complexity by institution of new transcriptional domains can be seen very clearly in certain kinds of eggs that have a more or less invariant early cleavage pattern. This includes the eggs of most phyla, but excludes vertebrate eggs, cephalopod eggs, and most arthropod eggs (Davidson, 1990, 1991). Invariant early lineage just means that the apparatus, which sets up the cleavage planes with respect to the anisotropic organization of the egg, operates more or less similarly in every egg of the species, so that in every embryo the future fate of the descendants of most cells can be recognized by their position and ancestry (Type 1 embryonic specification; Davidson, 1990). The significance for us is that Type 1 embryogenesis provides mechanistically simple, and easily identified, cases of cell type-specification. The research "model system" species that develop in this way are *C. elegans*, sea urchins (most common are *S. purpuratus*, *Lytechinus variegates*, and *Paracentrotus lividus*), and ascidians (most usually *Ciona intestinalis* and *Halocynthia roretzi*). In Fig. 3.2 lineage and fate maps for these frequently used embryos are shown diagrammatically, for reference to what follows.

At the extreme are cases in which the invariantly inherited, localized components of given early embryonic cells are not only necessary but sufficient to produce a

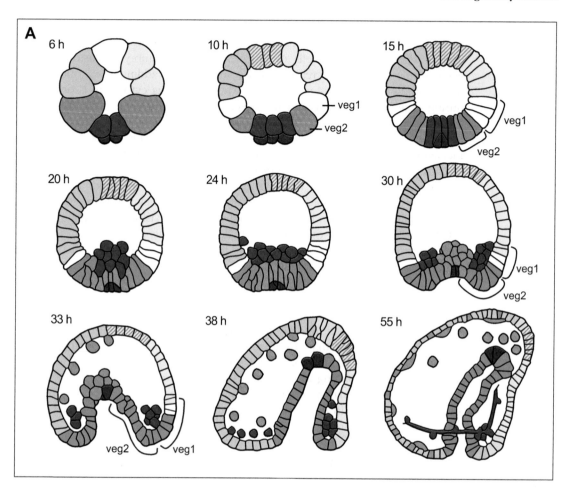

FIGURE 3.2 Lineage and fate maps in sea urchin, nematode, and ascidian embryos. These are three Type 1 embryos commonly used for research. (A) The sea urchin *Strongylocentrotus purpuratus*, color-coded tracings from photomicrographs (from A. Ransick and E. Davidson). veg$_1$ and veg$_2$ are rings of 8 cells each, arising from their parental cells at the horizontal 6th cleavage. From veg$_1$ derives ecto-derm plus (mainly) hindgut endoderm; and from veg$_2$ nonskeletogenic (secondary) mesenchyme (meso-dermal cell types) plus gut endoderm. Skeletogenic mesenchyme lineage, red; endoderm, blue; secondary mesenchyme, violet; oral ectoderm, yellow; apical oral ectoderm, hatched yellow; aboral ectoderm, green; unspecified cells, white. 6 and 10 h, cleavage stages; 15 h, blastula stage; 20 and 24 h, mesenchyme blastula; 30, 33, 38 h, gastrula stages; 55 h, late gastrula or "prism" stage. (B) The nematode *Caenorhabditis elegans*. (B1) Lineage to 26-cell stage. Names of some sister blastomeres are included for early cleavages in posterior half. Lineages in which all progeny give rise to a single cell type are indicated in color (data from Sulston *et al.*, 1983). (B2) Diagram of 8-cell embryo, from ventral side.

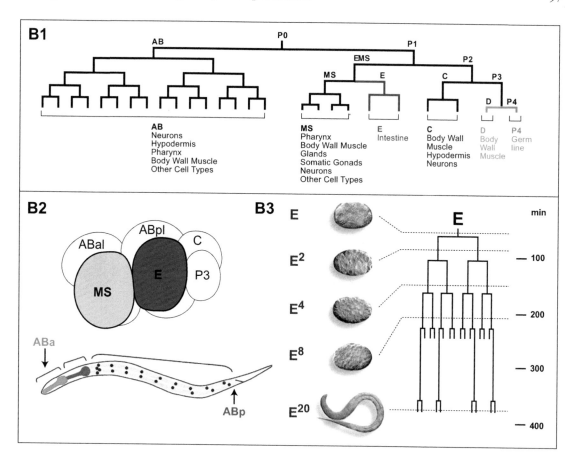

The E gut founder cell (red) and its sister blastomere, MS (gray), are shaded (from Goldstein, 1995). (B3) Lineage and disposition in embryo and larva of successive E-cell progeny (green), which entirely constitute the gut (Leung *et al.*, 1999). To the left is a drawing of a larva indicating in color the position of the gut (lavender), with individual gut cell nuclei indicated (20 E-cell stage) and pharynx, with anterior and posterior portions differently colored. The anterior pharynx derives from the ABa (anterior AB daughter cell) blastomere, the sister cell of which, Abp, migrates to the rectum, and the posterior pharynx derives from the MS blastomere (from Maduro and Rothman, 2002). (C) Ascidian embryo (the portions of embryonic lineages shown are similar in *Ciona* and *Halocynthia*). (C1) Drawing of 8-cell embryo from the side, blastomere names indicated: a, anterior; p, posterior (from Satoh *et al.*, 1996, after Conklin, 1905). (C2) 32- and 64-cell stages, ventral views (from Nishida, 2002a). Blastomere designations are given for right side in both drawings. In the 32-cell stage embryo the autonomously specified endoderm is shown in yellow; at the 64-cell stage the sister cells are indicated by short horizontal lines on the left side of the embryo. Color indicates that segregation of embryonic fates is

(Continues)

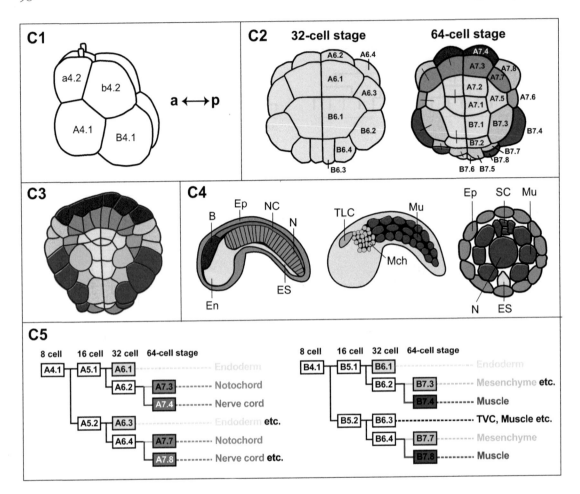

FIGURE 3.2 (continued)

complete, that is, all progeny will express the indicated fates: red, tail muscle; tan, notochord; dark green, epidermis; dark lavender, "nerve cord" (actually glial cells and axons only); turquoise, "trunk lateral cells" (set aside cells that give rise to adult blood cells; Hirano and Nishida, 1997). Gray cells remain unspecified. (C3) 110-cell gastrula stage, additional mesenchyme fate now specified, light green. (C4) Cell types and parts of larva, same color code as in C2 and C3. Left, sagittal midline section; center, same, but lateral to midline; right. Cross section: En, gut endoderm; ES, endodermal strand (gives rise to adult gut); B, brain; Ep, epidermis; NC, nerve cord; N, notochord; TLC, trunk lateral cells; Mu, tail muscle; Mch, mesenchyme (C3–C4, from Satoh, 1996). (C5) A4.1 and B4.1 lineages, and segregation of cell fates (refer to C1), same color coding as above (from Nishida, 2002a). TVC, trunk ventral cells (also adult set aside cells).

given state of differentiation in the lineal descendants of these cells. The mechanism begins with their inheritance of localized components which affect gene regulation, as in the diagram of Fig. 3.1G. Such cell lineages are said to be "autonomously specified." There are many well-known examples (for reviews, Davidson, 1990, 2001), for instance the skeletogenic (micromere) lineages of the sea urchin egg (Fig. 3.2A), or the muscle cell and endodermal cell lineages of the ascidian egg (Fig. 3.2C). But "autonomously specified" does not mean independent or uncommunicative. We have learned that often such autonomously specified embryonic cell lineages no sooner establish their unique zygotic regulatory state than they begin emitting signals to adjacent cells, producing further tiers of early specified cell lineages descendant from the blastomeres that receive these signals. Just to take the examples mentioned, the micromere lineages of the sea urchin embryo at 4th–6th cleavage emit a signal to adjacent blastomeres that is required to permit their specification as endomesodermal precursors. At 7th–9th cleavage, the micromeres express a Delta signal to adjacent cells that is required for their specification as mesoderm (Fig. 3.2A; Davidson *et al.*, 1998; Davidson, 2001; Sweet *et al.*, 2002). In the ascidian, at 5th cleavage and thereafter the autonomously specified endoderm lineages (Fig. 3.2C) inductively signal adjacent cells (via FGF ligands) to assume notochord or mesenchyme fate (Imai *et al.*, 2002; Nishida, 2002a; Miya and Nishida, 2003; Tokuoka *et al.*, 2004). These are striking examples because they demonstrate cleavage stage signaling on the part of the precociously specified autonomous cell lineages of very early embryos.

As knowledge of model system embryos has expanded, more and more specific, cleavage-stage interblastomere interactions have been revealed. No longer is it remarkable for some given signaling interaction to be shown to be required for an embryonic specification to occur; instead it would be remarkable if a specification occurred that is not signal dependent, other than in the relatively few confined cases of autonomous specification. We now have global discussions of proposed and demonstrated signaling interactions for many different kinds of embryo that develop in very different ways: *Xenopus* (Harland, 2004; Houston and Wylie, 2004), zebrafish and chick (Schier, 2001; Fraser and Stern, 2004; Kimelman and Bjornson, 2004; Stern, 2004), sea urchins (Davidson *et al.*, 1998), *C. elegans* (Schnabel and Preiss, 1997), and ascidians (Nishida, 2002a,b).

To make sense in terms of regulatory mechanism out of the maze of early embryonic signaling phenomenology, the following points are useful:

(*i*). The initial stages of signaling in embryos may utilize maternally encoded, anisotropically localized ligands to produce given regulatory states in adjacent recipient blastomeres. Two clear examples in *C. elegans* are expression at the 4-cell stage of a maternal ligand for the Notch receptor (Apx1) to signal from P2 to the posterior AB cell (cf. Fig. 3.2B), thereby initiating in it regulatory events that specify posterior as opposed to anterior cell fates within this clone; and the use of maternal Wnt ligands to signal from P2 to the adjacent EMS cell, thereby initiating a polarization that triggers an endoderm-specific gene regulatory network in the E daughter cell (Mickey *et al.*, 1996; Newman-Smith and Rothman, 1998; Maduro and Rothman, 2002; Maduro *et al.*, 2002).

(*ii*). A frequent result of early specification functions in the pregastrular embryo is the transcriptional activation of genes encoding emitted ligands (and sometimes molecules which bind to them and modulate their availability or activity), in confined spatial domains. On the other hand, the transcription factors which mediate the effects of the signal in the recipient cells are usually globally distributed and often of maternal origin. The effects of the signal are to activate or mobilize these transcription factors in localized spatial domains, e.g., by causing them to enter the nucleus. As an illustration, in sea urchin embryos the Notch receptor and Su(H) transcriptional mediator of N signaling are maternal and global in the pregastrular embryo, while the Delta ligand gene is specifically transcribed in skeletogenic cells and later in other cells (Sherwood and McClay, 1997; Oliveri *et al.*, 2002; Sweet *et al.*, 2002; Revilla-i-Domingo *et al.*, 2004). The *Xenopus* egg provides another example, this in a context where there is not an invariant embryonic cell lineage, but there is a maternally localized mRNA encoding a transcription factor, VegT, mentioned above (Fig. 3.1E). Among the initial transcriptional targets of this regulator are genes encoding the "Xnr" (Tgfβ-class) signaling ligands, which potentiate mesodermal specification in the overlying cells (Kofron *et al.*, 1999; Harland, 2004; this linkage is discussed in detail in Chapter 4).

(*iii*). Given signal systems are utilized and re-utilized to install diverse regulatory states in the same embryo; that is they cannot be assumed to be intrinsically dedicated to any specific "inductive" function. This is at root because their downstream transcriptional inputs are incorporated in diverse *cis*-regulatory designs. As specific illustrations of this point, in sea urchin embryos the N/Su(H) signaling system is first used in the spatial control of pigment cell regulatory genes (Ransick and Davidson, 2006) but then, in a second non-overlapping and independent phase of N signaling, in the control of regulatory genes in the endoderm (Walton *et al.*, 2006). Similarly in ascidians the Fgf 9/16/20 signaling system is required in the conditional specification of three different kinds of mesenchymal cell types, wherein its transcriptional input is incorporated into three different gene regulatory networks, controlling expression of three different sets of genes (Tokuoka *et al.*, 2004).

Territories

A useful concept in considering the regulatory construction of the pregastrular embryo is that of the "territory" (Davidson, 1989). Territories in this sense can be defined as multicellular (or multinuclear) domains of the embryo in which a given regulatory state is expressed. Territories are transient, functional way stations to the differentiated parts of the later embryo or larva. They are essentially of regulatory significance. They will usually be further subdivided into more refined, regionally specified domains in postgastrular development. Progressive territorial maps indicate the spatial relations of the domains of transcriptional regulatory state in different embryos, so that the possible contributions of early anisotropies and of interblastomere signaling may be directly considered. Territories are almost always polyclonal, though there are occasional exceptions such as the gut lineage of *C. elegans*, which is a single clone of cells descendant from the E founder blastomere (Fig. 3.2B). Polyclonal territories may include a fixed or variable set of cell lineages, depending on whether the animal has a fixed or variable (i.e., from embryo to embryo) cleavage. Before gastrulation (as that event is usually defined for diverse embryonic forms), there is little cell migration, except in amniote embryos, in particular chick (Stern, 2004). Thus the progression of pregastrular territories as development proceeds in general indicates directly the pattern of early regulatory states, not confounded by patterns of individual cell migration. Specific examples of territorial regulatory states are taken up in the following: Their importance is that they provide the initial bases for analysis of the gene networks which encode the genomic regulatory instructions for the embryonic beginnings of the animal body plan.

The concept of embryonic territory as a polycellular or polynuclear regulatory state domain is more general than is the mode of pregastrular specification. Such territories are established as the endpoint of both autonomous and signal-dependent specification processes, whether the territories are of more or less invariant cell lineage as in Type 1 embryonic processes or, on the other hand, they are never of the same exact lineage twice, as in vertebrate and some other large embryos. The process of territory formation along the dorsal/ventral axis in the syncytial *Drosophila* embryo illustrates this concept in relief. Here there are no discrete cells at the stage when spatial domains of regulatory state are established, and so there are no blastomere lineages, variant or invariant, to be specified either autonomously or conditionally. Instead, appropriate sets of regulatory genes are activated in coherent bands of appropriately positioned nuclei of no particular lineage relation to one another. As reviewed in Chapter 2, the decreasing concentrations of the Dorsal transcription factor are interpreted differentially by the diverse *cis*-regulatory designs of the modular control units governing the expression of these regulatory genes, many of which encode spatial repressors that set expression boundaries (Stathopoulos and Levine, 2005). These genes are hooked up together in a gene regulatory network, the organization of which establishes the respective spatial domains of expression (discussed explicitly in Chapter 4). For present purposes, the overall point is that the territories of regulatory gene expression set up in the syncytial stage end up defining precursor domains, which following cellularization, become the fields of cells from which form the heart, the CNS, the body wall mesoderm, and other embryonic parts that become apparent only after gastrulation (Levine and Davidson, 2005). Following cellularization, of course, the same kinds of progressive regional specification, mediated by local transcription of genes encoding signaling ligands, are required for construction of these parts in the *Drosophila* embryo as is seen in other postgastrular developmental processes.

To summarize, listed below are some principles of early development that devolve from the idea of territories as transient spatial regulatory states:

(*i*). Territorial regulatory states define functionally the domains from which the postgastrular parts of the embryo form by installing differential patterns of regulatory gene expression.

(*ii*). Territorial regulatory states are initially defined in three different ways: by autonomous blastomere specification; by conditional blastomere specification in response to interblastomere signals (and often both kinds of input are utilized); or in syncytia, by use of autonomously inherited anisotropic inputs and repressive gene networks. The comparative variety of early embryonic developmental mode in bilaterians can be reduced to these ways, which all converge on construction of territories.

(*iii*). Since territories foreshadow the domains of body (or larval body) parts, and since these are phylogenetic characters, global territorial topography is comparatively more conserved within clades than is mode of pregastrular specification, as long observed (Raff, 1996). For example, different arthropod groups use in different measure and different ways all three modes of pregastrular specification, but nonetheless, they generate similar territorial patterns of gene expression at gastrulation.

(*iv*). The developmental process by which territorial regulatory states are installed has the fundamental property that it is a one-way street. The underlying mechanisms are fully revealed only at the level of the gene regulatory networks that control the process. As we shall see in Chapter 4, the explanation lies in intranuclear regulatory network architecture, i.e., transcriptional installation of positive and negative feedback circuits of various kinds and in intercellular signaling "lockdowns." These include transcriptional expression of, and requirement for, "community effect" signaling, which reinforces extant regulatory states within territories and repressive signaling across territorial boundaries once they are established.

COMPARATIVE VIEW OF EMBRYONIC PROCESSES, AND THE ORGANIZATION OF THE UNDERLYING GENE REGULATORY NETWORKS

The organization of the pertinent gene regulatory network explains the organization of developmental processes, and this is a commutative relation: We should be able to learn something of the architecture of these networks from the architecture of the developmental process, as well as the reverse. Here we view the basic regulatory structure of embryonic body plan development comparatively, always a revealing exercise when considering the mechanisms by which bilaterians do business. There emerge the guidelines required to interpret the diverse structures of the gene regulatory networks considered in the next Chapter.

The analysis is shown in Fig. 3.3. Three different kinds of embryonic process are treated, and the color coding indicates four modes of overall gene regulatory network organization. Almost all forms of bilaterian embryogenesis begin by use of very early anisotropies and interblastomere signaling to establish their initial territorial regulatory states, as above. But then their strategies diverge markedly, driven by diverse forms of zygotic control circuitry. We begin with the form of embryogenesis that implies, and does in fact use, the most simply organized of all developmental gene regulatory networks.

Shallow Gene Regulatory Networks of Type 1 Embryos

A suite of coincident characteristics differentiates the Type 1 embryonic process from other embryonic forms. These include the following: (*i*) the egg is usually very small (<200 μm) and there are usually only a few hundred cells at gastrulation (in contrast to *Xenopus* and *Drosophila* eggs, for example; Fig. 3.3A, C, D); (*ii*) the zygotic genomes are activated at once after fertilization, and there is no transcriptionally quiescent period, as there is in vertebrate eggs; (*iii*) there is no migration of cells until after the whole embryo has been divided up into a mosaic of specified, territorial regulatory states, again in contrast to the major CNS and mesodermal components in vertebrate embryos, for example, in which both the allotment of cells to these domains and their regulatory states depend on their prior migratory behavior; (*iv*) there usually are to be found some autonomously specified cell lineages, which often are used as the source of signals for conditional specification of adjacent blastomeres (as exemplified in the specific sea urchin, *C. elegans*, and ascidian cases noted above); and (*v*) these embryos proceed as directly as possible to the expression of terminal differentiation genes, so their mode of development can be summarized as "direct cell type specification" (Davidson, 2001). In Fig. 3.3A Type 1 embryogenesis is parsed into its identifiable stages. In Fig. 3.3B are indicated the implied gene regulatory network functions of the portion of the control circuitry that extend from the initial installation of territorial zygotic regulatory state to the expression of differentiated gene batteries all over the embryo. Type 1 embryos were identified above by their invariant early cell lineage, and while there is probably no embryo with an invariant lineage that does not work as in Fig. 3.3A, the use of lineage is rather a "convenient" mode of operation that is allowed by other more basic features of such embryos, than it is a fundamental aspect of developmental mechanism. It is convenient in the sense that the cleavage plane positioning devices can be utilized as a means to install given regulatory states in the right place in the embryo, which otherwise has to be done by other spatial localization mechanisms.

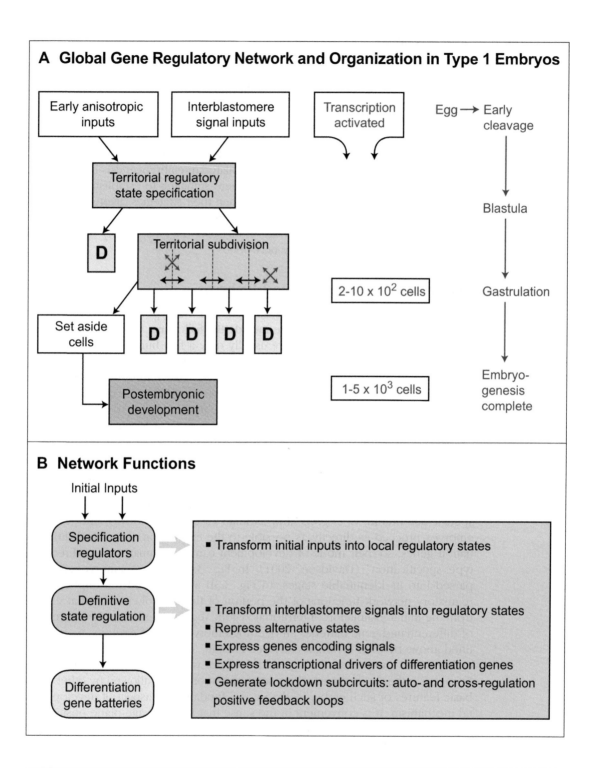

FIGURE 3.3 For legend see page 106.

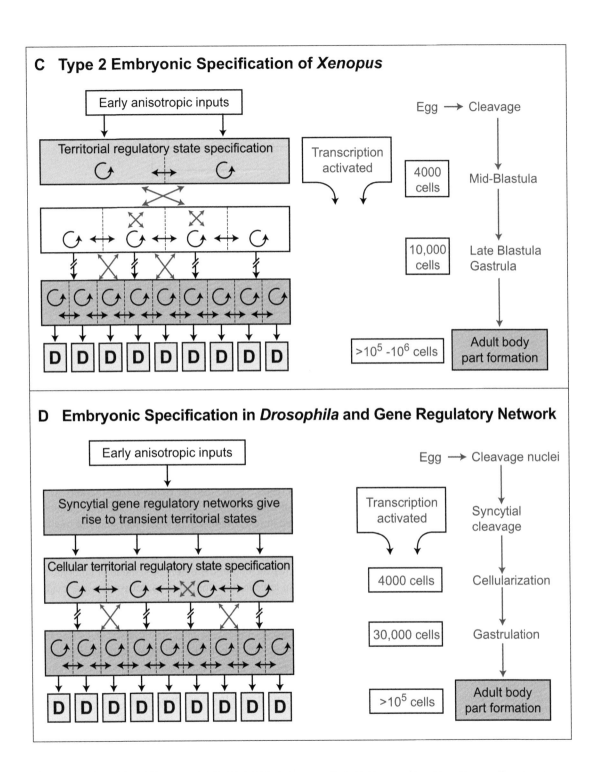

FIGURE 3.3 (continued)

(Continues)

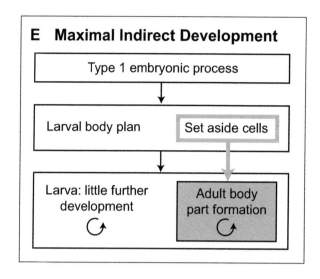

FIGURE 3.3 Diverse modes of development: sequential strategies for specification of spatial regulatory states. In A, C, D, sequences of embryonic functions are indicated in horizontal boxes on left; time proceeds from top to bottom in each drawing. Cell migrations are symbolized as crossed red arrows between or within embryonic territories. Inter-territory signaling is indicated by double-headed arrows, cell replication by reflexed curved arrows, progressive territorial subdivision by sequences of vertically bounded boxes (in respect to generation of regulatory states, this is the basic process in body part formation; see text). The outcome of development of the body plan in each case is ultimately to give rise to a spatial mosaic of differentiated cell types, symbolized as yellow boxes marked "D." In red, to the right, are shown the respective times of transcriptional activation and approximate cell numbers, and the embryological stages. As discussed in text, there are (at least) four kinds of mechanism resulting in establishment of regulatory states in development. (A) Type 1 embryo mode of direct cell type specification, indicated in turquoise. (B) Regulatory functions required of underlying gene network to execute the processes summarized in (A). (C) Vertebrate (here *Xenopus*) mode of embryonic specification in the context of massive cell migration, indicated in white (we do not yet know enough to characterize the structure of the underlying gene regulatory networks); the terminal establishment of the regulatory domains of adult body parts (see text) is in lavender. (D) *Drosophila* mode of syncytial regulatory control, indicated in tan. After cellularization this type of control system gives rise to a set of early specification functions that directly produces differentiated cell types (turquoise stripe) e.g., in heart, plus the complex series of progressive specifications needed for formation of many adult body parts (lavender). (E) Diagrammatic description of relation between larval and adult body plan in the extreme case of maximum indirect development (see text), as for example, in sea urchins.

This is an embryonic process that, as directly as possible, generates a confined set of differentiated cellular domains out of a small anisotropic egg equipped with canonical cleavage devices. That process implies a certain kind of gene regulatory network organization. In gene regulatory networks the differentiation gene batteries represent the terminal or peripheral class of interactions (cf. Chapter 1). Internal and upstream interactions at regulatory genes impinge directly on other genes of the network, but this is not so at the periphery, where reside the protein coding genes that directly encode the properties of differentiated cell types. There any given element of the gene regulatory network terminates. The gene regulatory network of the embryo begins its operation with the activation of the zygotic gene regulatory apparatus, in early cleavage. Thus, as Fig. 3.3A, B illustrates, in Type 1 embryos we can encompass both the beginning and the end of network elements, and this is not yet true of any other form of embryonic gene regulatory network. Furthermore, there are not many steps separating the beginning from the end.

Examples are considered in detail in the predecessor to this book (Davidson, 2001), which show how specific differentiation genes are activated soon after and in some cases even well before gastrulation in Type 1 embryos. Among these are biomineralization genes in the skeletogenic lineages of sea urchins, expressed even during the blastula stage, and cytoskeletal and other aboral ectoderm-specific genes (see Fig. 3.2A). To this may be added genes encoding pigment synthesis enzymes also expressed during blastula stage in prospective mesoderm cells (Calestani *et al.*, 2003), additional biomineralization genes (Oliveri *et al.*, 2003; Oliveri and Davidson, 2004), and gut-specific genes expressed in early gastrula (Rast *et al.*, 2002). An intestine cell-specific esterase gene is expressed in *C. elegans* as early as the 4 E-cell stage, which is illustrated in Fig. 3.2B (Edgar and McGhee, 1986; for discussion, Davidson, 2001). In ascidians (see Fig. 3.2C) muscle contractile protein genes are expressed in the autonomously specified tail muscle lineages after 5th and 6th cleavages (Satou *et al.*, 1995); many notochord-specific genes, e.g., a particular tropomyosin, are expressed at gastrulation or soon after in this conditionally specified lineage (Di Gregorio and Levine, 1999; Hotta *et al.*, 1999). To these examples may be added various genes expressed late in cleavage in precursors of mesenchyme cells (Tokuoka *et al.*, 2004), and many other genes, as a result of a genomic survey of embryonically expressed genes from *Ciona* (Fujiwara *et al.*, 2002). All this evidence supports the conclusion that direct cell type-specification marked by early expression of differentiation genes, long in advance of the morphogenesis of the structures in which they will be utilized, is anything but a peculiarity of a few genes, or of a few unusual cell lineages. Rather, as proposed earlier (Davidson, 1990, 2001), this is a basic mode of development in Type 1 embryos. Many differentiation genes that will be expressed in given cell types are indeed brought on line only later; multiple postgastrular specifications occur as well in Type 1 embryos (two tiers of such specifications are shown in Fig. 3.3A). But in some specific regulatory domains these embryos afford the remarkable opportunity to identify and analyze the gene regulatory networks that control major aspects of embryonic development from the beginning all the way to the peripheral end.

Taking a bird's eye view, Fig. 3.3A, B suggests that typically the Type 1 regulatory apparatus must minimally consist first, of an initial "territorial specification" component which provides a transcriptional interpretation of the cytoplasmic spatial cues and the interblastomere signals with which it is confronted soon after fertilization; and second, a component that functions to generate the definitive territorial regulatory states and maintain them in proper respect to one another (these regulatory machines are indicated by the blue color in the diagrams in Fig. 3.3A, B). Thereupon the regulatory machinery for operating differentiation gene batteries is called into play in each territory (yellow in these diagrams). The structure of differentiation gene batteries was briefly discussed in Chapter 1, and is not further considered here. But, as indicated in Fig. 3.3B, there are at least five separate kinds of function involved in formulating and operating the spatially related mosaic of territorial regulatory states, which is what the Type 1 embryo becomes (the general import of the fate maps of Fig. 3.2). These are the functional requirements that must be met by the causal interactions included in the underlying regulatory network. They are all illustrated explicitly in the network models for Type 1 embryos shown in the next Chapter.

Type 1 embryos are small and simple because they postpone to postembryonic phases of development a lot of the complex regulatory work required to build adult body parts. In this sense all Type 1 embryos develop somewhat indirectly, since their immediate role is to generate larval forms composed of differentiated, functional cellular structures, which they use in one manner or another to support their existence during the postembryonic development of their remaining adult body parts. The "degree of indirectness" is measured by the extent to which the parts formed in the embryo are retained and utilized in the adult. In *C. elegans*, essentially all the internal larval body plan is retained in the minute adult form, but in the first stage larva, of the total 558 cells there are 42 postembryonic blast cells (not including germline cells), which are set aside from the embryonic specification and differentiation functions, and their descendants subsequently form additional structures, including the hypodermis, some neurons, and sexual and external excretory organs (Sulston *et al.*, 1983). The *Ciona* adult is a large and anatomically complex animal that retains only rudiments of some head structures from the chordate-like "tadpole" larva which the embryo generates. The adult is built largely from larval set aside cells (Hirano and Nishida, 1997, 2000; Tomioka *et al.*, 2002). But this adult form is an evolutionary excursion, since the chordate tadpole is homologous to all other chordates, while the adult form is peculiar to its clade, and since also there are other ascidian forms (the Larvaceae) in which the tadpole is the adult form. The most extreme form of indirect development (maximum indirect development) is that seen in most sea urchins and in other echinoderms (though also in many other entirely unrelated clades; Peterson *et al.*, 2000). Most structures of the microscopically scaled sea urchin larva have no future at all in the adult body plan, and are thrown away at metamorphosis.

Furthermore, development of the adult body from a specific population of embryonic set aside cells occurs in these animals not sequentially with respect to the larval stage, but during it, within the body of the larva while it feeds (Fig. 3.2E; for reviews, Hyman, 1955; Pearse and Cameron, 1991; Peterson *et al.*, 1997). These distinctions aside, the conclusion is that in this type of development, formation of at least some, if not most, adult body parts occurs separately and by different "rules of procedure" from formation of the embryo/larva. Adult body part formation, represented in violet in the diagrams of Fig. 3.3, is controlled by a different kind of network architecture, as summarized explicitly below and illustrated in the postembryonic regulatory network fragments reviewed in Chapter 4.

More Complex Forms of Embryonic Process

In vertebrate embryos the order of processes is reversed with respect to Type 1 embryos, in that terminally differentiated states arise relatively much later in the process. This has immediate consequences for what is to be expected of the underlying gene regulatory network (Fig. 3.3C). Considering the *Xenopus* embryo as an example, several features jump out at once. First, the embryo is comparatively enormous in terms of the number of cells, compared to those just considered. In consequence of this basic parameter, it cannot utilize the fundamental stratagem of transcriptional specification of spatially defined cell lineages by intercellular signaling early in cleavage. Thus the embryonic genomes are not transcriptionally active at all until the mid-blastula stage, when there are about 4000 cells. At this point the nucleus to cytoplasm ratio finally achieves a value such that the level of transcription products can quickly affect the properties of the cells. Since before this the embryonic cells have no functional significance as generators and interpreters of transcriptional regulatory state, it is not surprising that after the initial cleavages, the cleavage planes are variable egg to egg, as is the cell lineage of the embryo in all its territorial domains (reviewed by Davidson, 1986, 1990; "Type 2 embryogenesis"). As noted above, there are essential anisotropies loaded into the egg, e.g., the vegetally localized *vegT* maternal mRNA (Fig. 3.1E) and the cytoskeletal apparatus that results in localization of the β-catenin transcriptional cofactor on the future dorsal side. When transcription resumes these are used to produce the first local regulatory states and to cause the local transcriptional expression of signals (e.g., the genes encoding the Xnr ligands, expressed under VegT control; see above), thus specifying other neighboring regulatory states. But then there intervenes a massive inter-territorial migration of cells (crossed arrows in Fig. 3.3C), and as in all vertebrate embryos at comparable stages, the regulatory state of any given cell depends partly on where it is and what neighboring cells are emitting what signals to it.

This is of course a very potent way of doing embryonic development, for it removes all the constraints of Type 1 specification *in situ*: any (very large) number of cells can play; they can be of any lineage so long as they are (relatively) unspecified; and morphological as well as purely regulatory domains can be established at the same time by the migrating, multiplying, cell populations. Note that in Fig. 3.3C the progressive further subdivision of the new territories formulated during gastrulation is symbolized (e.g., the various domains of the neural plate, the anterior endoderm, the paraxial mesoderm, etc.), always accompanied by further cell movement and cell division. In each region the process sooner or later merges into the adult body part formation mechanism, following the point when the domain of the embryo uniquely giving rise to a terminal structure such as the hindbrain or heart is specified. But the underlying organization of this elaborate, flexible, multistage developmental process is not transparent. As we shall see in the following Chapter, it is the beginning and ending portions of the diagram in Fig. 3.3C that we occasionally know something of at the level of the genomic regulatory code for formation of vertebrate embryos, i.e., the initial events of the mid-blastula to early gastrula on the one hand; and certain of the events of specific adult body part formation on the other. But unlike the case with Type 1 embryos, the connections between the initial and the ultimate mechanisms are as yet not accessible, as in entering a foyer that leads to a large hall where populations circulate and mix and then distribute through multiple exits, compared to, say, a narrow walk-in closet with a single exit. Nonetheless, we know that the whole process is genomically programmed and is specifically controlled through reproducible and invariant sequences of transcriptional expression of signaling systems. These function to cause installation of transient regulatory states, indicated experimentally as the regional appearance of given mRNAs encoding transcription factors.

The pregastrular regulatory design of the *Drosophila* embryo is again quite different in organization from either of the foregoing (Fig. 3.3D; Fig. 3.4). The embryo remains syncytial through the 13th cell cycle, the nuclei densely and evenly distributed in a single layer just beneath the surface of the large egg (see Fig. 3.4A), and cellularization then occurs with the inward extension of cell membranes throughout the surface to form the cellular blastoderm. But by then there has already occurred a succession of nuclear regulatory states, which before cellularization has blocked out a spatial coordinate system that serves as the regulatory basis for the immediately postgastrular formation of major elements of the body plan. Along the A/P axis, by the time of cellularization, every few nuclei of each future metameric unit express a certain set of transcription factors, repetitively in each metameric repeat of the thorax and abdomen (for review of expression patterns, Poustelnikova *et al.,* 2004; Ochoa-Espinosa *et al.,* 2005). The future head and posterior regions have also been specified so as to express different sets of transcription factors. In the D/V direction the future domains of the mesoderm, the neurogenic ectoderm, the heart, and a mid-dorsal structure, the amnioserosa, are differentially specified and display subdivided regulatory states, foreshadowing different future fates (see Fig. 3.4B). As detailed above (Chapter 2), the fundamental spatial regulatory functions are executed by the *cis*-regulatory modules that control the genes, which together constitute the regulatory states of each

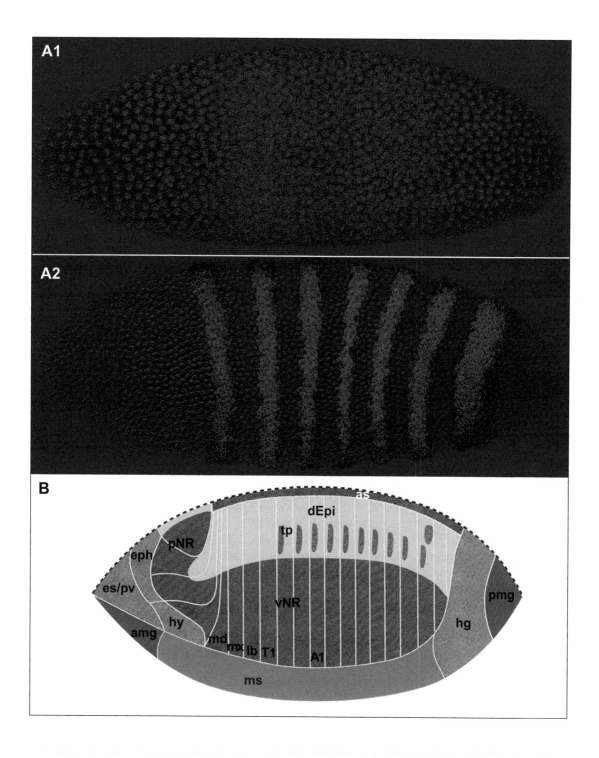

FIGURE 3.4 For legend see pages 112–113.

(Continues)

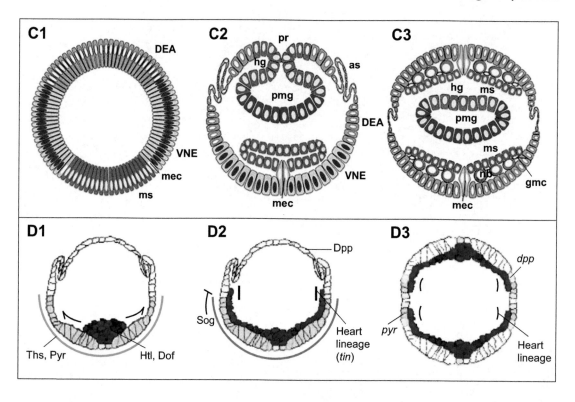

FIGURE 3.4 Syncytial and early cellular development in *Drosophila*. (A) Late syncytial stage and initiation of cellularization. (A1) Cycle 13 *Drosophila* embryo stained with DAPI to reveal nuclei (blue), and hybridized with a fluorescent probe for mRNA of the *ftz* pair rule gene (magenta). Two broad stripes of nuclei express this gene. (A2) Cycle 14 embryo, stained as in A. Cell walls extend down toward the interior of the egg, but have not yet sealed off individual cells. Expression of *ftz* has resolved into seven distinct stripes, though the borders remain rough until after cellularization is complete (images provided by D. Kosman and W. McGinnis). (B) Fate map of the *Drosophila* embryo blastoderm. From anterior (left) to posterior, abbreviations are es/pv, esophagus/proventriculus; amg, anterior midgut; eph, epipharynx; hy, hypopharynx; pNR, procephalic neurogenic region; tp, tracheal pits; dEpi, dorsal epidermis; as, amnioserosa; vNR ventral neurogenic region; md, mandibular segment; mx, maxillary segment; lb, labial segment; T1, first thoracic segment; ms, mesoderm; A1, first abdominal segment; hg, hindgut; pmg, posterior midgut (from Hartenstein, 1993). (C) Cellular fate maps and invaginations of the blastoderm (after Hartenstein, 1993). (C1) Blastula just after cellularization: green, mesodermal fate; violet, neurogenic fate; gray, epidermal fate. Abbreviations, in addition to those also used in B, are DEA, dorsal epidermis; VNE, ventral neurogenic ectoderm fate; mec, mesectoderm (midline) fate. (C2) Gastrula, posterior section showing gut and mesoderm invaginations (color coding and abbreviations as in B); pr, proctodaeum. (C3) Later gastrulation, germ band now reflexed to extend forward at

domain in the blastoderm. These respond initially to the Dorsal gradient (cf. Fig. 2.6), but soon to the products of other zygotically expressed regulatory genes. Thus they function as the nodes of a zygotic gene regulatory network, which is considered explicitly below, in Chapter 4. This network reveals unique features symbolized by the orange color in Fig. 3.3D; its character changes sharply upon cellularization. In the syncytial period there can be no intercellular signaling, and the network is built for continuing forward progression in regulatory state. Syncytial stage networks include no feedback loops or other stabilization devices, and domain boundaries are set by expression of spatial repressors in adjacent territories in both A/P and D/V regulatory systems (Levine and Davidson, 2005; Ochoa-Espinosa *et al.*, 2005). At cellularization everything begins to change. In the D/V network there now are expressed signaling ligands, and various feedbacks and cross regulations are set up similar to those seen in other developmental gene regulatory networks. As illustrated in Fig. 3.4C, during gastrulation, which shortly ensues, single cell thick layers of given blastodermal regions invaginate to form the future mesoderm and gut, and individual immigrant cells become the neuroblasts that produce the trunk CNS. There follow territorial subdivision and further cell division, with formation of additional layers of cells (see Hartenstein, 1993), and then the processes of adult body part formation and differentiation. Some such processes, e.g., heart formation, are becoming understood, as discussed below.

The intent of Fig. 3.3 is mainly to summarize comparatively the organization of three diverse forms of embryonic development and of the expected underlying networks. In addition, we see at a glance some items of evolutionary interest. First, there are very diverse ways of encoding the pregastrular and gastrular processes of development, some much more amenable to immediate genome

the top of the embryo to form a dorsal mirror image in the section with the ventral aspect. Neuroblasts (nb) and glial mother cells (gmc) are shown ingressing (from Hartenstein, 1993). (D) Migration of mesoderm following invagination (from Stathopoulos and Levine, 2005). (D1) Mesodermal cells (brown) spread laterally within the blastoderm wall in a response mediated by genes encoding the Fgf receptor and ancillary signal transduction proteins that they express (Htl, Dof), which when activated result in a transcriptional readout causing motile behavior. The Fgf ligands, generated by the cells of the blastoderm wall (blue line), are the products of the *ths* and *pyr* genes (Stathopoulos *et al.*, 2004). (D2) The dorsal-most mesodermal cells encounter a Dpp signal generated by the cells shown in white, which is prevented from spreading downward by the Dpp-binding protein Sog, generated by the cells shown in blue (cf. Fig. 3.4D for design of *sog* cis-regulatory module, which explains the expression of this gene in these cells). The mesodermal cells receiving the Dpp signal activate the *tinman* gene (*tin*) and become heart precursors (short black lines; see text). (D3) Germ band extension, showing domains of expression of *pyr, dpp,* and *tin* genes.

level analysis than others. But all roads lead eventually to the same kind of mechanism, adult body part formation and widespread regional installation of cell differentiation. In the predecessor to this book (Davidson, 2001) this process was aptly termed the "Secret of the Bilaterians."

REGULATORY STATE AND NETWORK CIRCUITRY IN BUILDING ADULT BODY PARTS

Steps in the Postembryonic Developmental Process

The developmental process by which animal body parts are built is controlled by a particular sequential mode of regulatory state deployments (Davidson, 1993, 2001). The starting points are defined by the spatial coordinate system erected by the end of the embryonic process. Although body parts are many and various, there are basic commonalities in the steps of the process by which they are formed. The steps of this process, in the basic same terms as our consideration of embryonic process, are

- (*i*). Installation of an initial "body part-specific" regulatory state in a field of cells, the progeny of which will constitute the body part; this is called the "progenitor field."
- (*ii*). Use of progenitor field regulators to provide inputs that, when integrated at the *cis*-regulatory level with those from signaling across the boundaries of the field, newly subregionalize it. The result is to produce additional new regulatory states within it.
- (*iii*). Execution of typical sets of subregional regulatory functions that enable each subregion to do the following: communicate to adjacent subregions, drive the process forward, exclude alternative subregional regulatory states, and activate downstream regulatory genes. These will operate cell cycle control functions, morphogenetic cell shape and movement functions, and differentiation gene battery control functions. The developmental "subroutines" summarized in (*ii*) and (*iii*) may often be deployed recursively, on finer and finer scales up to the most detailed level of species-specific, morphogenetic sculpting of the body part. Organization of body part-specific regulatory states is an essential aspect of the animal's evolutionary genomic heritage, for the animal is the sum of its body parts.

In Chapter 4 we encounter explicit examples of gene regulatory networks for aspects of adult body part formation, "aspects" because in no case do we as yet possess a complete, network level genomic program for construction of an adult bilaterian body part from start to finish. The examples we have produce recurrent illustrations of certain kinds of regulatory subcircuit devices, rather like the "chips" integrated into electronic circuit boards. As in this metaphor, such devices are used in the construction of all different kinds of body part, to execute the functions of (*ii*) and (*iii*) above. Here are some explicit examples: to receive signaling information, one of a small set of signal transduction subcircuits terminates in the *cis*-regulatory module of a gene encoding a subregional transcription factor, and this factor serves to provide another input into the network; to "communicate" to adjacent subregions, genes encoding signaling ligands are always part of subregional networks; to drive the process forward, cross-regulatory positive loops are commonly employed; to exclude alternative regulatory states, genes are included in subregional network architecture that encode repressors which engage the *cis*-regulatory modules of regulators expressed in adjacent subregions. These kinds of subcircuit features, not particular transcription factors or signal biochemistries, are what is functionally significant, as players of diverse identity fulfill each role.

The Progenitor Field Concept

The networks for diverse body parts incorporate similar "chip-like" subcircuit components, but what is it that defines body part-specific gene regulatory network architecture? This is a profound question, given that a basic homology in certain definitive body parts is apparent all across the Bilateria; for example, in their two-ended guts, their brains, their mesodermal body walls, their peripheral nervous systems, and so forth. On the other hand, each of these classes of body part is formed in a distinct way from the others. The issue lies at the heart of the problem of the evolution of bilaterian body plans, the subject of Chapter 5 of this book. But we can begin here to deal with the more confined question of which features of body part network architecture are body part-specific. The answer comes in three stages: comparative observations across the bilaterians suggest that the regulatory specification of progenitor fields tends to be body part-specific, certain complex circuit elements deployed just downstream of progenitor field specification may exist that are body part specific, and of course the differentiation gene batteries called into play at the end of each body part formation process are also at least in part specific. We focus now on the first of these, i.e., on the initial step in the mechanism of adult body part development.

As reviewed earlier and illustrated for diverse cases (Davidson, 2001), the telltale experimental index of progenitor field specification is expression of given transcription factors in a patch of cells that, as development proceeds, is revealed to contain the lineal progenitors of a future structure or organ, our "body part." Below we see for example after example that the initial transcription factors, in combination with other, usually spatially less confined factors, provide the inputs at the top of the gene regulatory network for development of that body part. Thus the specification of the progenitor field by institution of expression of the initial defining regulatory genes is a required function, and mutation or interference with the expression of these regulatory genes cancels the program for development of that body part (for review, Davidson, 2001). Understanding that the products of these genes provide the early inputs into a regional gene regulatory network for a specific developmental operation enables us to escape the mental error of regarding them as "master genes" that "make" the body part. They are necessary, but only the whole gene network is sufficient.

Figure 3.5A provides a canonical example of the progenitor field concept. Here we see a diagram of the parts of a *Drosophila* leg (Fig. 3.5A1), and the corresponding

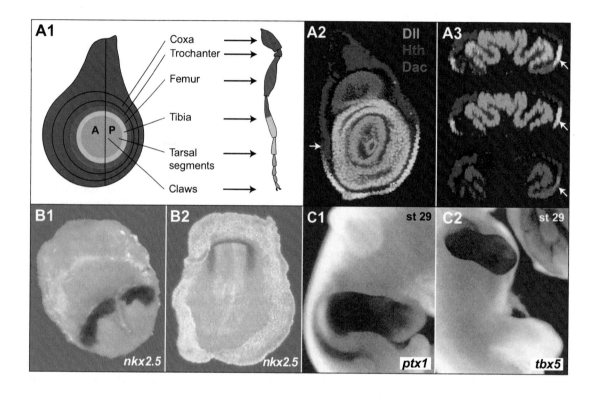

FIGURE 3.5 Progenitor fields and their subdivision in the formation of various adult bilaterian body parts. (A) The parts of the *Drosophila* leg. (A1) Diagram of leg imaginal disc, left, indicating the concentric map of regions destined to produce the morphological parts of the mature leg, illustrated in the drawing to the right. The most distal part derives from the central region where the distalless (*dll*) gene is expressed, and the "shoulder" or coxa from the most peripheral regions, which express *extradenticle* (*exd*) and *homothorax* (*hth*) genes (from Lecuit and Cohen, 1997). (A2) Immunocytological displays of expression of regulatory genes in the 3rd instar disc, color coded as indicated. *hth* and *dll* overlap appears yellow; *dachshund* (*dac*) and *dll* overlap appears light blue. (A3) Three images of disc shown in A2, optical cross sections at arrow. In top, expression of all three regulatory genes is shown; in middle,

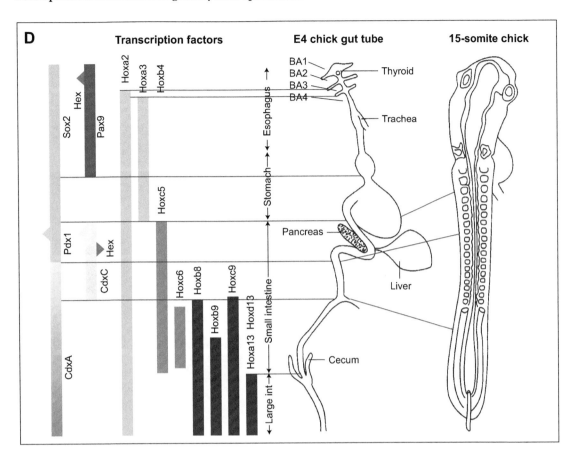

expression of *hth* and *dll* only; in bottom expression of *hth* and *dac* only. The spatial distribution of regulatory states indicated by these gene expression patterns exactly indicates the map of the morphological leg subparts that will derive from the disc (A2,3 from Wu and Cohen, 1999). (B) Expression of *nkx2.5* regulatory gene, visualized by *in situ* hybridization, marks the amniote heart progenitor field (from Harvey, 1996). (B1) Mouse, 7.5 day embryo; (B2) chick, stage 6. (C) Hindlimb and forelimb regulatory states (from Logan *et al.*, 1998. (C1) Expression of *pitx* regulatory gene in mouse embryo hindlimb bud; (C2) expression of *tbx5* regulatory gene in forelimb bud. Expression of these genes is necessary and sufficient to trigger downstream development of forelimb and hindlimb, respectively (Logan *et al.*, 1998; Rodriguez-Esteban *et al.*, 1999; Takeuchi *et al.*, 2003). (D) Regional regulatory states foreshadow the subparts of the gut endoderm (from Grapin-Botton and Melton, 2000; see also Beck *et al.*, 2000). Domains of transcription factor expression in the 15-somite embryo are shown on left, mapped onto the gut tube of the chick (center) and the 15-somite embryo as a whole on right. Pink triangles mark expression of *hex* gene in progenitor fields for thyroid (top) and for liver (center);

(Continues)

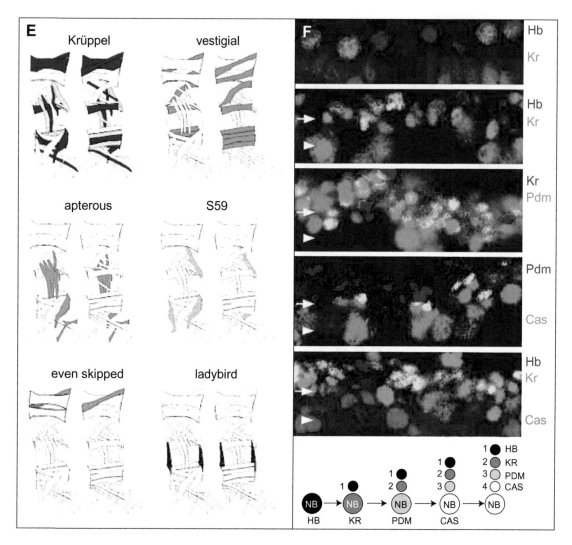

FIGURE 3.5 (continued)

blue triangle expression of *pdx* in progenitor field for pancreas. (E) Expression of diverse transcription factors in individual abdominal muscles in *Drosophila* (from Baylies *et al.,* 1998). Expression patterns are superimposed on diagrams of the individual muscles seen from external (left) and internal (right) views. The different regulatory states that mark these muscles are established during the development of the muscles from myoblasts, and are combinatorial in organization. (F) Expression of four different transcription factors, as indicated by color code, in neuroblasts giving rise to four classes of neuron in development of the ventral CNS of *Drosophila* (from Isshiki *et al.,* 2001). Expression is

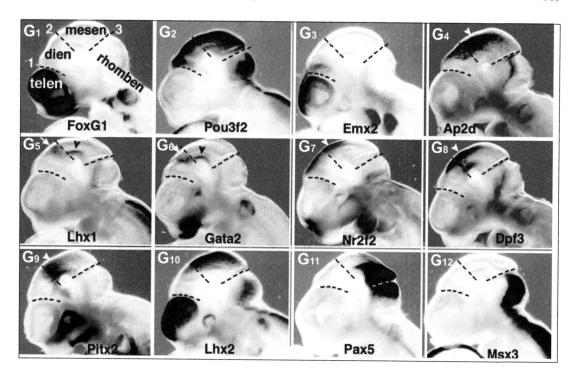

visualized by immunocytology. The neuroblasts are formed by successive divisions of the immigrant ganglion mother cell (Fig. 3.4C3). (G) Regulatory subdivision of the mammalian brain (from Gray *et al.,* 2004). *In situ* hybridizations are shown for 12 different transcription factor mRNAs, as indicated in each panel, in day 10.5 mouse embryos. Each is expressed in a different future morphological domain or subdomain of the developing brain. Abbreviations (G1): telen, telencephalon (cerebral cortex); dien, diencephalon (hypothalamus and thalamus); rhomben, rhombencephalon (hindbrain and cerebellum); mesen, mesencephalon (midbrain). Dashed lines indicate the boundaries of these regions.

domains of the leg imaginal disc from which these respective parts will later arise, defined by patterns of transcription factor gene expression (Lecuit and Cohen, 1997). These patterns represent the regulatory states that are required to set in train the development of each leg part.

Progenitor field specification in Bilateria occurs in canonical locations in respect to the body plan: brains form at the anterior end, hearts in the paraxial mesoderm, the intestine is posterior to the pharynx, in vertebrates the appendages occur at branchial and pelvic levels of the trunk and so forth. These locations are defined with respect to embryonic spatial coordinates. As an output of the late embryonic regulatory system, genes encoding signal ligands are expressed in given domains of the embryo, and the *cis*-regulatory modules controlling expression of progenitor field regulators respond directly to these signal inputs. For example, the expression of the *tinman* gene in *Drosophila* defines the progenitor field for the future heart with respect to the design of the embryo (Fig. 3.4D2), as expression of its ortholog, *nkx2.5*, defines the heart progenitor field in vertebrates (Fig. 3.5B). The *tinman cis*-regulatory module controlling this phase of expression is motivated by transcription factors (Mad, Medea) which are activated in response to Dpp signaling (Azpiazu and Frasch, 1993; Bodmer, 1993; Yin *et al.*, 1997), and this is the signal, transcriptionally expressed early on in the dorsal portion of the embryonic blastoderm, which designates the dorsally migrating bilateral strips of mesoderm where the heart will form (Fig. 3.4D). An additional embryonic spatial input is expression of the Twist transcription factor, which is expressed from the beginning in embryonic mesoderm: so these embryonic inputs in combination specify that the heart progenitor field will start out in the dorsal mesoderm of the embryo body plan.

In progenitor field specification the cells of the progenitor field must be in a (relatively) unspecified state before expression of the relevant regulatory gene, and the location of the progenitor field should be malleable, depending on where the signal is expressed. Both have often been shown to be true. This is the general import of observations recurrent in the long history of vertebrate experimental embryology, in which transplantation of tissue (which proves to be the source of the signals requisite for progenitor field specification) was found to cause ectopic body part development. Modern studies in the chick, for example, have established that ectopic presentation of Fgf signal ligands can induce ectopic limb formation anywhere in the lateral region between the normal limbs (Isaac *et al.*, 2000; for review, Tickle, 2004). As illustrated in Fig. 3.5C, the forelimb and hindlimb buds are distinguished by different transcriptional regulatory states (i.e., expression of *ptx1* and *tbx5*, respectively), though earlier both express *hoxc9* and other posterior group *hox* genes (Nelson *et al.*, 1996; Cohn *et al.*, 1997; Logan *et al.*, 1998).

Whatever its evolutionary significance, a deep and complex issue taken up in Chapter 5, there is remarkable conservation in the identity of the transcription factors that set up given progenitor fields across the bilaterians. Throughout the bilaterians the *nkx* homeodomain genes are used generally in early mesodermal specification events (Jagla *et al.*, 2001), including heart specification (Fig. 3.5B), *brachyury* and certain *forkhead* genes are expressed in hindgut progenitor fields (Wu and Lengyel, 1998; Lengyel and Iwaki, 2002), *otx* is expressed at the anterior end of bilaterian nervous systems, and so forth. A minimal conclusion is that regulatory linkages between the embryonic spatial regulatory apparatus, and the control systems of some genes defining the major progenitor fields of the body plan, are ancient and evolutionarily robust. Whether viewed historically or mechanistically, the apparatus controlling progenitor field specification is a fundamental and definitive aspect of the genomic regulatory code for the formation of bilaterian body parts.

Subdivision of the Progenitor Field Regulatory State

Multiple transient regulatory states separate the expression of regulators in the initial stages of progenitor field specification from the expression of controllers of cell type-specific differentiation gene batteries. Thus diverse cell types, and distinct morphological domains, will emerge from the initial progenitor field (reviewed in Davidson, 2001); for instance, think of the many discrete cell types that compose the heart and limbs developing from the progenitor fields illustrated, respectively, in Fig. 3.5A and Fig. 3.5B (even discounting the immigrant muscle and neuronal cell types that invade the growing limb from the dorsal axis). These many cell types are the output of regulatory states that are put into place only at the terminal steps of the progenitor field subdivision process. The fine scale regulatory expression patterns set up at the terminal stages of body part formation are of particular significance, because it is they which mark the actual developmental components of the finished body part, its internal modular structure. Usually these more advanced regulatory gene expression patterns agree with the intuition of the classical morphologist. Even so, as illustrated in the latter parts of Fig. 3.5, the detailed spatial precision of regulatory state programming in the later subdivision stages of adult body part development can be nothing short of overwhelming.

We see in Fig. 3.5D that the functional organs derived from the vertebrate gut tube are all functionally indicated, long in advance of their morphogenetic formation, by unique transcriptional regulatory codes. While aspects of this code are more or less conserved throughout the vertebrates (Chapter 5), note that in nonvertebrate chordates such as amphioxus (let alone the ascidian tadpole) the structures deriving from these transient regulatory subdivisions (large and small intestine, liver, pancreas, etc.), are lacking. Thus the developmental programs that follow the initial subdivision functions seen in Fig. 3.5D are evolutionary build-ons with respect to the basic chordate body plan. Proceeding to a finer level of specification, Fig. 3.5E shows the amazing precision of the regulatory functions by which every muscle group in the abdominal larval body wall of *Drosophila* is specified (Baylies *et al.*, 1998). This fate/space subdivision of course follows the earlier subdivision of the mesoderm into domains that will produce somatic muscle as opposed to heart or visceral mesoderm (Azpiazu *et al.*, 1996; Fig. 3.4), which is in turn downstream of the initial mechanism of embryonic mesoderm specification (Fig. 3.4D). Similarly, the neuroblasts that give rise to the *Drosophila* ventral CNS produce a variety of neuronal and glial cell types, and as illustrated in Fig. 3.5F, specification of each of these follows from expression of fate-specific regulatory genes. Again, this is a fine scale subdivision called into play following earlier events by which the neurogenic territory is initially specified and then divided into midline, medial, and lateral domains (Cornell and Von Ohlen, 2000; Stathopoulos and Levine, 2005; further discussed in Chapters 4 and 5). Finally, illustrating a middle stage of regulatory subdivision, we see in Fig. 3.5G a hint of the control process employed in specification of the major regions of the developing mammalian brain: The expression of 12 different transcription factors is illustrated, each in a unique subregion of what we name as anatomical components of the forming brain. This evidence presages the complexity of the underlying control system, which is yet to be experimentally revealed, for these expression patterns are but the observed, intermediate outputs of an encoded gene regulatory network.

BODY PLAN DEVELOPMENT AND THE ORGANISM

The homologous body parts of bilaterians differ clade to clade in their structures, sizes, shapes, and capacities. Developmentally earlier changes affect morphology and function more importantly. However, differences among organisms expand the further into the developmental process we look, and are greatest for the terminal phases. This last class of differences reflects changes in the spatial subdivision of regulatory states at the stage that determines where differentiation gene batteries are to be deployed. Examples include placement of bristles on the legs of various Drosophilid species, which depends on *hox* gene transcription domains, and control of bone length in bird forelimbs, which is also *hox* gene-dependent (reviewed by Davidson, 2001), but these examples can be multiplied many fold. The general conclusion is that the later portions of the process of regulatory field subdivision during adult body part formation are evolutionarily most flexible, the significance of which we explore in Chapter 5. In contrast, the establishment of the progenitor field at the correct embryonic address and the formation of the basic, spatial organization of the body part are evolutionarily much more conservative. This is a general property of the adult body part formation mechanism, which in itself predicts that there will emerge distinct architectural features associated with the respective aspects of the underlying gene regulatory networks.

Development of the body plan, that is, the process of embryogenesis and the processes of adult body part formation, are to be thought of essentially as the ultimate readout of genomically encoded systems for setting up sequences of spatial regulatory states. The embryonic and postembryonic regulatory systems operating at various levels and in various organisms have distinguishing features, which directly illuminate how they work, the subject of this Chapter. It is changes in these encoded systems for spatial subdivision that produce what we refer to as evolution of morphology, the generation of novelty, and the diversity in the body plans of descendants from common ancestors. But there are other aspects of development as well: for mammals some of these are of vital importance, such as the continuous "post-body plan" developmental renewal of our immune systems, and "regenerative" developmental repair processes that rely on dedicated set aside stem cells. The rules are undoubtedly different in some respect for these processes, for they often occur in single, migratory, pluripotential cells and they cannot utilize the tightly controlled, stage-by-stage intercellular architecture that is typically the source of the signals by which transcriptional subdivision of regulatory fields is oriented in body part formation. "Post-body plan" developmental processes are prominent in the vertebrates, and whether or how extensively they occur elsewhere is little known.

It is often said that development and evolution are two sides of the same coin. That is because mechanistic explanation of both leads directly to the heritable genomic regulatory code. These mechanisms, the subject of this book, are specifically those by which development and evolution of the bilaterian body plan likely take place. They are the causal explanations of the heritable morphological forms of animal life. But the development of the body plan provides only the physical basis of the potentialities of the species. The details of post-body plan development in each animal, its physiological responses, and its interactions with its environment make up a broader biological world that extends beyond the primary existence of its genetically encoded body plan.

CHAPTER 4

Gene Regulatory Networks for Development: What They Are, How They Work, and What They Mean

GENERAL STRUCTURAL PROPERTIES OF DEVELOPMENTAL GENE REGULATORY NETWORKS

Considering development a process of regulatory state specification, as in the last Chapter, leads seamlessly into the domain of genomic control logic, where the conditional assembly of developmental regulatory states is encoded. But unlike in Chapter 2, in which the focus was the individual *cis*-regulatory module and its functions as a multiple input logic processor, our scale and scope must here become system-wide. We have seen that regulatory state is the output of networks of *cis*-regulatory modules and the control genes they operate. The dimensions of a given gene network analysis depend on the number of regulatory states in space and time the developmental process entails, as in the diagrams of Fig. 3.3. Whatever their extent, however, developmental gene regulatory networks have an internal structure, in that they are composed of diverse kinds of modular parts and connections among these parts. Here "modular" takes on a simple functional meaning: it is used to denote small subsets of genes within the overall network that together execute given "jobs," e.g., to operate a certain differentiation gene battery, or to transduce an extracellular signal into a certain regulatory state. In what follows, sets of regulatory genes that execute modular functions are usually referred to as constituting "subcircuits" of the network, because as we shall shortly see they are "wired together" within the subcircuit by their gene regulatory interactions. Just as the target site inputs of an individual *cis*-regulatory module are integrated to generate novel outputs according to its genomic design, so the outputs of these subcircuits are integrated to generate logic outputs which depend on their organization, that is, their wiring architecture. Thereby we come to the most important imperative of the symbolic apparatus by which gene regulatory networks must be portrayed and interpreted. Since at each of the constituent *cis*-regulatory modules of the network the inputs are specified by the target site DNA sequence, the most fundamental of the things a network presentation has to do is to make explicit these genomically encoded inputs. The inputs themselves consist of transcription factors generated at other nodes of the network. Thus portrayed, the sum of the relationships that link given regulatory genes with their target *cis*-regulatory modules not only provide an immediate and unequivocal map of how each regulatory state is encoded in the genome, but also an immediate and unequivocal set of objects for experimental test. The inputs at each *cis*-regulatory module in the network can be authenticated, like any *cis*-regulatory predictions, by isolation of the module from the genome and causal analysis of their function by mutation. Authenticated in this manner, such networks differ from standard genetic diagrams of epistatic relationships in the crucial respect that they depict direct regulatory relationships encoded in target site DNA, rather than a mix of direct and indirect relationships that cannot necessarily be related to, or interpreted in terms of, genomic sequence. The term "node" in a gene regulatory network is used as shorthand for a gene encoding a given transcription factor plus its relevant *cis*-regulatory control module(s). The most important requirement of gene regulatory network presentation is that it precisely specify both the *cis*-regulatory inputs at each node, and the source of these inputs at other nodes (Arnone and Davidson, 1997; Bolouri and Davidson, 2002; Longabaugh *et al.*, 2005).

Overall Network Structure

Developmental gene regulatory networks are composed fundamentally of genomic components, that is, the genes and their relevant *cis*-regulatory modules at its nodes; and of regulatory state components, that is, the transcription factors which provide regulatory inputs into these modules. As an aid in this discussion, a brief glossary of the special use of terms in this area is given in Fig. 4.1A. In the form of presentation used in the following (already encountered in Fig. 1.3), network inputs are portrayed diagrammatically by vertical arrows (positive inputs) or by vertical barred stems (negative inputs) impinging on horizontal lines representing the genomic components. The outputs, indicated in the diagrams as bent arrows, represent both the mRNA and the encoded protein (e.g., Fig. 4.1B, C). This is a valid argument if it is the case that if transcribed, the RNA will be processed, exported to the cytoplasm, and translated. While that is usually so, in certain cases there are additional regulatory interventions downstream of transcription, most prominently interference with mRNA translation or survival by microRNAs which bind to sequence elements within target messages. In such cases there are required additional notations which separately gate the transcriptional expression of a regulatory gene *per se* and the function of its protein product (a specific example is included among the networks reviewed below).

Since life is cyclical, there will always be a prior regulatory state. In respect to any given developmental network, the initial state is defined as the sum of the inputs into the most upstream genes in that network. For example, in early development, where a gene regulatory network might begin with the first zygotic regulatory genes to be activated, the initial state would consist of the stored maternal regulators which are required to activate these genes in whatever spatial domain the network concerns. The terminal components of any given developmental gene regulatory network are genes the outputs of which do not affect other genes in the network. As summarized in Fig. 3.3 and ancillary discussion, in adult body part formation, or in Type 1 embryos, the natural terminal subcircuits are those encoding differentiation gene batteries (the internal organization of which was taken up in Chapter 1). For a gene regulatory network to be complete in its power to explain development, it should extend all the way to the expression of the relevant sets of cell type-specific proteins. The initial regulatory state and the terminal elements of the network can be thought of as together constituting the periphery of the network. The rest, the whole central apparatus of the network, consists of nodes occupied by genes which encode transcription factors or components of signaling systems: The basic role of all these genes is to generate spatial and temporal controls on the transcriptional expression of other genes.

The functional connections between the outputs of given regulatory genes and their target *cis*-regulatory modules at other nodes of the network are sometimes known as its "linkages." If correctly known, the linkages provide an exact map

A TERMINOLOGY AND GLOSSARY

General designations:

Authentication of network: experimental test of network linkages at *cis*-regulatory level.

Completeness of network: fraction of regulatory genes required for control of spatial patterns of gene expression that is explicitly included in a given network.

Inputs: transcription factors binding to a given *cis*-regulatory module at a node of the network.

Network linkage: functional connection between genes encoding transcription factors and the target *cis*-regulatory modules that these factors bind to and regulate.

Node of network: gene and relevant *cis*-regulatory modules which receive inputs from elsewhere in network, and provide outputs destined to targets elsewhere in network.

Output: RNA (or protein) product of a gene at a node of the network.

Periphery of network: initial inputs received from a prior developmental regulatory state which animate the most upstream nodes of a given network; or the final outputs of a developmental network; i.e., expression of differentiation gene batteries.

Process diagram: "bird's eye" network image that indicates relative positions of network modules in context of developmental parts and/or stages.

View from the genome (VFG): network image showing all *cis*-regulatory linkages summed over all regulatory states.

View from the nucleus (VFN): network image indicating those linkages active in the regulatory state that obtains in a given nucleus at a given time.

Parts of a developmental gene regulatory network:

Differentiation gene batteries: sets of genes that respond to a common set of cell-type specific regulators, which encode at the protein level the functional and structural properties of that cell type.

Input and output linkages: *cis*-regulatory controls on network subcircuits which either switch them on or repress them in given developmental situations.

Kernels: conserved subcircuit consisting of regulatory genes which interact with one another and which are dedicated to a specific developmental function.

"Plug-ins": common subcircuits that are utilized for many different developmental functions.

Subcircuits: small set of genes and *cis*-regulatory elements functionally linked so as to execute a given developmental"job."

FIGURE 4.1 For legend see pages 131–133.

B SUBCIRCUITS THAT DO DEVELOPMENTAL JOBS

B1 Subdivision of territory by subcircuit AND logic

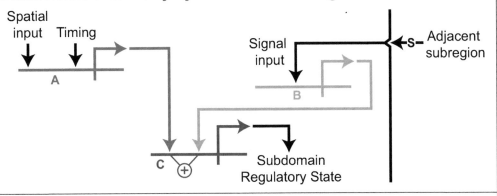

B2 Transformation of transient spatial regulatory input into stable regulatory state

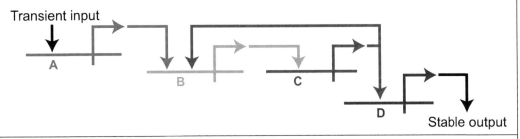

B3 Eventual specification of confined spatial expression beginning with broad domain

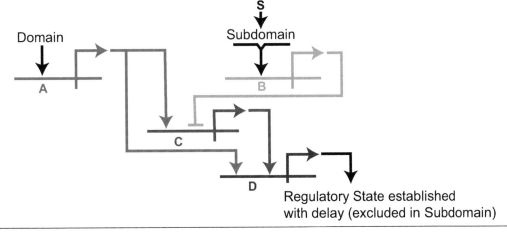

FIGURE 4.1 (continued)

(Continues)

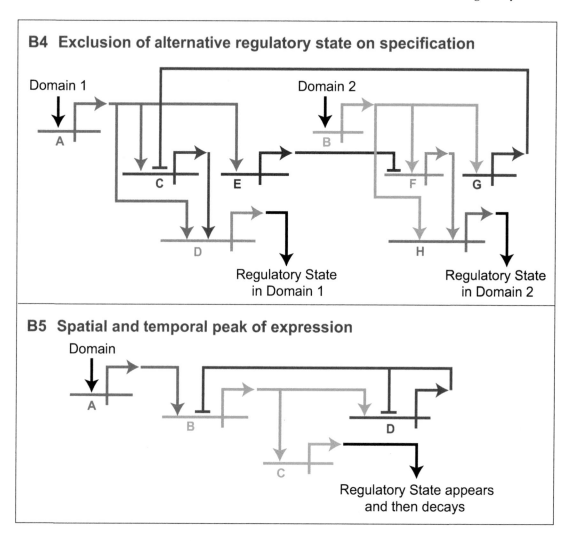

B4 Exclusion of alternative regulatory state on specification

Domain 1

Domain 2

A

B

C

E

F

G

D

Regulatory State
in Domain 1

H

Regulatory State
in Domain 2

B5 Spatial and temporal peak of expression

Domain

A

B

C

D

Regulatory State appears
and then decays

FIGURE 4.1 Fundamentals of developmental gene regulatory networks: glossary, diagrammatic symbolism, illustrative subcircuits, levels of analysis. (A) Glossary of useful concepts: summary definitions of phrases and terms as used in text. (B1—B6) Examples of network subcircuits. In each the output is an alteration of regulatory state, as indicated in black at the bottom of each panel, and in each case the biological consequence is to accomplish a given developmental "job," as indicated in red. Genes are color-coded, named with capital letters, and their relevant *cis*-regulatory modules are indicated as horizontal lines; the broken arrows denote the transcription of the gene. All genes in these diagrams encode transcription factors. The outputs of these genes, i.e., the protein factors

C PROCESS DIAGRAM AND NETWORK

C1 Process Diagram

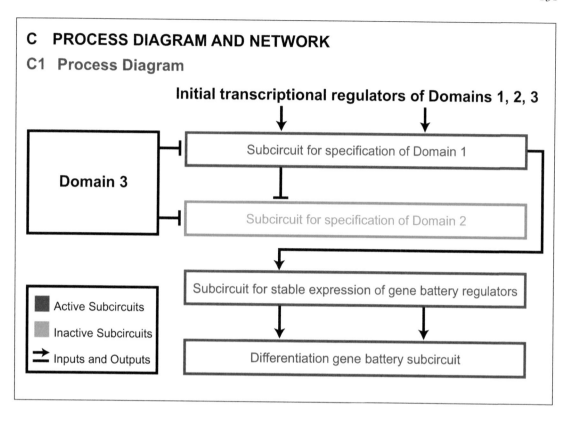

translated from their mRNA, are inputs to the *cis*-regulatory modules of other genes in the subcircuits, as indicated by the arrows extending from one gene to another, colored as the gene from which the output derives. Inputs are positive (activation) if they end in arrows; negative (repression) if they end in bars. Initial inputs from outside the subcircuits, signals from other cells (S), and subcircuit outputs, are shown in black. The circle containing "+" in B1 is an example of a designation of a logic function executed by the *cis*-regulatory module, here an "AND" logic operation (see Chapter 2). The diagrammatic symbolism is in general that utilized in the BioTapestry software for presentation of developmental gene regulatory networks (Longabaugh *et al.*, 2005; http://labs.systemsbiology.net/bolouri/software/BioTapestry/), and this same form of presentation is used throughout this book. (C) Levels of analysis. (C1) Process diagram, in which general functions required by the biology are considered as modular "black box" components that the network must include, though their internal wiring is unknown or not explicit. In this example a spatial domain of a developing system (Domain 1) is required to achieve a particular regulatory state. An exclusion function (e.g., see part B4 of this Figure) produced by Domain1 forbids the specification of Domain 2 to occur in Domain 1, while both the Domain 1 and 2

(Continues)

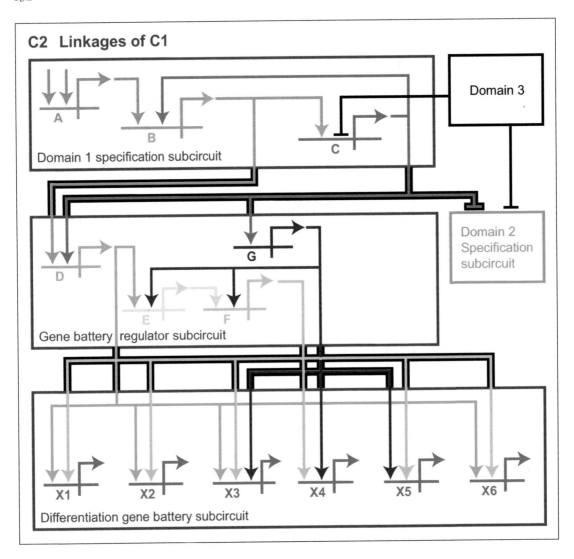

FIGURE 4.1 (continued)

states are excluded from Domain 3. Having generated the Domain 1 regulatory state (i.e., become specified), a stable regulatory apparatus for controlling a differentiation gene battery must be set up; and then the batteries including the differentiation genes themselves must be expressed. (C2) Network circuitry, explicitly indicating the linkages and therefore the underlying genomic regulatory code which executes the modular functions of (C1); that is, the circuitry which explains what lies within the black boxes of the Process Diagram and why these functions emerge as their outputs.

of the causal role of each regulatory gene in the network, and they show the inputs which are integrated at each *cis*-regulatory module to control the time and place of expression of the gene that it governs. Given the particular designs of the target *cis*-regulatory modules (cf. Chapter 2), this information suffices to explain why the regulatory and signaling genes in the internal domains of the network are differentially expressed where they are at successive stages. Therefore they explain how the developmental process is controlled. Furthermore, and most essentially, the explanation is in the direct terms of genomic input target sites. Thus the gene regulatory network provides the "transformation function" that relates the inherited, static, regulatory code in the DNA sequence to the progressive process of spatial gene regulation in development.

The causal importance of the inputs within a network is usually established initially by some form of knockout or loss of function experiments in which given factors are removed from the regulatory state either by genetic means, or by intervention in the molecular biology of expression of particular genes. The most common means of effecting the requisitely specific "trans-perturbations" of regulatory gene expression involve introduction of sequence-specific nucleic acids which target given mRNAs, e.g., "interference RNA" of various kinds (Novina and Sharp, 2004) or morpholino-substituted antisense oligonucleotides (which have enhanced stability compared to standard oligonucleotides (e.g., Nasevicius and Ekker, 2000; Heasman *et al.*, 2000). Other methods available either have undesirable pleiotropic effects or more boutique usefulness, limited to special circumstances. These include, for instance, introduction of mRNAs encoding fusions of the target transcription factor with obligate repressor protein domains (Jaynes and O'Farrell, 1991), or of mRNAs encoding versions of the factor lacking transcriptional activation or DNA binding domains, or mRNAs encoding immunological reagents that target the transcription factor (Bogarad *et al.*, 1998). Gain-of-function perturbations are also sometimes useful, as by injection of mRNA encoding transcription factors into eggs, forcing expression under an ectopically expressed regulatory system, or introduction of fusions between the DNA binding domain and an obligate activator domain (Davidson *et al.*, 2005). However, stratagems that involve overexpression of transcription factors often yield artifacts unless the intracellular concentrations are carefully controlled. But no matter how precisely done and gene specific are trans-perturbations, it is impossible by these means alone to exclude indirect effects and demonstrate a direct input except by *cis*-regulatory analysis (i.e., to distinguish between the case that Gene A affects Gene B and Gene B then affects Gene C, from the case that Gene A directly affects Gene C).

Outputs of each subcircuit are marked with black outlines. The specification subcircuit of Domain 1 includes a feedback loop which renders the output regulatory state independent of the initial inputs (parallel orange vertical arrows). A cross-regulating set of control genes is then activated and these generate the sets of drivers required by the differentiation genes at the bottom of the Figure.

If on the other hand an input target site in a nodal *cis*-regulatory module of a network is mutated, and as assayed in a gene transfer experiment the result is loss of the input function (cf. Chapter 2), the input is shown to be direct. Turning the argument around, a great strength of gene regulatory networks is that the linkages of which they are composed can all be experimentally authenticated by *cis*-regulatory measurements. In practice, if the time and place of expression of regulatory genes is accurately known and this knowledge is incorporated in the interpretation of trans-perturbation experiments, the majority of erroneous assignments of inputs that are actually indirect can be avoided *a priori*, and other essential aspects of network architecture inferred. For example, if Gene A is expressed in an overlapping time frame with Gene B, and knockout of Gene A function depresses the activity of Gene B, but these genes are expressed in different adjacent spatial domains, a signaling function downstream of Gene A and upstream of Gene B is indicated. These kinds of logic manipulations, and the organization of network architecture from epistatic relations within the same spatial domain, can increasingly be done by machine (e.g., Longabaugh *et al.*, 2005).

In addition to the authenticity of a gene regulatory network, that is, the extent to which it has been certified at the key nodes by *cis*-regulatory analysis, its completeness is an essential property. If it is incomplete, some spatial and temporal aspects of the expression of the genes in the network that depend on missing nodes and inputs will remain unexplained. But this could be something of a slippery slope. As we know from thorough examinations of a few *cis*-regulatory control systems, most notably that of the sea urchin *endo16* gene (Chapter 2), in order to explain every quantitative aspect of temporal regulatory output, occupancy of every target site must be considered. On the other hand, the causal logic of development depends on, and results in, spatially and temporally differential gene expression. Therefore those aspects of the input regulatory state which vary in space and stage must suffice to explain the qualitatively differential expression of the next tier of genes in the network. These inputs are none other than the "drivers" of our earlier discussion (Chapter 2; Yuh *et al.*, 1998, 2001; Istrail and Davidson, 2005). Other inputs that are constantly available can be regarded as necessary for *cis*-regulatory processing of the varying driver inputs, but they cannot account for the expression of a gene in one domain and not another, or at one time and not another. They are in that sense part of the invariant ambient machine for gene expression control; in *endo16*, as we recall, this class includes seven of the ten different species of input into Modules A and B (Fig. 2.3). Therefore, for most explanatory purposes, driver inputs alone suffice in a developmental gene regulatory network, together with the genes producing these drivers as their outputs.

In Chapter 3 the diverse phenomena of development were treated as transformations of regulatory state, but regulatory state throughout has meant driver regulatory state, the product of regulatory genes expressed differentially and variably. That is, development is basically encoded in the driver regulatory genome. This is the underlying reason that gene regulatory networks consisting of driver inputs and outputs and not all inputs can explain development (that is, not necessarily including the genes that encode ubiquitous and time-invariant transcription factors which are also utilized in the relevant *cis*-regulatory modules). The field has been feeling its way toward this simple resolution since the first evidence in the late 1950s and early 1960s suggesting that differential gene expression is causal for cell type-specific properties, and for embryonic development (e.g., see Davidson, 1968). The illuminating power of driver gene regulatory networks, in the most assorted and different developmental contexts, is illustrated in the various examples below. This is not to say that that their value will not be augmented when some day we can expand these networks to include every input, developmentally variable or not. Of course it will, but in a different direction, as we take up in a following discussion in this Section.

Gene Regulatory Networks from Different Perspectives: Subcircuits to Process Diagrams

A developmental gene regulatory network represents a physically real set of functional linkages, and hence it can be viewed from different angles, and at different focal lengths. One can zoom in on a particular *cis*-regulatory module, or zoom out to look at the organization of the whole process at such a distance that none of the detailed circuitry is visible and instead the only features in view are the relationships of the diverse subcircuits to one another in the context of the major domains of the developmental process. Termed a "process diagram" (Davidson *et al.*, 2002a), this is sometimes a very useful way to start building a network, and it generally provides an overview of the subcircuits that the network is composed of, and how they might be related to inter-domain signaling interactions. Or, since gene regulatory networks have both genomic and regulatory state components, they can be considered from either vantage point. A view of the network that considers at once all interactions occurring at all nodes of the network at all times and in all relevant spatial domains, is referred to as the "view from the genome" (VFG); a view that specifies only the interactions occurring in a given regulatory state, i.e., in a given nucleus at a given moment in time, as "the view from the nucleus" (VFN; Bolouri and Davidson, 2002). The VFG is that which provides the direct prediction of what driver target sites must be encoded in the DNA of the nodal *cis*-regulatory modules, since these structural features exist independently of their state of activity. In other words, the VFG shows the static organization of the regulatory genome underlying a given phase of development, including all the components, subcircuits, and linkages of which the network is composed. The VFN, on the other hand, is the developmental biologist's image of the network: It tells what is happening differentially in the spatial/temporal domains of the system, what interactions occur when and where, as a result of what inputs. Thus there is one VFG but potentially many VFNs for a given network, depending on the number of diverse regulatory states it accommodates.

The VFG is the best way to perceive the regulatory logic that emerges only at the network level. Subcircuits which carry out given developmental jobs often have successive phases of activity: for example, a subcircuit might be activated by a transient initial input and then set up a stable cross-regulatory relationship (of which we see several real instances below). No one VFN will convey its overall structure as will the subcircuit VFG. In Fig. 4.1B are five examples of network subcircuits. Every node in each of these examples includes a gene that encodes a transcription factor. These subcircuits would each execute a unit developmental function, as indicated. They are portrayed as VFGs so their structural relationships may be visualized at a glance even if their functions are generated sequentially. While, for the purposes of this Figure, these examples are all synthetic, their real-life resemblances will become apparent in the following. The main points here are that the unit functions, or the developmental "jobs" which subcircuits such as these execute cannot be perceived by looking at any one gene: the function is the property of the subcircuit. These functions grow out of the nature of the linkages between the particular *cis*-regulatory modules that they include. They carry out developmental processes that lie in the realm of biological observation: spatial territorial subdivision (Figs. 4.1B1 and B3), institution of regulatory state by setting up a stable feedback loop (Fig. 4.1B2), exclusion of alternative regulatory states following specification (Fig. 4.1B4), and generating temporal expression peaks (Fig. 4.1B5); the examples could be multiplied. Functions expressed in the currency of developmental processes require the information processing power afforded by network subcircuits, at the least. As the sub-circuits behave as modular components, networks have a "pebbly" kind of constitution. Their subcircuits, the regulatory "pebbles" that generate their developmental functions, are connected to one another by intermodular link-ages, which designate the inputs to each subcircuit, and determine how (or if) its output will be used.

The usefulness of the bird's eye, process diagram view of a network will now be less abstract. It designates the subcircuits in terms of their developmental unit functions without specifying them, whether as part of an *a priori* or an *a posteriori* analysis. As in the illustration in Fig. 4.1C, these developmental functions are transparent in the process diagram of Fig. 4.1C1 relative to the wiring diagram of Fig. 4.1C2. On the other hand, once subdivided by the conceptual exercise of generating the process diagram, the structure/function causality of the network subcircuit organization is easily absorbed because it has been modularized.

Qualitative Diversity of Network Components

The image of a modular, "pebbly" internal organization, different from a maze of equipotent connections lacking local functional features, suggests basic inhomo-geneity in network structure and function. But that is just the beginning. Gene regulatory networks are made up of qualitatively different kinds of parts, includ-ing different kinds of subcircuits, different kinds of regulatory devices external to these, and different peripheral elements. These inhomogeneities are significant for the study of development, as they account for different aspects of function; for any experimental attempt to reformulate given networks; and certainly for evolution, as we take up in Chapter 5.

At the outset let us distinguish the meaning of the term network "subcircuit" as used in this discussion from its meaning in other contexts where it is synonymous with network "motif." This usage figures prominently in many abstract discussions of the nature of network building-blocks (particularly in the context of yeast, bacteriological, and physiological networks, e.g., Lee *et al.*, 2002; Milo *et al.*, 2002; Balázsi *et al.*, 2005). A subcircuit as used here is a small set of particular, functionally linked regulatory genes which together execute a developmentally defined job. A motif is a small set of genes functionally linked in some particular way, and how they are linked, irrespective of the identity of these genes or the biological job they perform, provides the definition. For example, a positive feedback loop (Gene B activates Gene C; Gene C activates Gene B), as for instance in Fig. 4.1B2, is a motif. So is the "forward feed" motif seen in Figs. 4.1B3 and B4, and in Fig. 4.1C2 as well (e.g., in Fig. 4.1B3, Gene A activates Gene C; Gene C activates Gene D; Gene A also activates gene D). As we see below, these motifs occur frequently, and in many entirely unrelated developmental subcircuits. They contribute in a canonical way to the properties of the subcircuit. Given motifs have implications for the kinetics with which the subcircuit operates, and for other (developmentally more important) information processing functions: for instance in the forward feed motif the inputs into the *cis*-regulatory modules controlling Genes A and C, and their domains of expression are never the same, and the device is often used as a combinatorial means to sharpen the spatial expression of the target Gene D (see Fig. 4.1B3 or B4). Similarly the feedback motif not only stabilizes and amplifies regulatory state, but it also renders no longer needed the prior regulatory state which set it going at an earlier stage (e.g., Gene A input in Fig. 4.1B2). Motif properties are thus important, but in themselves they do not entirely describe the job the subcircuit does, even abstractly, as do the legends shown in red in each panel of Fig. 4.1B. The subcircuit always includes additional architecture than that of a canonical motif. Motifs cannot definitively be associated with any particular regulatory genes or signaling pathway, because each appears in many different subcircuits. Subcircuit analysis of gene regulatory networks, rather than exclusive motif analysis, displays the modular functional components of development at the level useful for relating the genomic regulatory code to the observable process of development.

Class distinctions can be made among different kinds of subcircuits (Davidson and Erwin, 2006). A very important class of subcircuit consists of sets of specific genes the products of which form functional cassettes that are utilized repeatedly in different developmental processes within the life cycle of given animals and in different animals. Their prevalence and prominence contributes largely to the powerful image of a shared bilaterian regulatory toolkit. Such cassettes are deployed in such a variety of ways that they seem to be inserted into the network wherever their function is useful; hence the name we apply to them, "plug-ins." Plug-ins cannot by definition be associated uniquely with any specific developmental role.

The most prominent examples are provided by the small set of intercellular signaling systems used in developmental specification processes: for example throughout the bilaterian world Hedgehog, TGFβ, Wnt, FGF, and Notch signaling cassettes are utilized and reutilized in diverse developmental contexts. The cassette includes in each case the genes (and their regulatory modules) encoding the ligands, the receptors, and the transcription factors which provide the immediate early responses used to transduce the intercellular signal at the transcription level in the cell expressing the receptor. In each of these signaling systems the cassettes are largely invariant in their orthologous molecular constitutions, as they appear in different bilaterians, but their target *cis*-regulatory modules occur in as great variety as the diversity of the jobs they do. Thus the distinction between the plug-in and the device into which it is plugged is not in the least obscure. Other kinds of plug-ins would be genes and their regulatory systems that encode transcription factors which form complexes and function jointly all across the bilaterians, for instance, *exd/hth* (*Drosophila*), and *meis/tale* (mouse). Note, however, that in the relevant cases the components of these transcriptional factor plug-ins are all expressed differentially, just as are the signaling plug-ins. They are not ubiquitously present components of the cellular biochemical machinery for transcription, like the Groucho corepressor or most TAFs or histone acetylases, etc. The importance is that their deployment must be programmed at the *cis*-regulatory level. Encoding their expression, thereby controlling the deployment of plug-ins, is a key function executed within the subcircuits of gene regulatory networks.

Other subcircuits are, in contrast, dedicated specifically to one given developmental function, the same function in different animals, functions such as gut or heart formation, gastrulation, patterning of appendages, and so forth. This is a main subject of Chapter 5, and for now as a place holder, we note only that there exist such gene network subcircuits. They are functionally essential in development, and in respect to their internal "wiring" they are conserved in evolution. They are termed the "kernels" of the network. In addition to dedication to given biological tasks, kernels have other definitive properties. The genes of which they are composed are linked to one another in multiple ways, directly or indirectly ("recursive wiring"); and deletion of function of any or many of the genes of the kernel therefore causes the same loss of function phenotype. Taking the oversimplified example in Fig. 4.1C2 as a heuristic illustration of a kernel, loss of either Gene B or C expression in this "specification subcircuit" will equally cause loss of expression of all the differentiation genes (X1–X6). The cross-regulatory subcircuit that controls the differentiation gene battery is also recursively wired and loss of any one of the gene functions would be equally catastrophic for the whole system, a force for conservation of the internal wiring. For at least the few available cases, these properties of dedicated developmental function, recursive wiring, common loss of function phenotype, and evolutionary conservation, suffice for the identification of kernels. That having been said, however, an overwhelming feature of the evidence thus far is its thinness. We as yet have available so little hard comparative developmental gene network data that it is not yet possible to infer how many or what kernels exist in bilaterian gene regulatory networks, beyond the few examples that establish their existence, and several other probable cases (Chapter 5). But as we see below, kernels are a particularly potent, significant, class of subcircuit.

Subcircuits have to be told what to do and what not to do; like all aspects of genomic developmental regulatory programming from the *cis*-regulatory module up, they operate conditionally on the inputs they receive from the ambient regulatory state. In the diagrams of Figs. 4.1B, C the inputs and outputs, with respect to the subcircuits, are shown in black. The purpose is merely to indicate that there are several biologically different kinds of input, though in a very general sense, at the gene regulatory level they all work more or less as switches, in conceptual terms positioned on the outside of the boxes that contain the subcircuits. There are two general kinds of input switches. Some transcription factors function positively as inputs when they bind their *cis*-regulatory target sites, either as initial inputs or the output products of upstream subcircuits, and in the absence of the positive input, the subcircuit is silent (e.g., in all the examples of Fig. 4.1B). Others function actively as silencers, which when they bind their *cis*-regulatory target sites repress expression at particular nodes of the network subcircuits. As discussed earlier (Chapter 2) many signaling systems act both ways: they are silencers in the absence of the signal and permissive or collaborative activators if the receptor receives the signal (Barolo and Posakony, 2002; Istrail and Davidson, 2005). External signal-mediated inputs are shown in Figs. 4.1B1, B3, affecting Gene B in each case. Except at the periphery the outputs of given subcircuits are the inputs of the next subcircuit (Fig. 4.1C2). Note in Fig. 4.1C that in domain 3 there is an external clamp on expression of either the domain 1 or domain 2 subcircuits, explicit in the diagram for domain 1. This kind of negative input device is very commonly used in development to delimit progenitor fields for given body parts, and we see many examples below. The significance of its place in the subcircuit wiring diagram of Fig. 4.1C2 is that these clamps do not build, specify, or generate regulatory state; rather, they merely prevent the operation of those subcircuits that do, as illustrated for domain 1. Input/output (I/O) devices that lead into and out of the subcircuits, including the external switch plug-ins which convey signaling data from adjacent cells, are to be considered a separate category of network parts. They are like a skein of "wires" connecting the subcircuit "pebbles" of the network.

Though in this discussion they appear mainly at the bushy perimeter of the gene regulatory network, the differentiation gene batteries contain the largest amount of protein coding information (see discussion of structure and complexity of differentiation gene batteries in Chapter 1; Fig. 1.5). Rarely do we know more than a sample of their constituents. But those we do know suffice to indicate that their *cis*-regulatory modules obey a small set of regulators. This in turn implies that in the hierarchical organization of developmental control systems (Fig. 3.3), there will exist subcircuits which immediately overlie each differentiation gene battery called into play, the function of which is to generate a relatively stable regulatory state controlling the deployment of differentiation genes. The requirement for stability of state means there will be feedbacks or cross-regulation, some form of recursive wiring, in the subcircuits executing this role. Such is the subcircuit in the lowest tier of regulatory genes in the diagram of Fig. 4.1C2, and as we soon shall see, such are the differentiation control subcircuits that have been derived experimentally.

Extensions

The driver gene regulatory networks considered below are essentially genome-based maps of the inherited control logic for given phases of development, in which large numbers of explanatory interrelations are conveyed by the network architecture. As such, developmental regulatory networks are an end in themselves, that is, if they can be made reasonably complete, and are authenticated at the *cis*-regulatory level. But since they convey the regulatory logic explicitly, they may also serve as the basis for extensions into two other kinds of "how it works" questions. These two other domains of inquiry indeed intersect: they are the actual information processing code deriving from the intrinsic design of the *cis*-regulatory nodes of the network; and the kinetics with which the network operates. We know just enough from modeling and a few experimentally based explorations to feel the strong scientific lure of extensions into these areas.

Suppose all the target sites in nodal *cis*-regulatory modules of a network were known and explicitly included, that is, those for transcription factors that contribute to processing of the driver inputs, in addition to those directly mediating the driver inputs. As we have earlier seen many of the functions executed by the factors other than the drivers *per se* are in the nature of conditional Boolean logic switches (Chapter 2; cf. Yuh *et al.*, 2001; Istrail and Davidson, 2005), while others act to modify quantitatively the driver inputs; for example, to amplify them. First and foremost, were all sites included, the network map would now give the complete regulatory genome for that phase of development. While the network relations of the drivers alone probably suffice to explain the developmental progression of regulatory states, a map that made explicit the complete regulatory genome would suffice to explain all of the relevant regulatory DNA sequence itself, in functional terms. This is not exactly the same objective as those of the developmental biologist, but it is an ultimately important one. The basic experimental requirement for extension to this level of knowledge is mutational analysis of every binding site, plus a logic analysis of how these sites work together to process whatever input information they confront (e.g., as in the *endo16* gene; Yuh *et al.*, 2001; Fig. 2.3). Given a quantitative description of the input regulatory state at any given moment, (i.e., the concentrations of the relevant transcription factors in a given nucleus at that time), the time course of change in regulatory state at that node, and given the primary DNA sequence, it would become possible to predict quantitatively the output at that node.

In order for the functions mediated at each site to be executed the sites have to be occupied. As briefly discussed in Chapter 2, occupancy depends on cooperativity and DNA-protein interaction constants as well as the factor concentrations (for statistical mechanical and equilibrium treatments, McGhee and von Hippel, 1974; Ackers *et al.*, 1982; Emerson *et al.*, 1985; von Hippel and Berg, 1986; Berg and von Hippel, 1987). In a comprehensive network analysis, whether these parameters or the occupancies were directly measured, or both, a network that included the complete modular regulatory genome components would bring to life the function of each node as a little machine which utilizes input regulatory factor concentrations according to the unique combinatorial logic program afforded by its design. The next big qualitative extension in developmental gene regulatory network analysis will be invention of ways to express nodal network functions in terms of logic transactions executed upon the inputs, so as to generate the spatial instructions inserted at that node, and then to integrate these regulatory outputs at the subcircuit and network system levels. Thus the quantitative output generated by each of these machines could be explicitly understood. At present we approximate the quantitative regulatory functions at network nodes by black boxes which process qualitative driver inputs by simple overall logic operations, e.g., AND logic or dominant repression logic. But even if they are correctly situated, and have the right inputs and outputs, such black boxes are only serving as place holders for the mechanistic understanding of their internal workings to come.

To proceed to the kinetics with which networks operate, a further consideration of events which occur on the DNA at the network nodes is required: this is the relation between *cis*-regulatory output and actual transcription rate. That is, ideally we would like to know for each node at any time, given the target site occupancies, what will be the rate of RNA transcript generation from that node? Mechanistically this relation will depend in detail on the nature of the activation functions stimulated by the particular *cis*-regulatory impetus received by that individual basal transcription apparatus, a difficult and complex matter, but there are escapes available. For one thing, it is relatively easy to measure transcriptional output directly and thus short-circuit this issue; or as done by Bolouri and Davidson (2003) a more or less mechanism-independent assumption can be made that transcriptional activity is proportional to driver site occupancy until it approaches the saturating absolute value of the maximum transcription rate characteristic of the system.

The dynamics with which networks operate depend basically on the parameters that determine the time required from the production of the primary transcript of a gene encoding a transcription factor to the production of the primary transcript of a target gene responding to that encoded factor. There are many such parameters, all to some extent system-dependent (including temperature); some rate limiting, some not; some more easily measured than others; some particular to given molecular species; and others more or less similar at a default value for a given cell type. For any given product, the relation between the time course of its accumulation and the relevant synthesis and turnover rate constants (over the period that they are constant) is given by a linear first order differential equation, assuming that a stochastic first order decay process is responsible for the turnover (e.g., Galau *et al.*, 1977), and these constants are conventionally extracted by fitting the measured product accumulation curves to the appropriate function. The same basic approach is followed in the case of cascades of gene expression where a regulatory gene is transcribed and its product is required for transcription of a downstream gene, which in turn causes transcription of a further downstream gene, except that the solution of the large linked systems of differential equations now required produces a need for extensive measurements (or assumptions) to constrain the many parameters that arise.

Though *a priori* models of these kinds can indeed be built, it must be said that they are not of much use in the context of typical developmental gene regulatory network architecture, because these networks do not usually consist only of linear gene expression cascades. Instead they include *cis*-regulatory repression, branch points, gates, positive and negative feedbacks, combinatorial inputs which may act differently from one another, switches, intercellular signaling, and so forth. Furthermore, the turnover rates of given transcription factors are key but highly variable parameters which directly determine the relationship between the activity of the gene encoding it and the amount of the factor present at any given time. Transcription factor turnover rates are difficult to measure, or at least relatively few such measurements exist. So the result is that only for certain special developmental situations are there computationally driven reconstructions of the operation dynamics of significant portions of any gene regulatory network. A good exemplar is a model for the anterior/posterior sequence of gap regulatory gene expression in the syncytial *Drosophila* embryo (Jaeger *et al.*, 2004). Here the special advantages for this kind of analysis are first, that the action happens in a single layer of nuclei at the surface of the embryo (cf. Fig. 3.4), much facilitating measurement of the rise and fall of given transcription factors by imaging methods; second, that because there are no cell walls the factors diffuse away from their sites of synthesis and so their momentary concentrations any time and place can be calculated by application of well-known diffusion functions together with the synthesis and decay rates. Nonetheless, comparison of model results to carefully timed series of quantitative factor distribution measurements, reveals that in order to explain the observed shifts in spatial distribution, *cis*-regulatory processing of the mutually repressive interactions of these genes is required in addition to the computed spatial factor concentrations (Jaeger *et al.*, 2004). A main useful outcome of these and other measurements and analyses (Nasiadka *et al.*, 2002) has been the determination of the relevant rates with which the factor concentrations change; that is, of the general rate of regulatory state change during a given developmental process. In the syncytial *Drosophila* embryo the time between successive states in a cascade of this network is a matter of 10-20 minutes; but this is again a system variable, and in the sea urchin embryo it is two to three hours (Bolouri and Davidson, 2003). In both cases the activation of the next gene long precedes attainment of steady state, and is dependent on the initial rate of factor production, not the steady state level (Bolouri and Davidson, 2003). Part of the absolute difference in rate between *Drosophila* and sea urchin systems is due to the difference in temperature, 24°C versus 15°C, though not all of it can be. But since animals of given clades live at many different temperatures, it is obvious that there is nothing of profound causal significance in the absolute values of these rates in any case.

In some ways the kinetic function of a developmental gene regulatory network bears the same relation to the regulatory DNA code that determines the architecture of the network, as does chromatin modification mediated by the transcription factors bound at *cis*-regulatory target sites to that same code (Chapter 1). Both are multistep processes required to put into biochemical effect the regulatory processes encoded in the genome; both follow downstream from, and are qualitatively specified by, sequence-specific interactions occurring at *cis*-regulatory DNA target sites. Siren-like, both also invite the mechanistically inclined to forget or take for granted the underlying causal regulatory logic, and focus exclusively on the details of its effectuation, confusing that with qualitative causality. The problem of resolving network kinetics falls into two conceptual domains: first, the relation over time between transcription factor inputs, site occupancies, and transcriptional outputs at network nodes, according to their complete *cis*-regulatory design; and second, the chain of subsequent off-the-DNA synthesis and turnover processes that intervene between transcription of a regulatory gene and occupancy of targets which its encoded factor recognizes. The last relates the kinetic operation of one node to that of its downstream targets, but the information specifying which are its downstream targets is resident in the architecture of the network, and is necessary to resolve the kinetics, rather than the reverse. The first of these conceptual domains would require comprehensive *cis*-regulatory network analysis of nodal information processing, as discussed above, but ultimately this will likely be the extension of current gene regulatory networks that will most deeply enhance our comprehension of the genomic regulatory code for development. For in development the multiple inputs at each node convey much spatial information, as well as controlling quantitative output over time; *cis*-regulatory information processing generates new spatial as well as temporal outputs (cf. Chapter 2) which are then fed into the operation of the network.

We turn now to present knowledge, and a diverse sample of experimentally derived developmental gene regulatory networks. Two of these examples, the sea urchin and *Drosophila* networks, are at real-life scale; that is, they are approaching completeness for the processes with which they deal, and each illuminates the program for its process of development in a way that can only emerge at the level of network architecture. Others, while they provide marvelous examples of network circuit design in respect to biological function, are rapidly being augmented, but are to be considered works in progress. It is the nature of gene regulatory networks that they are continuously being updated, as they become more complete and more intensively authenticated; the following examples are snapshots of current knowledge. So far we have examined in principle the unique value of the developmental gene regulatory network, as the bridge between the genome and the processes of development: now we are to use that bridge, and return to the problems of regulatory state specification laid out in Chapter 3. Only the scale is now that of real-life regulatory interactions; the players are the sets of genes that construct the real states; the contexts are particular; and the chain of developmental causality is rooted where it initiates, in the regulatory genome.

GENE REGULATORY NETWORKS FOR EMBRYONIC DEVELOPMENT

The Sea Urchin Endomesoderm Network

Embryonic development in indirectly developing sea urchins (here *Strongylocentrotus purpuratus*) is a canonical Type 1 process, in which the regulatory system early generates lineage-specific territories, and then immediately drives on toward the definition of embryo-wide cell type specification domains (territorial organization and developmental fates in these embryos are summarized in Fig. 3.2A). The general sequence of progressive regulatory functions underlying this type of development was given in Fig. 3.3A. In Fig. 4.2A we see these functions in the specific terms of the endomesodermal territories of the sea urchin embryo. This is a process diagram, as discussed above, such that each of the oval circles putatively contains a subcircuit of regulatory genes, activated early on either by anisotropic initial inputs, by intercellular cleavage stage signaling, or by inputs generated by the prior regulatory state in the respective territorial nuclei. Time, from early cleavage to the 24–30 h late blastula stage (at 30 h there are about 600 cells), proceeds from top to bottom. In each of the endomesodermal domains of Fig. 4.2 cell types arise in which differentiation genes have begun to be expressed by the end of the period to which the diagram pertains (indicated in oblong boxes at the bottom). The relevant territories are the skeletogenic mesenchyme domain, formed exclusively from descendants of four 5th cleavage "micromeres," the mesodermal territory, which gives rise to several cell types, though the only ones considered here are pigment cells; and the endodermal domain, which produces the gut. Most of the gut and the mesodermal cells derive from a single ring of eight, 6th cleavage cells ("veg$_2$"). The hindgut, however, is formed from descendants of their ("veg$_1$") sister cells toward the end of the morphological process of gut formation (Logan and McClay, 1997; Ransick and Davidson, 1998). Gastrulation begins only at about 30 h with the invagination of all veg$_2$ progeny. Before this, at 21–24 h, the 32 descendants of the skeletogenic micromeres individually invade the blastocoel. The necessary point of this chronology is that the process diagram, and the network of gene regulatory interactions that we will shortly focus upon, concern only the period of specification that terminates with the skeletogenic cells within the blastocoel, though they are not yet executing the secretion of skeletal biomineral. The future endoderm

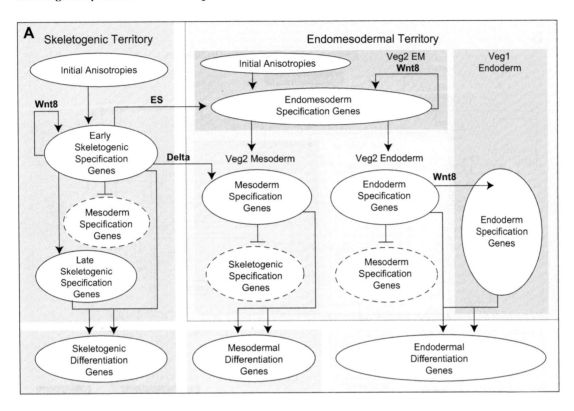

FIGURE 4.2 Network for specification of sea urchin embryo endomesoderm: process diagram, view from the genome, and *cis*-regulatory evidence of selected subcircuit functions. (A) Process diagram (modified from Revilla-i-Domingo and Davidson, 2003). The diagram and network in this Figure pertain to zygotic regulatory functions occurring in the cleavage and blastula stages, i.e., up to about 24–30 h, when gastrulation gets under way (see Fig. 3.2A for the territories and anatomy of the embryo). The color-coded areas of the process diagram represent the domains of the embryo as indicated; the relevant phenomenology of development is summarized in text. Black arrows labeled "ES," "Delta," and "Wnt8" indicate signaling interactions. ES is an early signal emitted from the skeletogenic domain (4th–6th cleavage micromeres) that is necessary and sufficient for endomesodermal specification to occur in adjacent cells, which functions by canceling out an otherwise repressive function (Ransick and Davidson, 1993, 1995; Oliveri *et al.*, 2003; Kenny *et al.*, 2003); see text for the Delta and Wnt8 signals. The linkages in the networks below reside within the oval areas of this diagram (cf. Fig. 4.1C); those areas surrounded by dashed lines are network subcircuits for alternative states which are excluded by repression in the domains shown. (B) The endomesoderm network, as a view from the genome. The network, initially presented by Davidson *et al.* (2002a,b), is continuously updated

(Continues)

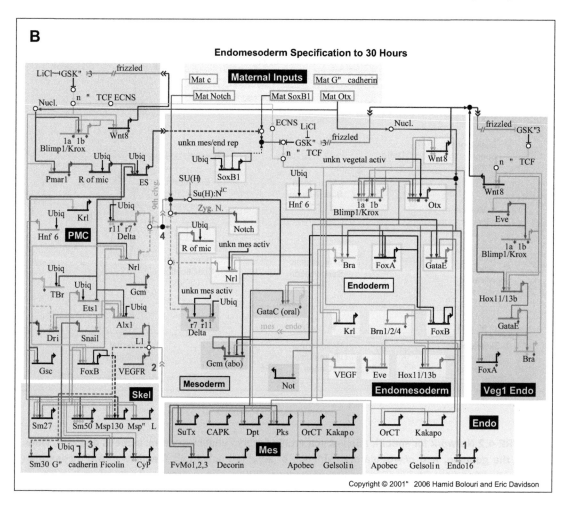

FIGURE 4.2 (continued)

at http://sugp.caltech.edu/endomes/, where views from the nuclei at any desired time and place and much underlying experimental data as well as a current list of supporting references are maintained. Diagrammatic symbolism (BioTapestry; Longabaugh *et al.*, 2005) is as in Fig. 4.1. Genes in the bottom rectangles encode differentiation proteins, of which only small samples are included; most remaining genes in the network encode transcription factors; a few encode components of signaling systems, as indicated by the names beneath the genes. From left to right the areas of the diagram represent interactions occurring in the skeletogenic domain (lavender); in the endoderm and the remainder of the mesodermal (i.e., nonskeletogenic) cell types (light green area: genes on blue backgrounds function ultimately in mesoderm, those on yellow backgrounds in endoderm); and in the late invaginating veg$_1$

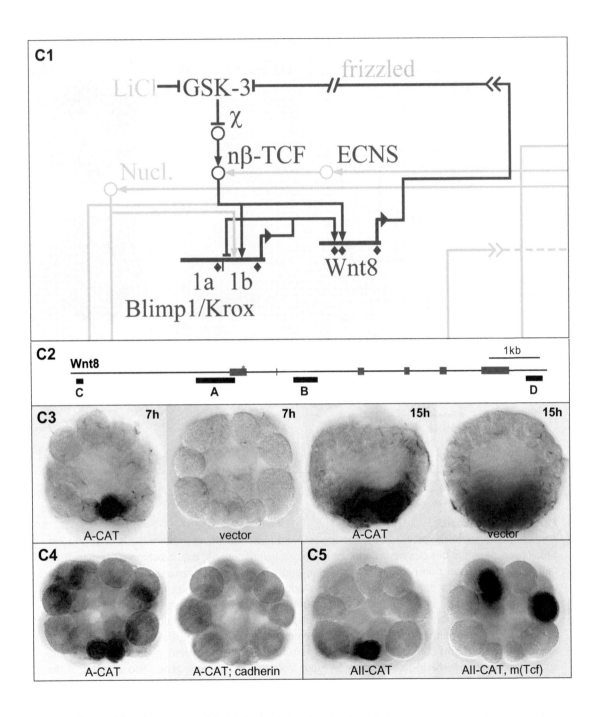

FIGURE 4.2 (continued)

(Continues)

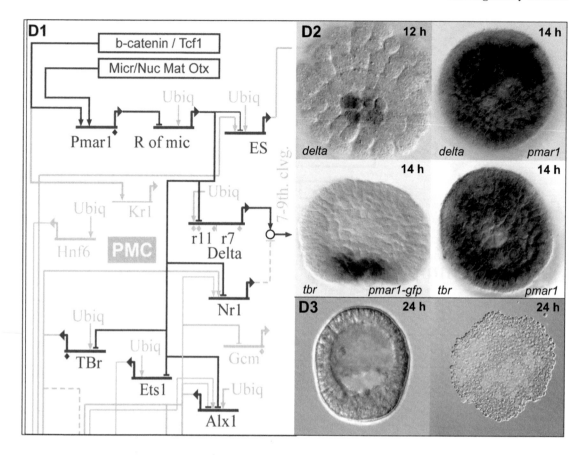

FIGURE 4.2 (continued)

endoderm (orange background). At the top are relevant maternal inputs (maternal mRNA or protein). Small open circles in this diagram represent cytoplasmic interactions, e.g., biochemical transactions among signal transduction pathway components, and small solid black circles represent junctions, where multiple sources of given inputs merge. In addition the diagram indicates nodes at which the *cis*-regulatory modules executing the functions specified in the network are known (red dots), and the role of individual target sites has been established by mutation (green dots). (C) *cis*-Regulatory control of the *wnt8* reinforcing loop. (C1) Enlargement of an area at the top of the network in (B), showing the *wnt8* and *blimp1/krox* genes. As described in text the *wnt8* gene is activated with nuclearization of maternal β-catenin, plus an input from the *blimp1/krox* gene, also a target of the β-catenin/Tcf1 input. The zygotically produced Wnt8 ligand is received in adjacent endomesoderm cells, where it drives further β-catenin nuclearization via canonical signal transduction machinery. Since both *wnt8* and *blimp1/krox* respond to the β-catenin/Tcf1 input, this functions as a positive reinforcing sub-circuit: all cells of the endomesodermal domain both express and respond to Wnt8. It is progressively turned off after some hours, however, in the sequence with which it was activated, due to repression

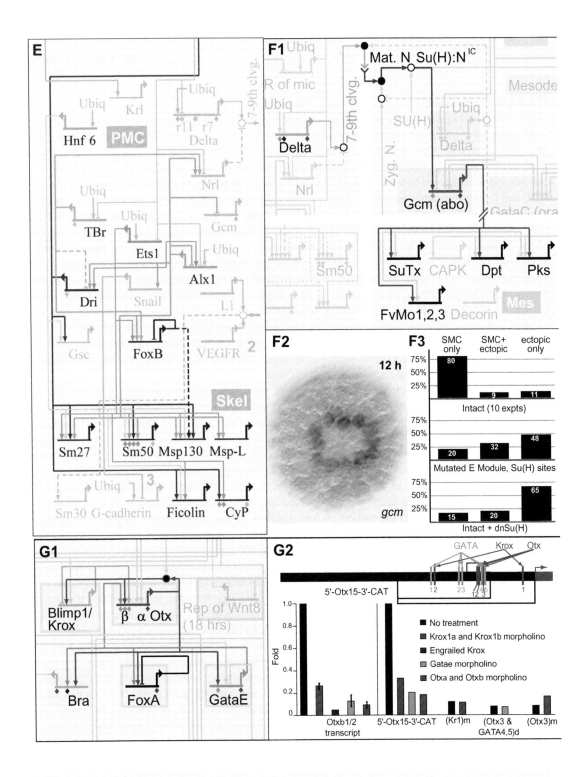

FIGURE 4.2 (continued)

(Continues)

FIGURE 4.2 (continued)

of the *blimp1/krox* gene by its own product (Livi and Davidson, 2006). The result is the cessation of *wnt8* expression and eventual disappearance of β-catenin, and of *blimp1/krox* mRNA, first in the skeletogenic domain, then the mesodermal domain, then the endodermal domain. (C2) Map of the *wnt8* gene, displaying exons (red; transcription is from left to right) and four genomic sequence patches highly conserved between *Strongylocentrotus purpuratus* and *Lytechinus variegatus*. Of these A, B, and C have important regulatory activity in the early embryo: A produces accurate early expression requiring the β-catenin/Tcf1 input; B boosts expression, but acts ectopically in isolation, control over which is exerted by the repressive effect of the Tcf1 interaction at sites in A in the absence of the β-catenin cofactor; and C contains closely linked Tcf1/Blimp1/krox sites and requires both inputs for function. (C3) Double *in situ* hybridization, displaying accurate expression of *wnt8A-CAT* expression construct ("A-CAT") 7 and 15 h after fertilization and injection into eggs. Endogenous *wnt8* mRNA is shown in yellow ("vector"), first in skeletogenic micromeres, then throughout the veg_2 domain (cf. Fig. 3.2A). CAT reporter mRNA is developed in purple. Incorporation of exogenous DNA is clonal and mosaic, though entirely random, and the construct expresses only within cells included in the normal domain of expression. (C4) Effect of removing β-catenin/Tcf1 input on the expression of the same transgene, by introduction of cadherin mRNA. This traps β-catenin in the egg cytoplasm. Control is on left, cadherin-treated egg on right; expression, again monitored by *in situ* hybridization, is totally abolished. (C5) Control of ectopic expression at Tcf1 sites: a construct containing a subregion of Module A is expressed accurately, left; but if its two Tcf1 sites are mutated, ectopic expression occurs, here seen in ectoderm cells. (C2–C5) are from Minokawa *et al.* (2005). (D) Control of skeletogenic micromere-specific regulatory state by a double repression network subcircuit. (D1) region of network in (B) highlighted in red to display the *pmar1* gene. In the micromere lineage the Pmar1 repressor represses the otherwise globally active *repressor of micromere* gene, allowing activation of the downstream targets under control of this system (Oliveri *et al.*, 2002; see text). The *pmar1* gene is activated specifically in micromeres by two localized inputs, nuclearized Otx and nuclearized β-catenin. (D2) Derepression by global introduction of *pmar1* mRNA of two of the *pmar1* targets, the genes encoding the Delta ligand and the Tbrain transcription factor. Normal expression, monitored by *in situ* hybridization, is shown on left, confined in both genes to the skeletogenic micromere domains, and the globally ectopic expression produced by injection of *pmar1* mRNA is shown on right (from Oliveri *et al.*, 2002). (D3) Response of *delta cis*-regulatory module to *pmar1* mRNA: accurate expression of *deltaR11-GFP* construct (see Fig. 2.2) in skeletogenic mesenchyme just after ingression into the blastocoel on left; widespread ectopic expression of same construct in an embryo bearing *pmar1* mRNA on right (from Revilla-i-Domingo *et al.*, 2004). (E) Skeletogenic gene battery (*sm27*, and *sm50* biomineralization genes; *msp130* and *msp130L* glycoprotein genes, and *cyclophilin* and *ficolin* genes) and its subcircuit of regulatory drivers. (F) *cis*-Regulatory transformation of the Delta signal into the pigment cell regulatory state. (F1) Network subcircuit, highlighting transduction of the Delta signal ligand by maternal Notch:Su(H) transcription factor, which activates the *gcm* gene. This in turn locks itself on, and drives expression of pigment cell differentiation genes which encode pigment synthesis enzymes (Calestani *et al.*, 2003). (F2) *In situ* hybridization displaying expression of *gcm*, exquisitely confined to the single ring of cells that receive the Delta signal from the skeletogenic cells that abut them within (from Ransick *et al.*, 2002). (F3) *cis*-Regulatory control of *gcm* response to Notch/Delta signaling. Spatial expression of *gcm-GFP* expression constructs, scored as percent of embryos expressing and mesoderm cells of veg$_2$ lineage still remain in the epithelial wall of the embryo, having not yet begun invagination. Thus except for the skeletogenic cells, no cellular migration takes place during the period considered (for modern summaries of the lineage and territorial organization of the embryo, see Davidson, 1989; Davidson *et al.*, 1998; Davidson *et al.*, 2002a). What has happened by 24–30 h is essentially only the spatial institution of territorial regulatory states (in the sense of Fig. 3.3A). But as the gene regulatory network in Fig. 4.2B impressively shows, these pregastrular endomesodermal specification functions require an enormous regulatory architecture, wired into the *cis*-regulatory modules of almost 50 genes.

Typically for Type 1 embryos, essential signaling interactions required for territorial specification occur among adjacent blastomeres in the fixed matrix of the mid-cleavage and early blastula stage sea urchin embryo. The two such interactions with which we are especially to be concerned are indicated in Fig. 4.2A. The first is a signal consisting of the Wnt8 ligand, which is synthesized and secreted by all endomesoderm cells and received by all endomesoderm cells, and that is required for continuing expression of multiple key endomesodermal regulatory genes (Wikramanayake *et al.*, 1998, 2004; Emily-Fenouille *et al.*, 1998; Logan *et al.*, 1999). The second signal consists of the Delta ligand, which is synthesized by the skeletogenic progenitor cells and received via the Notch receptor in the immediately adjacent veg$_2$ cells. Thereby these are transformed into precursors of other kinds of mesodermal cell, in particular, pigment cells (Sherwood and McClay, 1997, 1999, 2001; Sweet *et al.*, 1999).

correctly in "secondary mesenchyme" (smc), the mesodermal origin of the pigment cells, or elsewhere in the embryo (ectopic) or both (from Ransick and Davidson, 2006). The intact construct is compared to one in which the relevant Su(H) sites are mutated, which decreases correct expression sharply, and also permits extensive ectopic expression. At the bottom a similar effect can be seen from introducing the intact construct into eggs expressing an altered form of Su(H) protein lacking a DNA-binding domain, which nonproductively scavenges intracellular Notch. The ectopic expression in the presence of mutated Su(H), or mutated Su(H) target sites, is due to the failure of repression normally mediated by Su(H) when bound to *cis*-regulatory DNA in the absence of N protein signaling (i.e., here outside the prospective mesoderm). (G) *cis*-Regulatory control of a key component of the *otx-gatae* network kernel discussed in text. (G1) Structure of the kernel, highlighted, which features a positive feedback loop between *otx* and *gatae* genes. Its function is to lock in the endoderm regulatory state, to which both these factors are important contributors. (G2) *cis*-Regulatory experiments on the relevant module of the *otx* gene (from Yuh *et al.*, 2004). Target sites in the DNA sequence for the factors predicted from the network analysis are shown at top, in the context of an expression construct (5'*otx15*-3'-*CAT*). Normalized expression of endogenous *otx* gene assessed by QPCR measurement of mRNA is shown in left panel, when each of the three predicted inputs is removed as indicated. At right is expression of construct assessed by QPCR measurement of CAT mRNA, in the same perturbations, as indicated by the color code. Almost identical results occur in the presence of the normal inputs but only if the *cis*-regulatory target sites are mutated (black bars, m) as indicated (Kr, Blimp1/Krox site).

A nearly complete gene regulatory network for pregastrular endomesoderm specification in sea urchins is shown in Fig. 4.2B, as a view from the genome. The form and significance of the diagrammatic presentation are as discussed above, and the developmental biology to which it directly pertains is that summarized in the process diagram of Fig. 4.2A. The network is based on four sources of evidence: (*i*) the time course, (*ii*) the place(s) of expression of all the genes included, measured by quantitative PCR (QPCR) and *in situ* hybridization, respectively; (*iii*) a large-scale perturbation analysis in which expression of each gene was blocked (usually by application of morpholino-substituted antisense oligonucleotides) and the effects on level of expression of all other relevant genes in the network measured sensitively by QPCR; and then (*iv*) an extensive authentication by direct *cis*-regulatory analysis at the key nodes of the network. From the genome sequence of *S. purpuratus,* the species for which the network was derived, all the encoded transcription factors have been predicted, and their temporal and spatial patterns of expression in the embryo determined (Howard-Ashby *et al.*, 2006a,b; Materna *et al.*, 2006; Oliveri *et al.*, 2006a). A few regulatory genes were thus revealed to be expressed specifically in the endomesoderm before 24 h which had not been evident as players earlier, and which can now be linked into the architecture of the network. But these studies indicate that this network is unlikely to be far from complete, that is, lacking important causal drivers of the progression of regulatory states which it explains. The network is maintained and updated continuously at http://sugp.caltech.edu/endomes/, where can be found most of the underlying gene expression and perturbation data, as well as views from the nuclei at any desired time or location in the endomesoderm (for reviews and references, Davidson *et al.*, 2002a,b; Oliveri and Davidson, 2004; Levine and Davidson, 2005; and legend to Fig. 4.2B). In the following we delve into the subcircuits and functionalities of the enormous maze of genomically encoded interactions pictured in Fig. 4.2B, and revisit in concrete form many of the concepts treated abstractly in Chapter 3 and in the previous section of this Chapter. The network is based on experimental fact. Its essential meaning is in the end that it conveys the significance of relationships which transcend the particular properties of its components (here biochemical). We shall see that the network explicitly presents the genomic regulatory logic by which the specification functions of Fig. 4.2A are effected.

As outlined in Chapter 3, the earliest zygotic operations of a spatially localized nature in embryogenesis reflect *cis*-regulatory responses to the initial anisotropies presented in the cleavage-stage embryo. The sea urchin embryo affords a major case. This is the nuclearization of maternal β-catenin in all and only all cell lineages that will give rise to endomesodermal territories (Fig. 3.1F and ancillary discussion). The regulatory influence of this transcriptional cofactor (of the Tcf1 DNA-binding transcription factor) can now be traced directly into the *cis*-regulatory target sites of key endomesodermal regulatory genes (light blue lines in Fig. 4.2B). Among the genes specifically requiring direct β-catenin/Tcf1 inputs are *blimp1/krox* in the future endomesoderm; *pmar1* in the future skeletogenic domain; and *wnt8* in both. A second early anisotropy of regulatory significance for this network is the localization of maternal Otx transcription factor within the micromeres (4th cleavage parents of the cells which form the dedicated skeletogenic lineage; Fig. 3.2A). Right after these cells are born, both β-catenin (Wikramanayake *et al.*, 2004; Minokawa *et al.*, 2005) and Otx (Chuang *et al.*, 1996) are specifically concentrated in their nuclei, and both inputs are used for the transient activation of the *pmar1* specification gene exclusively in that lineage (Oliveri *et al.*, 2002). Direct experimental evidence adds the important point that the *cis*-regulatory system of the *pmar1* gene is the exclusive genomic response apparatus in the regulatory network for the skeletogenic lineage at which the β-catenin input is required. The evidence has the following structure (Oliveri *et al.*, 2003): First, ectopic introduction (at normal per cell levels) of *pmar1* mRNA, which in undisturbed embryos is expressed only in micromeres and their immediate descendants, suffices to turn any blastomere of the embryo into a skeletogenic cell, even future ectoderm cells which would never normally express skeletogenic genes. Furthermore, prevention of β-catenin nuclearization (by trapping it in the cytoplasm with excess cadherin) cancels transcriptional activation of the *pmar1* gene and skeletogenic lineage specification. However, this last effect is rescued by co-introduction of *pmar1* mRNA. Thus in a normal skeletogenic precursor overexpressing cadherin, or even in an injected ectoderm cell, *pmar1* mRNA alone suffices to short circuit the lack of β-catenin/Tcf1 input and cause transformation to skeletogenic fate. The gene regulatory network in Fig. 4.2B thus not only shows what the β-catenin anisotropy does in the skeletogenic territory at the gene regulatory level, but it also shows the only thing it does in this territory.

Figures 4.2C(1–5) illustrate different modular subcircuits of the endomesoderm network. It is educational to think about these examples both individually and together. They are all informative in regard to the main objective, to illuminate the regulatory code for specification so we can really understand why this embryo develops just as it does. But they also directly illustrate many of the particular principles addressed above: (*i*) that the *cis*-regulatory DNA sequence code provides the necessary and sufficient explanation for the primary developmental specification functions; (*ii*) that these unit biological functions emerge from the interactions of small sets of regulatory genes "wired" together in subcircuits; (*iii*) that signaling inputs act by directly affecting the activity of transcriptional regulatory subcircuits; and (*iv*) that there are indeed in Type 1 embryos direct regulatory links that extend across a relatively shallow informational landscape, all the way from the initial anisotropies of the early embryo to the territorial expression of differentiation gene batteries.

The first example is the *wnt8-blimp1/krox* subcircuit (Fig. 4.2A, B, C1–5). The idea is that following the anisotropic nuclearization of β-catenin in cleavage-stage skeletogenic and endomesodermal precursor cells, all the cells in these territories signal to one another ("community effect," Gurdon, 1988), and this signaling is necessary for continued endomesodermal and skeletogenic specification (Davidson *et al.*, 2002b). It is mediated by transcriptional expression of the *wnt8* gene throughout these territories but nowhere else. Reception of the Wnt8 ligand by neighboring cells causes further progressive nuclearization of β-catenin by the well-known canonical Wnt response pathway, which is operative in these embryos, and is necessary for endomesoderm specification (Emily-Fenouille *et al.*, 1998; Wikramanayake *et al.*, 1998, 2004). The very important effect is to transfer the initial dependence on the transient β-catenin input into a reinforcing forward drive loop that now (until it is canceled many hours later) depends on zygotic gene expression. The specific regulatory prediction of the network architecture is thus that the *cis*-regulatory control system of the *wnt8* gene (Fig. 4.2C2) will obligatorily utilize Tcf1 target sites, as well as target sites for the Blimp1/Krox factor. This last establishes a second interlocking, positively acting, loop, since the *blimp1/krox* gene is itself dependent on the β-catenin/Tcf1 input (Fig. 4.2C1). Experimental examination of the genomic structure and function of the *wnt8 cis*-regulatory system indeed reveals that the predicted inputs are required (Minokawa *et al.*, 2005; Fig. 4.2C4, 5), and more. Not only is the *wnt8* gene activated directly through specific *cis*-regulatory target sites by the β-catenin/Tcf1 and Blimp1/Krox inputs, respectively, but an additional function of the Tcf1 sites in one of its *cis*-regulatory modules is to repress ectopic transcription in the absence of the β-catenin/Tcf1 input (Fig. 4.2C5). The elegant little subcircuit mechanism portrayed in Fig. 4.2C1 is thus written directly in the genomic regulatory code.

A second example is shown in Fig. 4.2D1. Here we see the subcircuit that causes all the initial zygotic transcriptional functions particular to the skeletogenic lineage to be expressed only there. As we have just seen, the initial transducer of the maternal regulatory indices of this lineage is the *pmar1* gene, and the subcircuit consists of the set of genes which generates the primary, lineage-specific regulatory state, and also accounts for the expression of the Delta signal received by the adjacent future mesoderm cells (Fig. 4.2A). Rather surprisingly, the spatial specificity of these functions depends on a double repression. The *pmar1* gene encodes an obligatory repressor (Oliveri *et al.*, 2002), and it in turn represses a second, globally active zygotic gene also encoding a repressor, which keeps all the genes of this subcircuit off except in the founder cells of the skeletogenic lineage where *pmar1* is active. Thus wherever *pmar1* is expressed, the target genes of the global repressor such as *delta*, or *tbrain* (*tbr*), a regulatory gene required for skeletogenesis, are also expressed (Fig. 4.2D2, 3). To examine the genomic code for this subcircuit, the *cis*-regulatory module responsible for *delta* gene expression in the skeletogenic lineage of the cleavage stage embryo was isolated (Revilla-i-Domingo *et al.*, 2004; its identification by interspecific sequence comparison was illustrated above in Fig. 2.2A1, 2). Again we see that the functionality of the architecture, that is, of the circuitry causing expression to be confined to the micromeres, resides in *cis*-regulatory design at a node of the network: when introduced into an egg in which *pmar1* mRNA is being translated in all cells, the *delta cis*-regulatory expression construct, like the endogenous gene, also is expressed globally (Fig. 4.2D3).

In addition to emission of signals the role of the skeletogenic territory of the embryo is to proceed as directly as possible to its differentiated function, secretion of the calcite skeletal biomineral. Portions of the gene batteries performing this function are known, and as shown in Fig. 4.2E, they provide an excellent illustration of a peripheral gene regulatory network "part." These differentiation gene batteries are located in the network downstream of the *pmar1* specification subsystem, just as indicated in Fig. 4.2B, and are controlled by a partially cross-regulated set of genes which encode the transcription factors required by the differentiation genes (Makabe *et al.*, 1995; Oliveri *et al.*, 2002, 2006a; Otim *et al.*, 2004). Note that the latter genes fall into three sets with respect to the pairs of driver inputs they use. The essential point here is that the network explicitly traverses all the way across the process (Fig. 4.2D1, E), from the interpretation of the initial anisotropic cues, to skeletogenic lineage specification, to the expression of differentiation gene batteries. The concept that Type 1 embryos operate by means of "shallow" regulatory systems (Chapter 3) is displayed here in the concrete terms of network architecture.

Proceeding in the network diagrams of Fig. 4.2B and Fig. 4.2F1 sideways, along the link from the Delta ligand to its Notch (N) receptor in the prospective mesoderm cells, we see that the genomic way station where this signal input is transformed into a change in transcriptional regulatory state is the *gcm* gene (Fig. 4.2F2). The prediction is that the *cis*-regulatory system of this gene will respond to N signaling. That is, in more detail, the relevant *cis*-regulatory module should include sites for the transcription factor which mediates N signaling (the sea urchin orthologue of the Suppressor of Hairless factor), and these sites should be necessary both for expression of *gcm* in the mesoderm cells and for pigment cell specification downstream of this; also, as is sometimes reported for N signaling systems, the sites might be used as well for repression of the *gcm* target gene in ectopic locations in the absence of the N ligand (Morel and Schweisguth, 2000; Barolo and Posakony, 2002; Davidson *et al.*, 2002a). These predictions were all confirmed on experimental examination of the *gcm* *cis*-regulatory system (Ransick and Davidson, 2006; see Fig. 4.2F3). Note that the cases considered in Figs. 4.2C1 and 4.2F1 together include the two main endomesodermal signaling interactions revealed by experimental embryology, a completely different level of observation. The examples demonstrate exactly why these signals are produced where they are, and how the respective intercellular spatial interactions are processed into network inputs, according to the sequence-specific genomic regulatory code.

The concept of network kernel was introduced above, in considering the qualitative diversity of network parts. The sea urchin endomesodermal network affords at least one excellent example. This is the five-gene subcircuit highlighted in Fig. 4.2G1 (Yuh *et al.*, 2005). Briefly, this VFG diagram summarizes the sequence of events underlying gut formation, as follows: the *blimp1/krox* gene is activated in the embryonic endomesoderm early in cleavage, and resolves to the endoderm during blastula stage (Livi and Davidson, 2006). During this time it in turn contributes an input required for activation of a *cis*-regulatory module of the *otx* gene, which directs transcription of a particular *otx* mRNA form (β1/2-*otx* mRNA). This *otx* gene product auto-activates its own *cis*-regulatory module, and also activates the gene encoding the Gatae regulator, a major controller of endodermal functions. As shown in Fig. 4.2G1 two of these functions are mediated by expression of the *foxa* gene (see below), and of the *brachyury* (*bra*) gene, needed for gastrulation (Rast *et al.*, 2002; Peterson and McClay, 2003). Meanwhile the *gatae* gene product feeds back to provide a necessary input to the $\beta 1/2$-*otx cis*-regulatory module, rendering the Blimp1/Krox input no longer requisite, and indeed it is transient due to *blimp1/krox* autorepression (Livi and Davidson, 2006). The ultimate consequence is that the process of endoderm specification is driven inexorably forward, locked by feedback into generation of the definitive endodermal regulatory state defined by the presence in the same nuclei of Otx, Gatae, Bra, and Foxa drivers (Davidson *et al.*, 2002a). Interference with expression of any of these regulators causes loss of gut formation. The key DNA components in the subcircuit architecture are the *cis*-regulatory modules of the five genes of this kernel. Results are shown for the $\beta 1/2$-*otx cis*-regulatory module in Fig. 4.2C5. Here it can be seen that the only logic *modus operandi* that will fit the presumed function of this "processor," viz. AND logic, in fact obtains (cf. Chapter 2), both for the endogenous regulatory module and for an experimental expression construct driven by this module (Yuh *et al.*, 2005). As can be true only for an AND logic processor, interference with the "*trans*" input (i.e., by blocking translation of the respective factors) and "*cis*-interference" (i.e., by mutating the target sites for these proteins) have identical effects. The Gatae, Otx, and Blimp1/Krox inputs are all requisite for expression above a background level in the developing embryo.

The complex, recursive wiring of the kernel in Fig. 4.2C5 is not a quirk of recent sea urchin evolution. A very illuminating insight into its structure/function properties came from a comparison with the orthologous network subcircuit in a starfish (Hinman *et al.*, 2003). Echinoderms have a good fossil record, and so we know that the last common ancestor of starfish and sea urchins lived almost a half-billion years ago, just after or possibly even before the end of the Cambrian (Paul and Smith, 1988). Yet as Fig. 4.3 shows, all the key linkages in the sea urchin kernel are also present in the starfish, though other surrounding linkages are different. As taken up in Chapter 5, extreme evolutionary conservation of regulatory architecture such as displayed here is a definitive character of kernels. The several ways of looking at this, e.g., as a consequence of the lethality of interference with any one of its components, or of the recursivity of its wiring, all amount to the same thing: this subcircuit is an essential and inflexible modular element of the overall gene regulatory network.

Another property revealed by experimental examination of the sea urchin gene regulatory network is important to mention in passing, as it recurs as well in other of the networks we examine briefly below. This is what could be termed the "exclusion principle" (Oliveri and Davidson, 2006): as each state of specification is achieved in each domain of the embryo, alternative regulatory states are excluded. This may happen at many levels and in many ways, but the most fundamental is that directly encoded in the genomic regulatory architecture. It is the activation, as

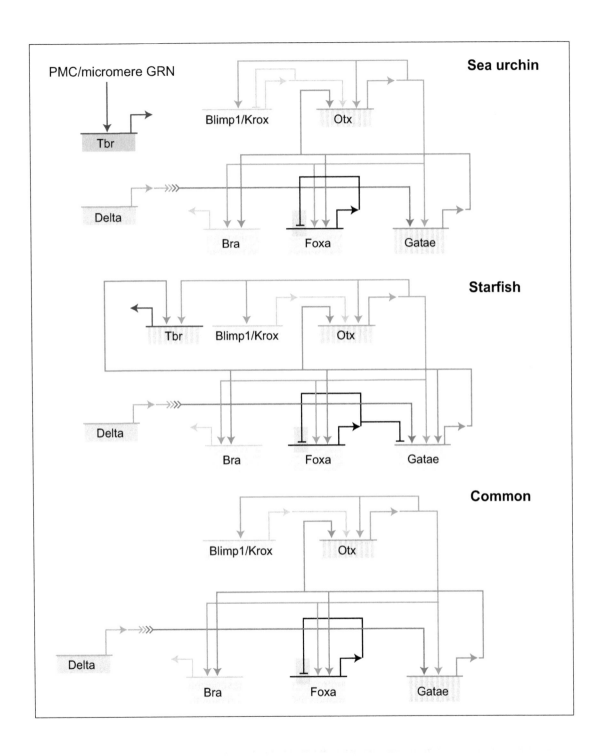

FIGURE 4.3 For legend see page 158.

(Continues)

part of the specification program, of genes encoding transcriptional repressors for regulatory genes essential to other modes of specification. Several examples are at hand: (*i*) In the endoderm the *foxa* gene is activated, as we have just seen, and one of its functions is to prevent expression of the *gcm* pigment cell subnetwork (Fig. 4.2F1), so that if *foxa* activation is arrested, the embryos turn up *gcm* expression and produce excess pigment cells at the expense of what should have been gut (Oliveri *et al.*, 2006b). (*ii*) In the mesodermal cells the skeletogenic pathway is specifically inactivated, so that if the *gcm* pigment cell specification system is turned on in these cells, they cease expression of skeletogenic functions (Damle and Davidson, 2006). (*iii*) The reverse is also true, so that if the mesodermal cells are forced to express skeletogenic regulators, they turn off the pigment cell program (Oliveri and Davidson, 2006). Additional examples are known elsewhere in the embryo (e.g., see Amore *et al.*, 2003), even though current evidence is still preliminary as to the extent of the exclusion effect. But it may be that every subnetwork that functions to set up a new regulatory specification state includes transcriptional repression of alternative states, particularly where these are the assigned properties of sister cells or of adjacent cells which have undergone some conditional specification process (Chapter 3). More broadly stated, an overall feature that almost glares forth from the sea urchin endomesoderm network is the presence of multiple forms of encoded device all of which act to ensure its correct forward progress: the intergenic feedback of its endoderm kernel, its several autoregulatory genes, its community effect signaling circuit, its cross-regulatory transcriptional subcircuits, its repressive regulatory exclusions. No wonder that

FIGURE 4.3 Detailed and extensive conservation of a kernel of the endomesoderm network. The top diagram shows essentially the same kernel of the *S. purpuratus* endomesoderm network as in Fig. 4.2G1, with the addition of the Delta input into the *gatae* gene in the endoderm (this signal is emitted from the adjacent mesodermal domain at late blastula-early gastrula stage; cf. Fig. 3.2A). The expression of the *tbrain* gene in the skeletogenic domain is also indicated (Fig. 4.2D). The middle panel shows the equivalent circuitry in a very distantly related echinoderm, the starfish *Asterina miniata* (Hinman *et al.*, 2003). Starfish and sea urchins have been evolving independently since a last common ancestor around the end of the Cambrian. At the bottom is reproduced the circuitry held exactly in common in the two species (modified from Davidson and Erwin, 2006). The kernel is entirely conserved, but there are significant peripheral differences, and many more not shown, that are farther removed from this location in the network. For example, in *A. miniata* the *tbrain* gene is expressed in the endomesoderm under control of the endomesodermal regulators Gatae and Otx, where it is required for endomesoderm formation, while in sea urchins this is a required skeletogenic gene not expressed and not needed in the developing endomesoderm; in *A. miniata*, *foxa* represses *gatae*, while it does not in the sea urchin; conversely the *blimp1/krox* gene represses itself in the sea urchin but does not in the starfish (Hinman *et al.*, 2003).

when sea urchin eggs are fertilized, virtually all of them develop properly and identically at more or less the same rate, almost impervious to their environment so long as they are not poisoned or cooked!

As we saw in Chapter 3, Type 1 embryos use similar regulatory algorithms, and so consideration of the sea urchin endomesoderm gene regulatory network will suffice for this class of embryonic process. In the two other major Type 1 experimental systems, the ascidian *Ciona* and *C. elegans* (Fig. 3.2), regulatory networks for several different embryonic processes are under study. In *Ciona* these include muscle cell specification downstream of the anisotropically localized maternal transcription factor Macho1 (Fig. 3.1D and ancillary discussion; Yagi *et al.*, 2004a, 2005), notochord specification (Yagi *et al.*, 2004b), and heart specification (Davidson *et al.*, 2005). In *C. elegans* the regulatory network underlying specification of EMS blastomere descendants (defining the progenitors of gut, posterior pharynx, among other cell types) has been subjected to genetic, *cis*-regulatory, and expression perturbation analyses (reviewed by Davidson, 2001; Maduro and Rothman, 2002; see Broitman-Maduro *et al.*, 2005; Maduro *et al.*, 2005; Witze *et al.*, 2006). A peculiarity of this system is the incidence of pairs of genes encoding Gata family factors, which probably cross-regulate, another kind of regulatory state lockdown device. All of these zygotic transcriptional networks for Type 1 embryonic processes display similar architectural features to those discussed in the context of the sea urchin endomesoderm network. Their common general characteristic is their genomically hardwired, direct progression from the initial interpretation of signaling and other spatial information in early cleavage, to the establishment of territorial regulatory states, to the rapid activation of differentiation gene batteries in the respective cell lineages.

Regulatory Gene Network for Specification of Mesoderm in the *Xenopus* Embryo

The *Xenopus* embryo is broadly representative of vertebrate embryos in its mode of development (Chapter 3; Fig. 3.5), in which the regulatory regionalization of the early territorial domains is followed by migration and complex spatial subdivision of growing cell populations, rather than by direct *in situ* differentiation as in Type 1 embryos. But it stands out in that we know more about the gene regulatory program which directs its initial territorial regionalization than we do for other vertebrate embryos. In Fig. 4.4 is reproduced a network diagram for specification of the mesodermal territories of this embryo (Koide *et al.*, 2005), based largely on *cis*-regulatory as well as perturbation evidence.

The earliest zygotic regulatory interactions are indicated in Fig. 4.4A1. They result in separation of future dorsal and ventrolateral mesodermal domains, the descendant cells of which will ultimately contribute to the entirely different body parts located, respectively, along the dorsal axis and surrounding the coelomic cavity of the larval animal. The regulatory state defining the dorsal domain includes the early expressed Siamois and Twin activators, and the Goosecoid (Gsc)

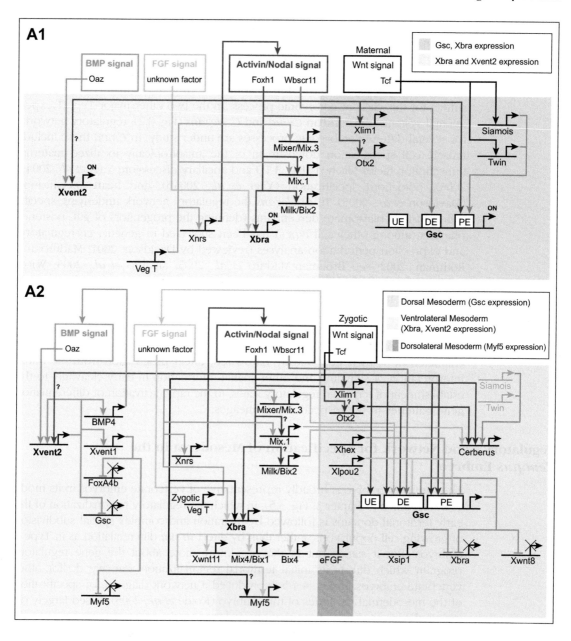

FIGURE 4.4 Gene regulatory network for specification of mesoderm in *Xenopus*, subcircuits, and examples of supporting *cis*-regulatory evidence. (A) Overall network based on *cis*-regulatory evidence

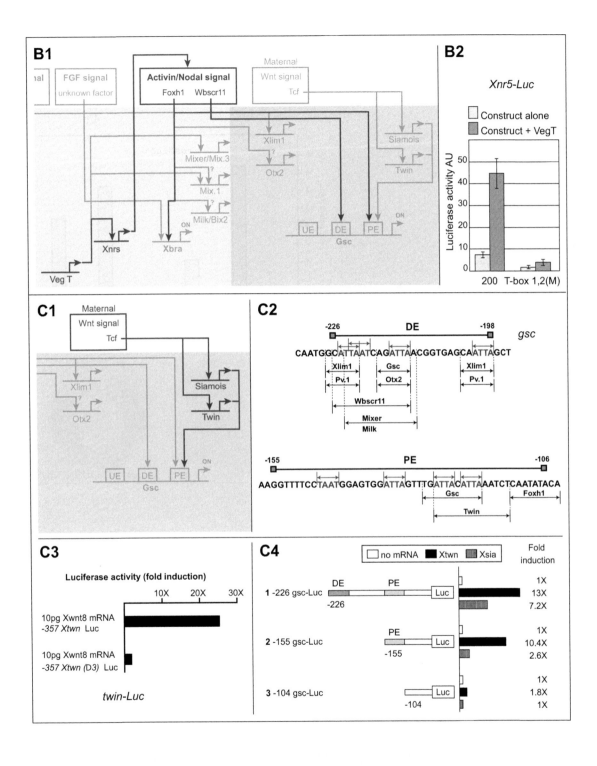

FIGURE 4.4 (continued)

(Continues)

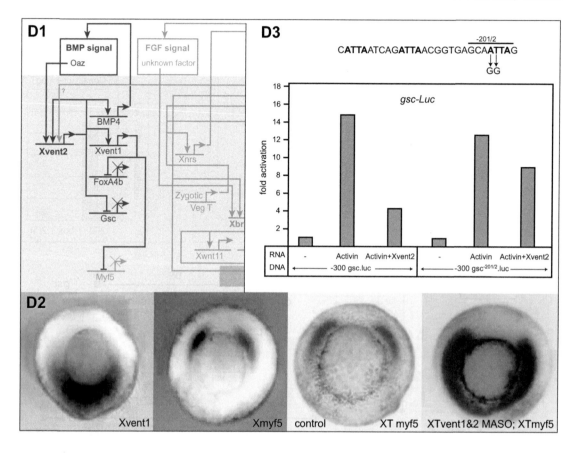

FIGURE 4.4 (continued)
of direct interactions as well as perturbation data, displayed as in Fig. 4.2 (after Koide *et al.*, 2005, where sources of evidence for each linkage in the network are listed). Genes shown encode transcription factors or signaling ligands as indicated. (A1) Initial interactions following reactivation of transcription at mid-blastula. (A2) Specification of dorsal, ventrolateral, and dorsolateral mesodermal domains in later blastula stages. (B) Subcircuit by which vegetal localization of maternal transcription factor VegT (Fig. 3.1) is transduced into the early zygotic program of gene expression. VegT causes transcription of genes encoding Xnr signaling ligands, contributing to activation of *goosecoid* (*gsc*) and *brachyury* (*xbra*) regulatory genes. (B1) Subcircuit from (A1) highlighted; (B2) activity of *cis*-regulatory luciferase expression construct from the *xnr-5* gene (Xnr5-luc). The construct was injected into one-cell *Xenopus* embryos, monitored by light output in absence and presence of additional injected VegT mRNA; either intact, or with the sites to which the VegT factor binds mutated (from Hilton *et al.*, 2003). The experiment proves that VegT directly activates the *xnr5 cis*-regulatory module. However, though not shown here, VegT also activates the *sox7* gene, which in turn contributes to *xnr* gene activation

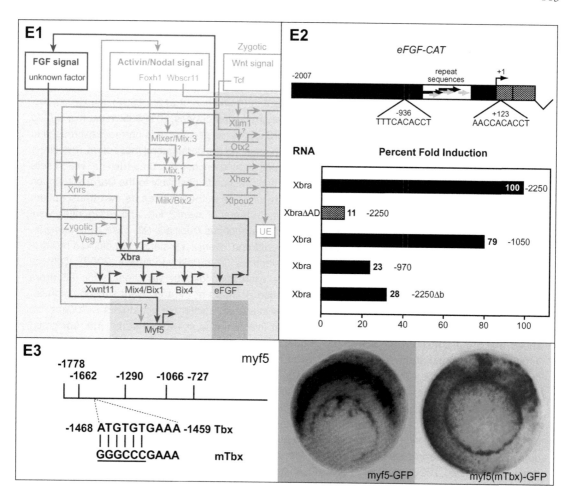

(Zhang *et al.*, 2005). An additional negative input into the *xnr5 cis*-regulatory system is provided by maternal FoxH1 factor, which prevents ectopic expression in vegetal cells (Kofron *et al.*, 2004). (C) Subcircuit showing transduction of Wnt signal on dorsal side of egg into initial regulatory state generated by expression of *siamois* and *twin* genes, thereby providing an input into the *gsc* gene *cis*-regulatory stem. (C1) Subcircuit from (A1) highlighted; (C2) sequences and target sites in the "distal" (DE) and "proximal" (PE) *cis*-regulatory modules of the *gsc* gene, including those interaction sites featured in (B1) and (C1) (from Koide *et al.*, 2005). (C3) Expression of *cis*-regulatory luciferase construct from the *xtwin* gene (*xtwn-luc*), in presence and absence of injected Wnt8 mRNA; and intact or with the Tcf1 sites mutated (–*357xtwn(D3)luc*). Constructs were injected into the animal pole ends of all

(Continues)

FIGURE 4.4 (continued)

four blastomeres at the four-cell stage, and at blastula stage animal caps were isolated, cultured 3 h, and assayed (from Laurent *et al.*, 1997). The experiment demonstrates direct response of *xtwin* to the signal-activated Tcf1 complex. (C4) *cis*-Regulatory experiments as in (C3) demonstrating that the PE of the *gsc* gene responds directly to Siamois and Twin factors (from Laurent *et al.*, 1997). (D) *xvent2* sub-circuit, including signaling feedback and alternative fate exclusion devices. (D1) Subcircuit, highlighted from (A2). In the ventrolateral mesodermal area *xvent2* represses *gsc*, which is normally expressed in the dorsal mesoderm, and *myf5*, which is normally expressed in the dorsolateral mesoderm. The direct positively responding targets of the *xvent2* gene include a gene encoding a Bmp4 ligand, which sets in train an intradomanic feedback loop, since *xvent2* itself responds positively to the Oaz transcription factor activated by Bmp signal reception. Note that *xvent2* also autostimulates its own transcription. (D2) Normal exclusive domains of expression of *xvent1*, a target of *xvent2*, left, and *xmyf5*, right, visualized by *in situ* hybridization; and normal expression of a *Xenopus tropicalis* (*Xt*) *myf5* gene (*xtmyf5*), left, and expression of same construct when both *xtvent1* and *xtvent2* translation are blocked by introduction of morpholino-substituted antisense oligonucleotides (from Polli and Amaya, 2002). (D3) *cis*-Regulatory demonstration that *xvent2* excludes *gsc* expression directly via sites in the PE (from Trindade *et al.*, 1999). Relevant *cis*-regulatory sequence is shown at the top, *xvent* sites in boldface, and −20/12 mutation destroying one of these sites indicated. Luciferase constructs (−*300gsc.luc*) were introduced and assayed as in (C3), except that injection was at two-cell stage. The bar graph shows that the *gsc* response to activin [i.e., nodal response; cf. (B)], is mediated by the PE present in the construct and that the introduction of *xvent2* mRNA represses construct output. However this does not happen in the −20/12 mutation. (E) *xbra* signal-mediated feedback subcircuit, and downstream regulatory genes. (E1) Subcircuit highlighted from (A2), in which *xbra* drives expression of a gene encoding an eFGF ligand, but the *xbra cis*-regulatory system is in turn activated in consequence of FGF signal reception: here is another intradomanic reinforcing signaling loop operating at the transcriptional level. The *mix4*, *bix4*, and *myf5* genes all encode transcription factors. (E2) *cis*-Regulatory domain of eFGF gene including *xbra* target sites, of which the distal site responds to Xbra factor (from Casey *et al.*, 1998). Control eFGF-CAT construct is shown at top. The expression constructs were injected into a marginal zone (tier C) blastomere at 32-cell stage together with the indicated RNA, and CAT activity measured at a later gastrula stage. XbraDAD mRNA lacks the region coding for the activation domain. The response of the constructs to *xbra* mRNA depends on presence of the distal site ("B"), but not the proximal site ("b") missing in −2250Db. (E3) Direct *cis*-regulatory control of *myf5* gene by Xbra factor (from Lin *et al.*, 2003). The T-box site which is the Xbra target is shown at left, and also a mutation of this site; a germline transgenic expressing a myf5-GFP construct, as assayed by *in situ* hybridization of GFP mRNA in center; and at right is a germline transgenic expressing the same construct except with the mutation in the T-box site, which has lost the ability to express the reporter.

repressor; that defining the ventrolateral domain includes the newly transcribed Xbra and Xvent2 factors. As discussed above (Chapter 3), and indicated in the network diagram, the anisotropically localized maternal VegT transcription factor kicks off this state specification process by activating genes encoding TGFβ factors (the Nodal-related factors, "Xnr's"). These signals are received by the adjacent cells and transduced via the maternal Foxh1 transcription factor (plus Smad factors) into positive transcriptional inputs to *xvent2*, *xbra*, and other genes, some of which are also direct targets of VegT (Tada *et al.*, 1998; Clements and Woodland, 2003). In the dorsal domain an additional input is combined with that stemming from the Nodal signal, viz., a Wnt signal-mediated Tcf1/β-catenin input, a localized gene regulatory function activated by an oriented cytoskeletal apparatus on the future dorsal side of the vegetal pole (cf. Chapter 3). Figure 4.4 shows that the Tcf1/β-catenin input is directly responsible for expression of *siamois* and *twin* genes (Lemaire *et al.*, 1995; Laurent *et al.*, 1997), and that their gene products are in turn integrated by AND logic together with the Foxh1/Smad input at the "PE" *cis*-regulatory module of the *gsc* gene. A second module of the *gsc* gene is activated via another Nodal signal transducer, the Wbscr11 factor (Watabe *et al.*, 1995; Ring *et al.*, 2002).

The chain of causality leading to the definition of the ventrolateral and dorsal regulatory states is encoded in the genome in the network subcircuits highlighted in Figs. 4.4B1 and C1 (the Nodal-stimulated and Wnt-stimulated subcircuits, respectively). Following are representative *cis*-regulatory demonstrations: in Fig. 4.4B2 evidence showing that the VegT input is required for expression of an *xnr5 cis*-regulatory module (Hilton *et al.*, 2003); in Fig. 4.4C2 the specific target sites where Foxh1 and Wbscr11 bind in the PE and DE *gsc* regulatory modules, respectively (Koide *et al.*, 2005); in Fig. 4.4C3 the demonstration that the *twin* gene *cis*-regulatory module requires the Wnt/β-catenin input; and in Fig. 4.4C4 evidence that Twin directly activates the *gsc* PE *cis*-regulatory module (Laurent *et al.*, 1997). Here we see once again that the network architecture of the initial part of the embryonic specification process devolves directly from the A's, C's, G's, and T's of the relevant *cis*-regulatory modules.

The whole of the above apparatus has as its biological function the genomic interpretation of the anisotropic initial inputs, VegT and Wnt, set up in the cytoplasm by the beginning of cleavage, and thereby the installation of the earliest zygotic nuclear regulatory states. What happens next is shown in Fig. 4.4A2. We note at once that the character of the circuitry now changes dramatically: first, many more regulatory genes become involved, including all the organizer specific genes indicated in the pink (i.e., the *gsc* expression) domain of Fig. 4.4A2; second, the process of further spatial subdivision begins (cf. Fig. 3.3), here the definition of the muscle-forming dorsolateral mesoderm; third, alternative fates are excluded by activation of domain-specific repressors throughout; and fourth, lockdown positive feedback loops make their appearance. Both alternative fate exclusion and positive feedback circuitry are illustrated in the subcircuit shown in Fig. 4.4D1. Thus expression of the *xvent2* gene leads to direct activation of a gene encoding a BMP4 ligand (Schuler-Metz *et al.*, 2000). This in turn drives further expression of *xvent2* in neighboring cells of the ventral mesoderm, producing a reinforcing feedback loop analogous to that operating via the Wnt8/β-catenin signal system in the sea urchin embryo (Fig. 4.2C). Two examples of alternative fate exclusion within this same subcircuit are illustrated in Figs. 4.4D2, 3, both depending on *xvent2* expression. The transcription factor encoded by this gene activates *xvent1*, and this gene which produces a repressor, directly blocks expression of the *myf5* muscle regulatory gene (Polli and Amaya, 2002). The normal dorsolateral domains of expression of that gene are thus bounded by the ventral domain of expression of *xvent1*, and if the *vent* genes are prevented from functioning, the expression of *myf5* spreads laterally and ventrally (Fig. 4.4D2). The *xvent1* gene acts as the "enforcer" of the ventrolateral mesoderm regulatory state, and another of its direct *cis*-regulatory targets is the *gsc* gene (Trindade *et al.*, 1999), expression of which is thus confined to the dorsal-most mesoderm (Fig. 4.4D1, D3). Gsc later acts as a reciprocal repressor, preventing expression in the dorsal-most mesoderm of the *xbra* regulatory gene, which in the ventrolateral domain is locked on by another feedback loop (Fig. 4.4E1). The key wiring in this loop is in the *cis*-regulatory module controlling expression of the eFGF gene which responds directly to the Xbra activator (Fig. 4.4E2; Casey *et al.*, 1998), while the *xbra cis*-regulatory module is in turn responsive to FGF signaling (Latinkić *et al.*, 1997). Xbra is a direct activator of the *myf5* regulatory gene (Lin *et al.*, 2003; Fig. 4.4E3), though it is allowed to function as such only on the dorsal side of the zone where it is expressed, as we saw above (Fig. 4.4D2).

The quality of the circuitry in Fig. 4.4A2 displays general similarities to that in the later phases of the endomesodermal specification network of the sea urchin: feedback lockdowns, alternative fate exclusions, cross-regulations abound. Yet the process starts very differently. We turn now to an early *Drosophila* network, which begins operation in an even more distinct mode, but then similarly resolves to an architectural character not unlike those just considered. A set of properties emerges, defined conveniently as kinds of subcircuits which at the level of network architecture seem always to be represented in the generic animal program for embryonic territorial specification.

Gene Regulatory Network Controlling Dorsal-Ventral Territorial Specification in the *Drosophila* Embryo

From a regulatory point of view the remarkable feature of the early *Drosophila* embryo is the accuracy and complexity of the pattern of spatial regulatory states established in the syncytial nuclei of the precellular blastoderm (cf. Fig. 3.4 and ancillary discussion). These patterning functions devolve from two fundamental and overlapping aspects of the genomic regulatory program. The primary program is encompassed in the diverse designs of individual *cis*-regulatory modules determining both the responsiveness of different genes to different levels of the Dorsal (Dl) transcription factor, and the identity of the other positively acting and repressive transcription factors providing inputs to these genes. Many of these *cis*-regulatory modules were considered in Chapter 2 (see Fig. 2.7C) as a canonical demonstration of encoded *cis*-regulatory structure/function relations. Only at the level of the network architecture, however, can we perceive the functional mechanism of the dorsal-ventral spatial patterning system, and not surprisingly it reveals a very special character directly related to the constraints of the syncytial nuclear system within which it operates. While it could be predicted long ago that spatial regulatory processes would somehow work very differently in this syncytial system (e.g., Davidson, 1990), the organizational features revealed by network analysis are striking in contrast to those displayed in the same territories in the same embryo within an hour and a half after cellularization begins. A major difference of course is the initiation of intercellular signaling as a regulatory device, but that is by no means the only difference.

The dorsal-ventral gene regulatory network is reproduced in Fig. 4.5A, encompassing both pre- and post-cellularization processes of state specification. The four light shaded areas on the left of the diagram represent domains of the precellular state. On the left is the extracellular domain where occurs the initial steps of the mechanism by which the Dl factor is nuclearized, beginning with generation in the follicle cells of the precursor of a ligand which, after processing in the ventral extracellular perivitelline space, engages the Toll receptor on the ventral surface of the egg, resulting in turn in the graded ventral-to-dorsal nuclearization within the egg of the maternal Dl factor (*op. cit.*, Chapter 2, here shown in pink). The network of *cis*-regulatory interactions in, respectively, the syncytial domains of the ventral mesoderm is shown in light blue; the ventral neurogenic ectoderm in yellow; and the dorsal ectoderm in light green. The more heavily shaded areas on the right represent the post-cellularization domains which during gastrulation (i.e., after 3 h postfertilization) produce definitive progenitors of major parts of the embryo. The dorsal-most region of the ectoderm initially defined by its expression of the *zen* regulatory gene (darker green) becomes dorsal epidermis and the extra-embryonic amnioserosa; the neurogenic ectoderm (dark tan) produces both epidermis and several distinct populations of invaginating trunk CNS cells (Fig. 3.4), viz. in central to lateral order the midline cells expressing the *sim* regulatory gene, the medial neuroblasts expressing the *vnd* regulatory gene, the intermediate neuroblasts expressing the *ind* regulatory gene, and the lateral neuroblasts expressing the *msh* regulatory gene (for review, Cornell and Von Ohlen, 2000). Also indicated at the right in Fig. 4.5 is the dorsal mesoderm (blue) where *tin* and *eve* are expressed, which gives rise to the heart and visceral mesoderm. Figure 4.5 is thus constructed to facilitate direct comparison of the architecture of the syncytial and the cellular domains of the network.

The syncytial portion of the network defines the regional causes of expression of genes encoding several repressors (Snail, Brk, Vnd, Ind), several activators [Twist, Singleminded (Sim), and Zen, in addition to Dl], and components of several signaling systems, which, however, cannot work until cell membranes appear in which the receptors and ancillary apparatus can be mounted. These include in

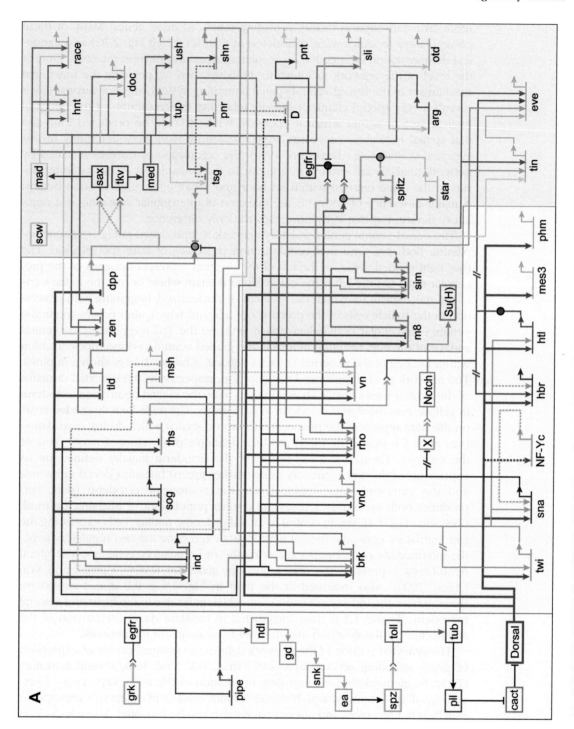

FIGURE 4.5 For legend see pages 169–171.

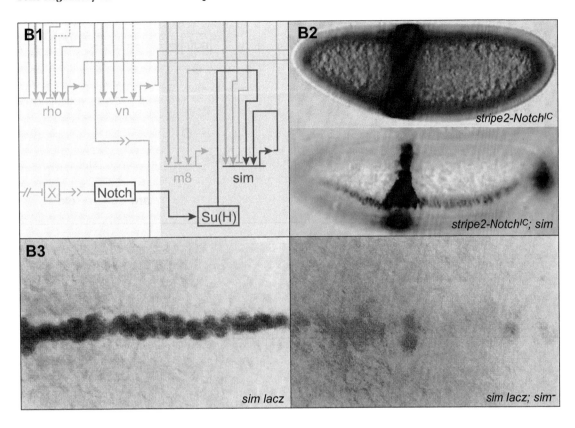

Gene regulatory network underlying dorsoventral specification in *Drosophila*; subcircuits and their functions. (A) Network, indicating three phases of the specification process (from Levine and Davidson, 2005). On left (pink area) is a summary of interactions leading to the ventral-to-dorsal gradient of nuclearized Dorsal transcription factor within the syncytial embryo (reviewed briefly above, in Chapters 2 and 3). These interactions begin in oogenesis in follicle cells, and following expression of the *pipe* gene in the ventral follicle cells, they occur in the perivitelline space between the egg membrane and the chorion. Their result is the proteolytic activation of the Spatzle (Spz) ligand, which is deposited in a graded cline on the ventral-to-lateral surface of the egg, and results in graded activation of the Toll receptor, generating the Dorsal nuclearization function. In the central region of the network (light green, dorsal ectoderm and dorsal neurogenic ectoderm; light yellow, ventral neurogenic ectoderm; light blue, ventral, mesodermal territory) are shown regulatory interactions among genes responding to the Dorsal input during the syncytial phase of development, from about 2 to 3 h postfertilization. At right (darker green, dorsal epidermis, dorsal neurogenic, and amnioserosal territories; tan, ventral epidermis and ventral neurogenic territories; darker blue, dorsal mesoderm, here heart, territory) are shown regulatory

(Continues)

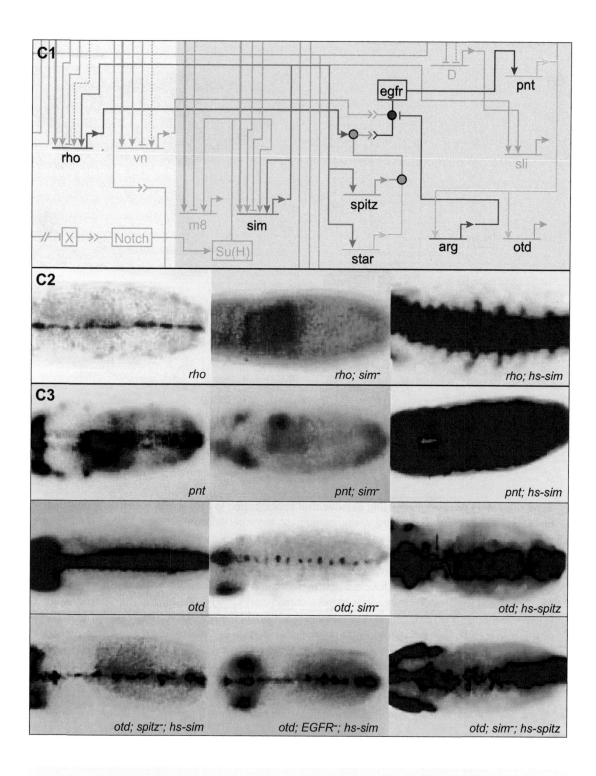

FIGURE 4.5 (continued)

the mesoderm the genes encoding the Htl FGF receptor and signal transduction protein Hbr; in the neurogenic ectoderm the *rho* and *vn* genes, which encode products mediating EGF signaling; in the dorsal ectoderm the *tld* and *sog* genes, the products of which spatially modulate the availability of the Dpp ligand; and the *ths* gene, which encodes an FGF ligand active when during gastrulation the mesoderm migrates dorsally and Ths comes in contact with the Htl receptor.

interactions occurring after cellularization and into gastrulation, up to about 5 h postfertilization (cf. Fig. 3.4). The network is based largely on evidence for direct *cis*-regulatory interactions, such as reviewed earlier in Chapter 2, as well as on observations on the effects on gene expression pattern of given mutations and of forced ectopic expression. Sources for each linkage are given in Levine and Davidson (2005). For roles of key genes in the developmental process see text and *ibid*. (B) *cis*-Regulatory control of *sim* expression following cellularization. (B1) Notch to *sim* subcircuit highlighted. (B2) Demonstration that Notch expression suffices for induction of *sim* in the ventral neurogenic ectoderm (from Cowden and Levine, 2002). At top, ectopic expression of the Notch intracellular domain [Notch(IC)]; this is the effector portion of the receptor which interacts with Su(H) transcription factor when the receptor is activated) is forced by introduction of an *eve stripe2-Notch(IC)* construct. Notch(IC) mRNA is visualized by *in situ* hybridization. Below, expression of endogenous *sim* gene is visualized similarly in embryo expressing this same construct. The *sim* probe reveals both the normal bilateral rows of *sim* positive cells at the dorsal edge of the mesoderm, but also ventral spurs of *sim* expression extending in a dorsal direction, where Notch(IC) is expressed under *eve* stripe 2 control. (B3) Autoregulation of *sim* (from Nambu *et al.*, 1991). A *sim-lacz* expression construct gives correct midline expression in midline cells following mesoderm invagination (left), but in *sim* mutants this expression is largely lost (right). (C) The *sim*-EGF signaling subcircuit. (C1) Subcircuit highlighted: following *sim* activation it drives expression of three target genes which collaborate to produce activated Spitz ligand in midline cells (*rho*, *star*, and *spitz* itself; see text). On reception by EGF receptor (EGFR) in neighboring cells, the regulatory gene *pointed* (*pnt*) is activated, in turn causing activation of the (in this context) epidermal regulatory gene *otd*, and of *argos*, the product of which blocks EGF signaling (and epidermal fate) more distally away from the midline (Golembo *et al.*, 1999; for quantitative model of Argos function in the control of Spitz activity range, see Reeves *et al.*, 2005). (C2) Demonstration that Sim is a driver of *rho* (from Chang *et al.*, 2001): visualization of *rho* expression in control embryos (left), *sim* mutant embryos (center), and *sim* overexpression embryos (under heat shock control) (right). (C3) *sim* control of *pnt* and *otd* expression (from Chang *et al.*, 2001). The top row shows *pnt* expression in control, *sim* mutant and *sim* overexpression embryos (cf. C2). The second row shows *otd* expression in control and *sim* mutant embryos, and in *sim* mutant embryos in which *spitz* is overexpressed under heat shock control. It follows from these results that *sim* is upstream of *pnt* and *otd*, but that it works on *otd* by means of causing *spitz* expression (cf. C1). The third row shows similarly that *otd* fails to respond to *sim* overexpression in *spitz* mutants (left), or in EGFR mutants (center), but that overexpression of *spitz* causes overexpression of *otd* even in *sim* mutants, all in accordance with the circuit shown in (C1).

These signaling components load the system for classes of region-specific interactions which occur only after cellularization, a unique transcriptional anticipation of communication states to come.

There are three dominant features of the syncytial portion of the network architecture. First, every gene shown responds directly at *cis*-regulatory target sites to Dl (these are about half of all probable Dl targets, as discussed in Chapter 2), and in the future neurogenic ectoderm they respond synergistically to the Twist activator as well, which is itself expressed in response to high levels of Dl. This is a classic "feed forward" relationship (see above), the function of which is here to sharpen the transcriptional activation mediated by given levels of Dl (*op. cit.*, Chapter 2). The Dl input is transient, however, and Twist/Dl activation pertains exclusively only during the syncytial phase: after cellularization these genes run on other inputs. In the dorsal ectoderm other activators than Twist collaborate with Dl (Fig. 2.6). Second, the boundaries of all the dorsal-ventral expression domains depend, by direct *cis*-regulatory interaction, on expression of repressors within the adjacent domains (the identities of some of the responsible repressors are not yet known). For example, in the syncytial portion of the network the Snail repressor can be seen to exclude expression of most of the genes from the mesodermal domain where Snail is expressed, also under Dl plus Twist control, and this sets the ventral boundaries of expression of the neurogenic ectoderm genes. Similarly, the Vnd repressor directly silences *ind* expression in its ventral neurogenic ectoderm domain (Von Ohlen and Doe, 2000), while Ind turns off the *msh* gene where Ind is expressed (Weiss *et al.*, 1998), initiating a set of adjacent median-to-lateral longitudinal stripes. So the first two characteristics of the architecture mean that in each dorsal-ventral domain certain genes are expressed that encode repressors, and these directly preclude expression of genes responding to them; or else the genes in the network are allowed to be active according to direct activating inputs from Dl plus partner(s). While the activation domains are to some extent set in space by the responsiveness to Dl concentrations and the partner factors, these domains exceed the observed expression boundaries, as shown by the expansion of the boundaries in the experimental absence of interactions with the repressors (see Chapter 2; an exception is the simplest and most upstream of the Dl responsive genes, Twist). That is all there is to it: no inter- or intragenic feedbacks; no lockdowns; no stabilization circuitry, either signal-dependent or -independent. This negative feature, so different from what we have been looking at, is the third general property of the syncytial stage network. It consists exclusively of univectorial inputs, arrows all running the same way in the diagram of Fig. 4.5A. These inputs direct the process up to the point permitted by the relatively simple patterns of activator presentation, and by the integrating capacities of the *cis*-regulatory modules, which specify the repressors to which each module will respond.

On cellularization multiple signaling pathways are activated, these immediately impinge upon the transcriptional control apparatus, and the quality of the architecture of the network of regulatory interactions controlling specification of the dorsal-ventral territories suddenly changes. This can be seen at a glance in Fig. 4.5A: the input arrows no longer all run in the same direction, more genes become involved, subcircuits are set up, and genes formerly under control of the syncytial system now receive inputs from the cellularized system (right to left arrows). Here we focus on the midline region of the neurogenic ectoderm as our example. The key player is the *sim* gene (Nambu *et al.*, 1991; Crews, 1998). Initially activated by Dl and Twist, and repressed by Snail in the future mesoderm, on cellularization the expression of this gene is confined to a single row of ventral neurogenic ectoderm nuclei on each side of the mesodermal strip by means of a cell membrane-bound Notch (N) ligand expressed in the boundary mesoderm cells. The N signal transducer Su(H) (as reviewed earlier) works as a toggle switch: in its liganded form it is required for normal activation of *sim*, but in the absence of N, that is, away from the boundaries, it represses *sim* expression (Morel and Schweisguth, 2000; Cowden and Levine, 2002). On gastrular invagination of the mesoderm, these two rows of cells are brought together and form the midline of the embryonic CNS (Fig. 3.4).

As shown in the subnetwork highlighted in Fig. 4.5B1, the *sim* gene then proceeds to lock itself on (Nambu *et al.*, 1991): experimental evidence demonstrating both the control of spatial expression of *sim* by N and the *cis*-regulatory autoactivation of *sim* is reproduced in Figs. 4.5B2, 3. A major function of the *sim* gene is to control the deployment of an EGF signaling system. Thus expression of several genes required for the production of active EGF ligands in the midline cells is controlled by the Sim transcription factor (subcircuit highlighted in Fig. 4.5C1). These *sim* target genes include *spitz*, encoding an EGF ligand; and *star* and *rho*, required for processing Spitz. Note that the circuitry is that of an indirect feed forward loop with Spitz activation as the downstream target: at the *cis*-regulatory level *sim* drives *rho* expression, utilizing a different *cis*-regulatory module than that initially responding to the Dl system (Ip *et al.*, 1992; Fig. 2.6) so that *rho* is now expressed under *sim* control in the midline cells, as demonstrated in the experiment shown in Fig. 4.5C2 (Chang *et al.*, 2001). Transcription of the *sim* gene also drives *star* and *spitz* expression (Golembo *et al.*, 1996; Lee *et al.*, 1999), but Rho (and Star) in turn drive the processing to active form of the Spitz protein, which is present more broadly, specifically in the midline.

The cells immediately adjacent to the midline will become ectoderm, and this spatial subdivision is encoded in the form of a transcriptional regulatory state installed by reception of the Spitz signal. A portion of the interesting circuitry by which this program operates is shown in the right side of the highlighted architecture in Fig. 4.5C1. A key regulatory gene activated by EGF signal transduction is *pointed* (*pnt*), and it in turn provides a required input to the *otd* gene, a transcriptional ectodermal regulator in these cells: the experiments in Fig. 4.5C3 (Chang *et al.*, 2001) show that spatial *pnt* expression depends on *sim* transcription, as does expression of *otd*, and that this dependence works through Spitz signaling.

Similar kinds of relationships obtain in the dorsal ectoderm, where spatial regulatory state specification is mediated by deployment of the Dpp signaling system.

This is indicated by the activation of the half dozen regulatory genes of the future amnioserosa and dorsal epidermis, shown at the upper right of the network in Fig. 4.5A (see legend).

What we see in the dorsal-ventral network architecture following cellularization throughout the *Drosophila* embryo is in essence similar to what we have encountered in the sea urchin and the *Xenopus* embryos: it would appear possible to designate a set of genomically encoded subcircuit network devices that are generally used for territorial specification in the development of all types of bilaterian embryo, just as implied in the "top down" analysis of Fig. 3.3. Returning to this schematic for a moment, we see that it correctly indicates that the process of territorial state specification is held in common in the three canonical forms of embryogenesis considered; the earlier occurring processes depend on the kind of embryo and the initial anisotropies with which it is equipped, and the following processes again differ. In the sea urchin embryo, at least in some territories, the network leads directly to cell-type differentiation (Fig. 4.2); in the *Xenopus* embryo the network we have (Fig. 4.4) controls events up to when massive cell migration is just beginning, and neither here nor for any vertebrate embryo is there yet available system level analysis of gene regulatory process for the gastrular and immediately postgastrular stages; in the *Drosophila* embryo what immediately follows the processes determined by the network of Fig. 4.5 is regional (larval) body part formation, always including much cell division, invaginations, and further spatial subdivisions. But the commonality of the programming devices used to drive territory state specification up to and the point where gastrulation begins here emerges as one of the underlying themes of embryonic process, a shared design feature of the bilaterian regulatory genome.

We can define the functional character of these devices in reasonably precise terms, as we have a growing number of similar examples, and soon no doubt will be able to build and install our own versions in living embryos. Once, by whatever means the initial territorial regulatory states are set up, an apparatus takes over the function of which is to install an amazing set of what we have loosely termed lockdown mechanisms that must at root be responsible for the inexorability, reproducibility, and flexible robustness of the early developmental process. These mechanisms include the following species of regulatory network subcircuit, all encountered in the foregoing examples, some in all of the examples: (*i*) signal-mediated positive feedback circuits, in which all the cells of a given territory participate, such that a regulatory gene active in each drives expression of a gene encoding a ligand, and the reception of this ligand in each (neighboring) cell results in enhanced expression of the same regulatory gene, and of course of its other downstream targets; (*ii*) direct positive cross-regulatory feedback loops among genes encoding territory-specific transcription factors; (*iii*) positive autoregulatory feedbacks that maintain such genes in an "on" state; (*iv*) intra-territorial expression of regulatory genes encoding repressors which preclude expression of alternative territorial regulatory states; and (*v*) feed-forward subcircuits in which genes that are pleiotropically important for the territory are synergistically driven both by an upstream regulator of that territory and by another regulatory gene also a target of the first, so that the target gene quantitatively and qualitatively utilizes the *cis*-regulatory inputs of both upstream regulators. In addition, territory-specification regulatory circuits of the embryo frequently and perhaps always express new signaling ligands which initiate further spatial subdivision. Embryonic territories that secrete intercellular signals affecting adjacent regulatory states used to be distinguished as "organizers," but it is likely that all newly specified embryonic territories do this, though the effects are more easily demonstrable with some than others. Since a common set of regulatory network subcircuits is utilized to similar ends in the pan-bilaterian process of pregastrular territorial state specification, treatment of the diverse processes of embryogenesis in these terms, that is, specifically as in Figs. 3.3, 4.2, 4.4, and 4.5, provides a view of early development which is at once more basic and more mechanistic than are all traditional approaches. Those approaches are exactly opposite in orientation: they begin with the usually unique anatomical transformations of the embryo, proceed to its prominent signaling interactions, and then may include some individual gene regulatory functions executed by given *cis*-regulatory modules active in the embryo (see any introductory text on development). An implication of this and the preceding Chapter is that general treatments of embryogenesis in bilaterians need to be redone in an entirely different way.

NETWORKS THAT CONTROL CONSTRUCTION OF COMPONENTS OF ADULT BODY PARTS

We see that the architectural forms of gene regulatory networks for development are diverse according to the job they must do: distinct kinds of networks set up syncytial spatial domains from those that do cell differentiation, etc. Adult body parts are many and various, and require control of morphological and growth functions as well as integration of prior with novel spatial regulatory states. So a first expectation is that a considerable variety of network architectures will underlie their developmental construction. As yet we know little of the architecture of networks for adult body part construction that can be taken as representative of the actual scale of these control systems, though very illuminating small subcircuits of these networks have been worked out for many of them. Among these are many *Drosophila* structures, including the amnioserosa (e.g., Reim *et al.*, 2003); wing coordinate system (Gómez-Skarmeta and Modollel, 1996; Yan *et al.*, 2004); wing veins (de Celis, 2003; Kölzer *et al.*, 2003; Lunde *et al.*, 2003; Crozatier *et al.*, 2004); heart (Reim *et al.*, 2005; Zaffran *et al.*, 2002); eye (Punzo *et al.*, 2002; Michaut *et al.*, 2003; Silver and Rebay, 2005); salivary gland (Zhou *et al.*, 2001); and hindgut (Lengyel and Iwaki, 2002; Johansen *et al.*, 2003). In vertebrates subnetworks and small circuits have been proposed for, among many other structures, the eye (Zuber *et al.*, 2003; Silver and Rebay, 2005); the neural crest (Meulemans and Bronner-Fraser, 2004); heart (Cripps and Olson, 2002; Davidson and Erwin, 2006); rhombomere specification (Davidson, 2001; Trümpel *et al.*, 2002); and pituitary (Dasen *et al.*, 1999; Scully and Rosenfeld, 2002).

Here we consider two cases in which the known network dimensions are such as to permit a more general analysis than is yet usually possible. These are a mammalian network leading from pancreas specification to β-cell differentiation; and a *C. elegans* network that controls the final specification of two particular neurons. These networks represent the control circuitry for different kinds of process, and they yield different kinds of take-home lesson. It is disappointing that many examples are not yet available to be adduced, as undoubtedly there soon will be, for there are tantalizing hints that a rich variety of hitherto unseen designs for regulatory logic processing will tumble forth when there are.

Network Organization in the Specification of Pancreatic β-Cells

Because of its immediate disease relevance, and the early realization that many forms of human diabetes devolve from mutations that affect the function of developmental regulatory genes, an enormous amount of evidence is available regarding the roles of such genes in the formation of the pancreas, and in the specification and differentiation of pancreatic cell types. Among recent efforts to formulate the linkages among these genes in the purview of a system-level transcriptional regulatory network are those of Jensen (2004), Brink (2003), Habener *et al.* (2005), and Servitja and Ferrer (2004). The portions of this network considered below, all too briefly, are organized essentially following Servitja and Ferrer (2004), and using many of the network linkages adduced by Jensen (2004), with some additional recent evidence. The pancreas forms upon installation of a specific regulatory state in a confined region of the gut, from dorsal and ventral "buds" which later fuse, and from the beginning each phase of specification and differentiation of each pancreatic cell type is characterized by expression of given sets of transcription factors. As always, the spatial placement of the progenitor field from which the pancreas arises depends on signaling from adjacent regions of the embryo, in particular notochord, mesenchyme, and aorta on the dorsal side; the subsequent subdivision of the pancreas into exocrine and endocrine, and other cell types depends on further signaling interactions generated by the operation of its gene regulatory network. Many different signaling systems are involved (for reviews see Grapin-Botton and Melton, 2000; Jensen, 2004). Here we focus not on the initial origins of the pancreas, but rather on the regulatory transactions leading from the specification of endocrine precursor cells in the pancreatic buds to ultimate formation of insulin-producing β-cells. A very important caveat is that information on the key *cis*-regulatory interactions of the genes involved is incomplete, and many genes known to be essential from knockouts in mice cannot even be placed in the network as yet; furthermore, data on each given gene has, in general, been obtained in different labs from data on other genes. Widely different procedures have been used in these various studies, including observations of specific gene expression in knockout mice, chip assays, and measurement of the effects of *cis*-regulatory mutations in gene transfer experiments carried out in developing animals, in cultured cell lines that express given sets of functions, or in differentiating embryonic stem cells. Those network linkages considered in the following, as illustrated in Fig. 4.6, are all supported by direct experimental *cis*-regulatory evidence obtained in gene transfer experiments (see legend for sources), as well as, in every case, by evidence obtained in targeted knockouts. Nonetheless, what is yet known of this very complex system cannot yet be regarded in the same light with respect to network authenticity or completeness, in the special senses defined earlier (Fig. 4.1A), as can the embryonic model system networks treated above.

The three subnetworks in Fig. 4.6 represent aspects of the circuitry controlling progression through three stages of this continuous process here arbitrarily separated (i.e., as VFNs). Our basic interest is the nature of the circuit architecture. The essential function of the subnetwork in Fig. 4.6A is to drive expression of the *pdx1* and of the *ngn3* genes. The *pdx1* gene is an early and essential regulator of the pancreatic progenitor field, activated directly or indirectly by HlxB9 and Hex factors in the dorsal and ventral buds, respectively, and directly by the foregut endoderm factor Hnf6 in the early buds (Jacquemin *et al.*, 2003). The *pdx1* regulatory gene does different things at different stages of pancreatic development, carrying out a succession of roles, as seen in its changing network linkages in the course of endocrine cell development. The *ngn3* gene executes a specific regulatory function in the generation of endocrine precursors, as we see in the following subnetworks. The dedicated subcircuit of Fig. 4.6A, which during specification of Ngn3+ precursor cells has the job of stepping up expression of both genes, has a curious structure. It appears to be driven by two little "motors" shown on the highlighted backgrounds in Fig. 4.6A, each consisting of a cross-regulatory positive feedback loop, i.e., the *hnf6-hnf1β* and the *foxa2-nr5a2* loops (the evidence for the latter is more extensive; see legend for references). The early and continuing Hnf6 input triggers both motors with an input assist from Gata6, another endodermal factor, into the *foxa2-nr5a2* "motor." The outputs from these "motors" can be thought of as the continuing generation of Hnf6 and Foxa2 factors, and both these outputs go to both target genes, *pdx1* and *ngn3*, functionally a rather symmetrical design. In terms of regulatory logic the circuit ensures that the *ngn3* and the *pdx1* genes receive quantitatively and qualitatively identical transcriptional inputs as the feedback "motors" ramp up, and indeed ectopic experimental expression of these two target genes suffices to produce ectopic endocrine specification (Grapin-Botton *et al.*, 2001). To gild the lily, *pdx1* also locks itself on by autoregulation.

In Fig. 4.6B we see how these two target genes become the drivers of the next, penultimate, set of regulatory genes, again by means of a curiously intertwined circuitry. The *pdx1* gene directly activates the *nkx6.1* gene, but it also activates the *pax4* gene which in turn also provides a positive input to the *nkx6.1* gene: a feed-forward loop of which *nkx6.1* is the distal target. The effector of the *ngn3* part of the subcircuit is its target, the *neuroD* regulatory gene, which encodes another bHLH transcription factor, and which may work together with Ngn3 or in sequence. In any case, NeuroD and Ngn3 activate the *nkx2.2* gene, and this

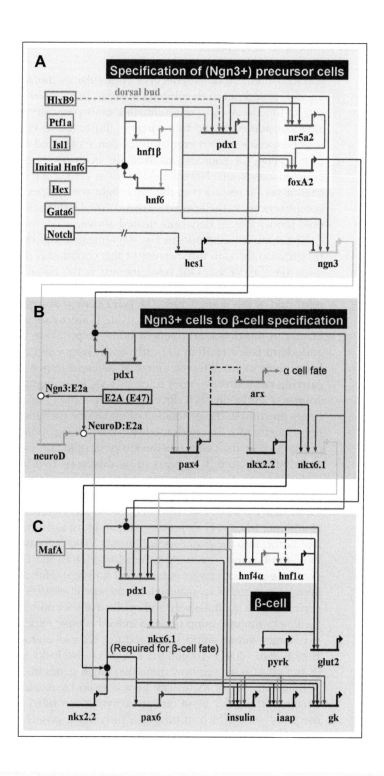

FIGURE 4.6 For legend see page 179.

in turn drives the same *nkx6.1* gene; while at the same time NeuroD and Ngn3 also activate *pax4* and hence also contribute to expression of *nkx6.1* by that route, another (two step) feed-forward loop targeting the same *nkx6.1* gene. One has the impression that the expression of this gene is rather important, and indeed the phenotype of *nkx6.1* knockout is drastic reduction of β-cells (Sander *et al.*, 2000). In terms of regulatory logic this circuitry acts as a tight coordinating device to ensure that its essential Nkx6.1 output is expressed in response to the pair of inputs deriving from the expression of *ngn3* and *pdx1* genes: the Pdx1 and NeuroD/Ngn3 inputs are integrated at the transiently expressed *pax4* gene, which then feeds into *nkx6.1*, while *nkx2.2* receives the same NeuroD/Ngn3 input as does *pax4*, and then passes it on to *nkx6.1*. There is a distinction in the circuit, however, in that in contrast, the *nkx2.2* gene is not directly or indirectly controlled by Pdx1 input, either in this or via the subcircuit of Fig. 4.6A; the *nkx6.1* gene integrates both regulatory pathways while *nkx2.2* reflects only the *ngn3* pathway. This might provide for possible different (here not specified) additional inputs, which could be important in the subsequent functions of Nkx2.2 and Nkx6.6 as direct drivers of β-cell differentiation genes.

The climax of this very tightly organized system is shown in Fig. 4.6C, which indicates the control circuitry of representative β-cell differentiation genes, including the most famous, the *insulin* gene. In addition to Nkx2.2 and NeuroD the inputs to this gene battery (*insulin*, *iapp*, and *glucokinase* genes) consist of Pdx1 and

Transcriptional regulatory network controlling development of mouse pancreatic β-cells, from specification to differentiation. The networks are built in BioTapestry (Longabaugh *et al.*, 2005); see Fig. 4.2A for symbolism. The networks for the three stages indicated (mostly after E13.5) are according to the general organization provided by Servitja and Ferrer (2004). Linkages shown are based on direct *cis*-regulatory analyses cited in this review, but with additional evidence from *cis*-regulatory studies cited in the reviews of Brink (2003), and Jensen (2004). Further evidence is from Wang *et al.*, (2004) (*pax4-nkx6.1* linkage and *nkx2.2-pax6* linkage); Iype *et al.*, (2004) (*nkx6.1* autoregulation); Gerrish *et al.*, (2004) (*foxa2-pdx1* linkage); Odom *et al.*, (2004) (*hnf4α-hnf1α* linkages); Haumaitre *et al.*, (2005) (hnf1β-hnf6 linkage); Itkin-Ansari *et al.*, (2005) (*neuroD1-nkx2.2* linkage). (A) Earlier β-cell specification stages, beginning with the pancreatic regulatory state represented by the transcription factors in rectangles at left. The output of this phase of the regulatory operation is stable expression of *ngn3* and *pdx1* in these cells. Highlighted areas include intergenic positive feedback loops. (B) Later specification of *ngn3* cells. The output of this part of the regulatory operation is the stable expression of many of the drivers of the differentiation gene batteries to be expressed *viz. neuroD, nkx2.2, nkx6.1, pdx1*. (C) Differentiation of β-cells. Additional drivers, *viz. pax6* and *mafA* are now required. An intergenic feedback loop is highlighted.

Pax6 factors, and in the case of the *insulin* gene, an additional factor, MafA. Note, however, the organization of the control apparatus: at the top is another cross-regulatory positive feedback loop (highlighted), the *hnf1α-hnf4α* "motor," this together with Foxa2 now provides the driver for high-level *pdx1* expression (plus *pdx1* autoregulation). In addition, *nkx6.1* autoregulates. The remainder of the circuitry consists largely of overlapping feed-forward loops. First, with respect to the *insulin* gene as the distal target, MafA drives *insulin*, but also the *pdx1* gene, while Pdx1 in turn is an activator of *insulin*; second, with respect to all three genes of the battery as distal targets, Nkx2.2 serves as a direct driver, but Nkx2.2 also drives *pax6*, and Pax6 in turn drives the differentiation genes; third, while Pax6 is a direct driver, it is also an activator of *pdx1*, and as we have seen, Pdx1 is in turn a driver of the differentiation genes.

The details aside, what shines forth in these subnetworks is their remarkable degree of recursive wiring: use of the same regulatory gene products in different combinations as inputs to other regulatory genes, as in feedback and feed-forward loops. Almost the whole of these networks consists of such recursive features. There are evolutionary implications, as we take up in the next Chapter, but also functional ones. For one thing, in the later phases of this developmental process this apparatus produces relatively high levels of the transcription factors (Servitja and Ferrer, 2004), and at appropriate times very high levels of secreted differentiation gene products. It is equipped with feedback step-up "motors" and autoregulatory lock-on devices, and the prevalence of feed-forward loops indicates frequent use of synergistic *cis*-regulatory mechanisms (cf. Chapter 2). For another, the recursive quality means that the same regulatory genes perform many different roles: the champion in this respect is *pdx1*, which in these diagrams over and over again provides inputs into diverse *cis*-regulatory modules, in different combinations. The different devices that control *pdx1* expression at various stages nail down this spatial and temporal developmental input into each of its differently functioning, target regulatory information processors. The developmental network for this adult body part component is thus composed of a heavily wired skein of programmed interactions in which many genes play multiple roles. This skein evidently provides a developmental framework of great stability. But, apart from the robust presentation of the developmental drivers explained by these networks, there will also be other inputs into *cis*-regulatory modules that function in differentiated β-cells, those responsible for endless minor variations in performance of the differentiation genes in response to ambient fluctuations of physiological significance. After all, the body part has not only to develop but also to function physiologically.

Cell Fate Exclusion by microRNAs in a Gene Regulatory Network Controlling Terminal Differentiation in *C. elegans* Taste Neurons

At the periphery of developmental regulatory networks lies apparatus that controls the enormous variety of differentiation gene batteries and the fine scale variations in terminal morphology which animals deploy. It is yet impossible to discern how much variety in network architecture underlies this peripheral functional diversity, but the following example suggests that many strange and wonderful system designs await discovery at this level. This example concerns two *C. elegans* neurons that express taste receptors, known as ASEL and ASER, i.e., left and right ASE neurons (Chang *et al.*, 2003). These are bilaterally symmetrical, morphologically alike cells, which are wired into the nervous system similarly, but they express different batteries of what are apparently chemoreceptor genes, and they respond to different ambient chemicals. The architecture of the gene regulatory network that maintains ASER function and excludes ASEL function in the right neuron, and vice versa in the left neuron, is reproduced in Fig. 4.7 (Johnston *et al.*, 2005). This is a driver gene regulatory network in the sense discussed above. It was worked out following isolation of classes of mutants in which both neurons express ASEL functions, or conversely ASER functions, or both lose functions. The exact effects on transcription of the other network genes were then determined in ASER and ASEL in the absence of expression of these individual genes, and when they were expressed ectopically, using transgenic GFP fusions (Chang *et al.*, 2003, 2004; Johnston *et al.*, 2005). A unique feature of this network is the key role played by two microRNAs. Each of these represses translation of a transcription factor required for one of the two cell fates, by binding to the 3′ UTR of the respective mRNA (Chang *et al.*, 2004; Johnston *et al.*, 2005). The network includes the regulatory functions by which transcription of the genes encoding these microRNAs is controlled, as well as the downstream regulation of at least a sample of the genes constituting the ASEL and ASER chemoreceptor differentiation gene batteries.

The fate exclusion circuitry is shown at the top of Fig. 4.7, and the mechanism directly controlling the differentiation genes in the lower part of the diagram. These genes are, in ASEL, the receptor *guanine cyclase* (*gcy*)-6 and -7 genes, and the *flp-20* and *flp-4* neuropeptide genes; and, in ASER, the *gcy-5* gene and the *ben-1* LDL receptor gene. The *cog-1* gene, which encodes a homeodomain transcription factor, is active in ASER, and is also equipped with a positive autoregulatory device, such as we have seen so often in this Chapter. The target of the *cog-1* gene in the network is the gene encoding the MIR-273 microRNA, which in ASER thus represses the translation of the mRNA encoding the Zn finger transcription factor Die-1. As shown in Fig. 4.7, Die-1 is directly or indirectly the driver of all the ASEL downstream genes in the network, and so the transcription of MIR-273 excludes expression of all these ASEL genes in ASER. In ASEL, conversely, the Die-1 factor activates transcription of a gene encoding the LSY-6 microRNA, and this blocks translation of the ASEL *cog-1* gene, excluding ASER fate in ASEL. Thus each cell operates an exclusionary lockdown mechanism mediated by a microRNA not transcribed in the other cell. There are further recursive features to this cell fate stabilization system: in ASEL a target of Die-1 is the *lim-6* homeodomain gene, which not only provides positive inputs into the *flp-20* and *flp-4* genes, while repressing the ASER *gcy-5* gene, but also apparently feeds back to further *die-1* and *lsy-6* transcription, reinforcing the ASEL regulatory state.

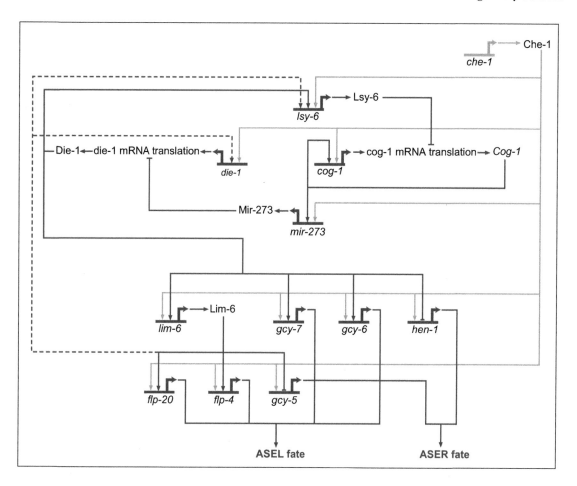

FIGURE 4.7 Gene regulatory network underlying differential gene expression in right and left ASE taste neurons in *C. elegans*. This Figure was kindly provided by O. Hobert, after Johnston *et al.*, (2005). Genes expressed in the right neuron, ASER, are shown in blue; genes expressed in the left neuron, ASEL, are shown in red (see text for gene functions). MicroRNAs are designated in capital letters, genes in lower case italics, and protein products in Roman with first letter only capitalized.

The Die-1 factor directly activates *gcy-6* and *gcy-7*, but it also represses the ASER gene *hen-1*.

The circuitry in Fig. 4.7 alternatively excludes either of two regulatory states but it is not to be regarded as a symmetrical "toggle switch" which might trip in an equivalent way in either direction. This can be seen in its structure and in the developmental process that leads to its installation. The structural asymmetry begins in the *cog-1* autoregulation, which suggests that the ASER fate might be dominant unless something else intervenes. In development, after the final mitosis in the lineage which gives rise to the two cells, they bilaterally express the same genes, including what are later the ASEL-specific *gcy-6, gcy-7, lim-6*, and *lsy-6* genes; but they also co-express the ASER regulator *cog-1* (Johnston *et al.*, 2005). The positively acting transcription factor produced by the *che-1* gene is required for the activation of these and the other relevant genes in both ASE neurons (Chang *et al.*, 2003; Fig. 4.7). But this early bilateral function phase is then superceded by a phase when both cells express the ASER regulator *cog-1* and the ASER marker *gcy-5*. An input external to the network shown must provide the mechanism by which the L/R fate assignment is installed. Since it is always the left neuron with respect to the rest of the animal which expresses the ASEL genes, an input external to the network shown must provide the mechanism by which the L/R fate assignment is installed. We can see from the network what happens once the assignment is made: in both cells the exclusion mechanisms now operate, in ASEL the expression of the differentiation genes is driven by Die-1 and the product of its target gene *lim-6*, and in ASER the differentiation genes may continue to be activated by Che-1 or additional regulators.

The unusual feature of this regulatory network circuit is its utilization of microRNAs to accomplish an irreversible exclusionary lockdown function. Negative control of protein concentration by microRNAs is observed commonly in many systems (reviews cited in Chapter 1), but while this is often regarded as an "alternative" control scenario with respect to conventional transcription regulation by means of DNA-protein interaction (i.e., interaction with transcription factors), we see here the opposite. A particularly illuminating feature of this network is that explicit in it is the *cis*-regulatory control of the genes encoding the LSY-6 and MIR-273 microRNAs, respectively, by the Die-1 and Cog-1 transcription factors. But why have microRNAs been incorporated in this system rather than the straight DNA-level transcriptional repressors that we have encountered over and over again in this Chapter in considering developmental network exclusion functions? Regulatory state exclusion is usually mediated by activation of genes encoding transcriptional repressors, which target the *cis*-regulatory apparatus of key regulatory genes driving alternative regulatory states. For example, we might imagine an ASEL/ASER circuit in which *cog-1* encodes a direct repressor of the *die-1* gene, thereby excluding its expression in ASER, while in ASEL *die-1* activates a gene encoding a repressor we could call Rep-L, which transcriptionally shuts down *cog-1*. This solution would require even less *cis*-regulatory programming than that in Fig. 4.7; i.e., the *cis*-regulatory modules controlling the *lsy-6* and *mir-273* genes versus that which would be needed to control the *rep-l* gene. One infelicitous aspect would be that the *cog-1* gene would have both to positively autoregulate itself and negatively regulate its target *die-1* gene, but such things are not unknown.

There is a much more general answer, however. It is emerging that there are many hundreds of microRNAs produced in animal genomes (Berezikov *et al.*, 2005), some relatively conserved and some clade-specific. On the other hand, the families of DNA-recognition domains that constitute the sequence-specific elements of transcription factors are in general pan-bilaterian; this includes repressors as well as activators. Furthermore, genes encoding transcription factors not only function pleiotropically but are often used over and over again in the course of development, which means that their target sites are present in multiple developmentally regulated genes. Like antisense oligonucleotides built in the laboratory to take out individual message species, however, microRNAs may easily target specific mRNA sequence elements, located in rapidly evolving noncoding regions of the transcript. The implications are first, that there could be less evolutionary constraint posed by prior useful interactions in the insertion of microRNA repression into a regulatory circuit, than in the insertion of a canonical transcriptional repressor, particularly considering the dominant behavior of many such repressors; and second, that these devices will be seen to occur most extensively at the periphery of gene regulatory networks, where the events of later development, and particularly the details of body part structure and function are programmed. A global study of the deployment of microRNAs in zebrafish development is particularly interesting in this connection (Wienholds *et al.*, 2005). Of 115 microRNAs studied, most also present in mammals (though not infrequently expressed differently), most were undetectable in zebrafish development until the embryo had become segmented, and they were not expressed strongly until after organogenesis is complete (96 h). The majority continues to be expressed in adults, furthermore, and by *in situ* hybridization, most are expressed in a very tissue-specific manner. Similar conclusions have been reached in studies of limb formation in mutant mice lacking the RNA processing enzyme required for making all microRNAs and by misexpression of specific microRNA (Harfe *et al.*, 2005; Hornstein *et al.*, 2005): here again the role of these regulators is found to be confined to the later phases of the developmental process and not to affect the complex developmental patterning underlying limb formation. Nor are patterning or function of *Drosophila* muscle affected by knockout of a conserved microRNA gene expressed specifically in muscle (Sokol and Ambros, 2005). Thus the prospective negative regulatory repertoire represented by these sequences is indeed often utilized during terminal differentiation and regulatory state stabilization, at the periphery of the network.

CONCLUDING REMARKS

The genomic sequences in which are embedded the code for animal development specify two different levels of informational transaction. The fundamental level is that of the individual, information-processing *cis*-regulatory module which responds conditionally to the multiple inputs which its combinatorial target sites define (Chapter 2). But developmental process is mediated by networks of these *cis*-regulatory modules, particularly those controlling expression of regulatory genes. The networks also process information, that is, the regulatory outputs generated at its nodes. The architecture of these networks, the topology of their linkages, is of course also specified by the target site sequences of the *cis*-regulatory modules which are at its nodes. The events which this architecture mandates represent the unfolding of the regulatory code for development. It is impossible to escape the image that the regulatory genome consists of a vast delocalized computational device. The network nodes are small, but resilient and flexibly responsive information processing machines, linked functionally into a compound organization. This organization of genomic computational devices, on an organismal scale the sum of many networks such as those considered in this Chapter, is what is encoded in the regulatory genome. It is what makes animal development possible: its enormous capacity to specify changing spatial and temporal expression of many thousands of genes over the days, months, and sometimes years that development requires.

The regulatory genome is itself the product of evolutionary process. But there follows from this obvious relation a conceptual price: evolution must be thought about in the terms of that which it has generated directly. Construction and alteration of developmental gene regulatory networks is the output of evolutionary process which accounts for animal body plans. Therefore we have to think about evolution in ways that take directly into account the structure/function properties of these networks, as they are now finally emerging.

CHAPTER 5

Gene Regulatory Networks: The Roots of Causality and Diversity in Animal Evolution

EVOLUTIONARY IMPLICATIONS OF THE STRUCTURE AND FUNCTION OF GENE REGULATORY NETWORKS

The concept that evolutionary change in body plans must devolve from change in the developmental regulatory program, since the body plan is the product of the developmental regulatory program, is where we began (Chapter 1). But that is only the familiar gateway. Recognition that the causal genomic underpinning of development can be represented in gene regulatory networks leads to precise ideas of how evolutionary change occurs: what kinds of changes occur, what are their diverse consequences, and what are their dynamics. The major take-home lesson of Chapter 3 was that development is fundamentally a process of change in spatial regulatory state. That of Chapter 4 was that this specific process is programmed by large networks of regulatory gene interactions, which are wired up according to the DNA sequence at their *cis*-regulatory nodes, and which are constructed of diverse kinds of subcircuits. Now let us think of the "time derivatives" of this control circuitry, that is, the processes of change over evolutionary time in the program by which successive regulatory states are imposed in the developing organism, caused by genomic sequence changes in the wiring or architecture of gene regulatory networks. The reward will be a predictive consideration of what kinds of evolutionary effects might devolve from different kinds of change in network linkages; conversely, we can use the record of life on this planet as a guide to what kinds of change in developmental regulatory networks are and are not likely to occur.

At the end of this Chapter, in considering animal origins, we return to a theme which recurs throughout this book, the concept of regulatory information processing. Chapter 2 was devoted to conditional *cis*-regulatory information processing, and Chapter 4 to the conditional information processing emergent at the network system level. Reflections on this theme gave rise to the concept of the developmental regulatory genome as a vast delocalized computer, the internal *cis*-regulatory nodes of which are hooked up through their functional *cis-trans* interactions. These arguments have powerful evolutionary corollaries. For in the end, the "derivatives" that in the sense above produce the process of evolution, are all changes in the DNA machinery for information processing.

The Parts of Gene Regulatory Networks, and the Qualities of Evolutionary Change

Changes in given functional linkages of gene regulatory networks occur at the DNA level by alteration of the *cis*-regulatory sequence defining transcription factor target sites. "Alterations" here might mean mutation, transposition, deletion, insertion, repetition, generation of novel sequence, and so forth. While such alterations may occur in any *cis*-regulatory module, and from the bird's eye view of individual modules may all look similar, they will have fundamentally different effects depending on where in the structure of the network they occur. In evolutionary terms this is the great example of how the properties in which one is interested must be perceived at the organizational level to which they pertain. The other side of this coin is that a focus exclusively at the level of change in the A's, C's, G's, and T's of the individual module, rather than on the level of change in the network architecture, may obliterate its real significance. For there are two dominant relations between given changes in network architecture and evolutionary process, both of which depend powerfully upon where in the structure of a developmental regulatory network these changes occur (Davidson and Erwin, 2006). The first is the relation between the type of network part affected by the change and the kind of alteration in body plan that might result. The second is the relation between the type of network part and the likelihood of change therein, which depends on the mechanistic constraints precluding those changes.

In Chapter 4, four distinct classes of network subcircuit were defined and briefly discussed: (*i*) differentiation gene batteries (including their immediate regulatory controllers; see Fig. 1.5 for canonical examples of gene battery regulatory structure); (*ii*) promiscuously used, invariant little subcircuits, such as pan-bilaterian signal transduction systems, which are termed "plug-ins" (Davidson and Erwin, 2006), because the same such systems are plugged in to a countless variety of networks and network locations; (*iii*) input/output (I/O) devices that act as switches on other network subcircuits, in that they either allow them to produce their outputs in a given circumstance or preclude their activity, i.e., either pass or block their inputs; and (*iv*) highly conserved, rigidly and recursively wired subcircuits which initiate specification of fields from which particular body parts arise, and which we refer to as the "kernels" of developmental gene regulatory networks. Figure 4.3 provided the example of a kernel for endoderm specification that has been conserved between starfish and sea urchins virtually unchanged for half a billion years (Hinman *et al.*, 2003). Davidson and Erwin (2006) concluded that the order of this parts list for networks roughly parallels the ease and frequency of evolutionary change (lability) in these diverse kinds of subcircuit, in that differentiation gene batteries are the most malleable and change frequently by several diverse paths, while at the other extreme, kernels are enormously resistant to change. As a graphic illustration of this parts list, in Fig. 5.1 the sea urchin endomesoderm network is reproduced from Fig. 4.2 with examples of the different kinds of parts highlighted. The evolutionary lability of each class of network part differs, and the evolutionary consequences of change in each class are distinct.

Differentiation genes are uniquely positioned in regulatory network structure, in that they reside at the periphery of the network. Unlike all of the genes that constitute the internal structure of the network, their products do not have the function of controlling other genes. They do not produce regulatory states, nor do they execute regulatory pattern formation. They are never participants in the wiring designs discussed in Chapter 4, which are characteristic of the internal architecture of developmental regulatory networks; for example, feedback and exclusion circuitry. The peripheral position of differentiation gene batteries in regulatory networks has the immediate consequence of freeing them from the functional constraints

Endomesoderm Specification to 30 Hours

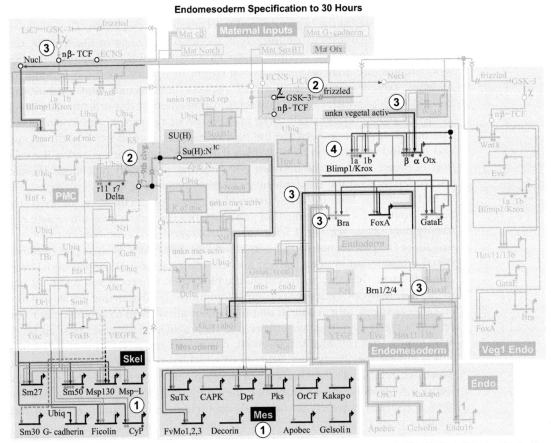

FIGURE 5.1 Illustrations of different classes of parts in a gene regulatory network. The sea urchin endomesoderm network is reproduced from Fig. 4.2, but grayed out so as to illustrate by colored-highlight various types of subcircuit or linkage: (1) differentiation gene batteries, here skeletogenic gene batteries (Fig. 4.2E), and pigment cell gene battery; (2) plug-ins, here the Wnt8-Tcf1/β-catenin subcircuit (Fig. 4.2C1), and the Delta-Su(H)-Notch signaling systems (Fig. 4.2F1); (3) examples of I/O linkages, here the maternal inputs into the *pmar1* gene (Fig. 4.2 D1), and into the *otxα* transcription unit, the repressive output from the *foxa* gene to the *gcm* gene, the output from the *bra* gene to cell motility genes of the gastrulating endoderm, and the outputs of the kernel shown in (4); (4) kernel, the same as in Figs. 4.2G1 and 4.3, here seen in the context of the whole network.

on change in their linkages, which for regulatory genes arise from cascading effects on expression of other genes, according to the network architecture. For this basic reason, change in differentiation gene batteries may occur endlessly. They change in many ways: by alterations in protein coding sequences which affect the functional character of their products; by alterations in *cis*-regulatory sequence which create new target sites for the regulators of these batteries, thus causing additional protein coding genes to be added to pre-existing gene batteries; or similarly, by degradation of target sites to cause loss of given genes from these batteries. The rates or intensities of expression of differentiation genes are also subject to posttranscriptional modulation, as well as to modulation at the transcriptional level. This last can occur by target site divergence in *cis*-regulatory modules regulating individual structural genes, or in the modules regulating the battery controllers. Differentiation gene batteries may contain tens or hundreds of genes, and one of the most powerful modes of evolutionary change in which they participate is their redeployment. The consequence of such redeployment is mobilization of the tens or hundreds of genes in the battery to a new developmental address. One need only think of the variation in developmental deployment of muscle differentiation gene batteries in the variously formed body walls and appendages of even such closely related animals as the amniote vertebrates (birds, reptiles, and mammals).

The constraints on evolutionary change in differentiation gene battery function are directly subject to selection, since they produce in detail most of the phenotypic functionalities with which the animal confronts its environment. When looked at in this way, classical microevolutionary Darwinian theory, which treats evolutionary change as a continuous process of small incremental variations of selective value, can be seen to be a suitable vehicle for treatment of changes in content, expression, and deployment of differentiation gene batteries (for review, see Davidson and Erwin, 2006). At the taxonomic level, differences produced in the terminal phases of development by evolutionary redeployment of the peripheral differentiation gene batteries of developmental gene regulatory networks are what distinguish congeneric species and confamilial genera. From the gene regulatory network perspective much of what causes species-specific differences can be said to be the consequence of microevolutionary alterations of differentiation gene battery function. These differences arise from internal changes in the membership of these batteries as enumerated above, or from changes in their deployment, which means alterations in their linkage into the immediately upstream developmental processes. Deployment changes occur at the *cis*-regulatory level of the gene battery controllers.

Another kind of change that occurs very easily at low taxonomic levels is change in body part size (think of domestic dogs, which are all recently diverged members belonging to the same species; Parker *et al.*, 2004). In a gene regulatory network, body part size is controlled by transcription factors expressed in the subcircuits that determine the patterning of given developmental fields, which directly regulate the downstream activity of cell cycle control genes such as the *cyclin* genes (e.g., Zhu and Rosenfeld, 2004; cf. Chapter 3).

But if microevolutionary treatment of continuous change fits reasonably with the underlying molecular biology of developmental operations at the network periphery, by the same token it must fail for evolutionary alteration of architecture in the internal domains of gene regulatory networks. For here the rules are all different: the network structure is hierarchical, so changes in more upstream regions may have large effects and not incremental small ones. Some kinds of network parts change at entirely different rates than others, and in general the consequences of change are qualitatively very diverse, depending on where they occur in the network structure. Differentiation gene batteries do not make body plans. What this means at root is that traditional microevolutionary theory is not useable for treatment of the molecular mechanisms by which evolution of the animal body plan has occurred. The body plan is encoded in the internal regions of the gene regulatory network, which during development determine the progression of spatial regulatory states, not the business of differentiation gene batteries.

At the opposite extreme from the continuously evolving differentiation gene batteries are the kernels of developmental gene regulatory networks (Chapter 4), exemplified above by the conserved endoderm specification subcircuit of Fig. 4.3. It is not yet known how generally there occur subcircuits in bilaterian developmental gene regulatory networks which fulfill the definitive criteria for kernels, a matter we take up in the following section. But because of their remarkable properties, even a few demonstrated or highly probable examples of kernels, such as might prove or strongly imply their existence, must change our views of body plan evolution. The definitive criteria for this species of network subcircuit are as follows: (*i*) The subcircuit consists of regulatory genes that share inputs from one another so that they are joined by multiple *cis*-regulatory linkages (the "recursive wiring" of Chapter 4). (*ii*) The subcircuit operates at the initial phase of regional regulatory state specification required in the formation of a given body part, e.g., in defining the progenitor field for that part. (*iii*) The subcircuit is dedicated to this particular function, and the same genes wired together in the same architecture (i.e., the same subcircuit) are not utilized for any other developmental function, though some of the genes of the subcircuit may be included in combinations with different other genes in other subcircuits. (*iv*) If any of the genes of the subcircuit are prevented from functioning, the consequence is failure of the body part to develop. A glance back at the kernel in Fig. 4.3 illustrates many of these points. Thus, with respect to the first, all five genes encode transcription factors, and their *cis*-regulatory modules are linked recursively in a feedback loop such that the *otx* gene receives inputs from three of the five genes, and the same is true of the *foxa* gene, while the *bra* and *gatae* genes are each activated by two inputs from within the subcircuit. With respect to the second criterion, this kernel is at the top of the endoderm specification process, and with respect to the fourth, interruption of expression of any one of its genes causes the same catastrophe, failure of the gut to form at all (see Chapter 4 for details and references). While the third criterion, exclusive dedication of the kernel to this function, could formally be validated only were all the developmental network architecture of the whole organism throughout its life cycle known, it is certainly true just from the patterns of gene expression that it cannot be used anywhere else during embryogenesis.

The conservation of almost every *cis*-regulatory linkage in the complex kernel of Fig. 4.3, for over half a billion years of independent evolution since the last common ancestor of starfish and sea urchins, is a shocking observation. It stands in contrast to the differences displayed by the surrounding parts of the starfish and sea urchin networks that have been compared (Hinman *et al.*, 2003; V. Hinman and Davidson, unpublished data). As we take up in the following, other classes of network subcircuit are characteristically labile. One can see on the basis of their definitive properties why kernels would be extremely conserved: given their properties, neither gene nor *cis*-regulatory input could easily be removed from them on pain of the gross failure of the body part to develop. If extreme conservation thus indeed follows from their definitive internal characteristics, kernels could provide an answer to what is perhaps the largest unsolved problem in bilaterian evolution (Davidson and Erwin, 2006). This could be why the basic body plans of the Bilateria as a whole, and of bilaterian phyla and superphyla (deuterostomes, chordates, panarthropods, etc.) have remained essentially unchanged at least since the Early Cambrian, 520 million years ago, during which most of them appear unmistakably in the fossil record (Chen, 2004; Valentine, 2004).

Upper level taxonomic groups such as phyla are essentially defined by the presence and fundamental design of their major body parts. If the initial developmental spatial organization of these parts is directed by network kernels, the mechanistic reason that no new phylum or superphylum level body parts have appeared since the Cambrian emerges directly from the intrinsically conservative properties of kernels: once assembled, if the lineage of animals depending on the use of these body parts itself was to survive, the kernels could not be taken apart and some time in the future redone a different way; they could only be added on to downstream. For example, the through gut with anterior mouth and posterior anus is a basic shared character of Bilateria. Once formulated, the kernels that organize the regulatory states required for developmental formulation of the through gut and its ends could never again be dispensed with. Subsequent clade-specific developmental variations in the ensuing developmental morphogenesis of the gut would have been constrained only by the requirement that they receive their initial inputs from the kernels that set up the progenitor fields for the gut and its anterior and posterior parts. Kernels could define properties shared by all bilaterians, properties of superphyla, or properties of individual phyla, a nested set (Davidson and Erwin, 2006). But since bilaterian superphyla and bilaterian phyla can be recognized in Early Cambrian assemblages, all members of this set have evidently been with us for at least 520 million years. The shared characters specific to each phylum or superphylum, properties for which the regulatory wiring arose prior to the Early Cambrian, would be those of which the initial stages of development are programmed by network kernels. During the time since, however, the fossil record and modern anatomy reveal enormous morphological variation within phyla; that is, potent alterations of the underlying developmental gene regulatory networks have been installed, differently in the different classes and orders of which the phyla are composed. We may ask, then what are the components of gene regulatory networks that drive subphyletic developmental diversification, and what are the malleable classes of network linkage?

The answers are based on a wealth of comparative observations, unorganized and serendipitous, but a wealth nevertheless. In the terms of the network parts considered in Chapter 4, subphyletic changes in body plan, great and small, have been driven by countless events of redeployment of signaling subcircuits and other "plug-ins," and by a general and prevalent process that amounts to installation of switches on the "outside" of spatial specification subcircuits; that is, by alteration of many kinds of input and output linkages in the developmental regulatory networks that control body part formation. New subcircuits that control spatial elaborations of simpler body part predecessors have also arisen (though not often); for example, in the developmental programs responsible for the major organs that all vertebrate chordates share.

The signaling cassettes that bilaterians utilize for development provide either an intradomanic or an interdomanic spatial input every time a regional subdivision occurs and a new regulatory state is established. We saw the canonical use of signaling inputs of both kinds in diverse processes of embryogenesis in Chapter 3, and many explicit linkages of signaling inputs into developmental gene regulatory networks in Chapter 4. As development proceeds, the deployment of signaling plug-ins becomes finer, so as to effect ever finer spatial subdivisions; for example, in the morphogenesis of terminal structures such as teeth (Jernvall and Thesleff, 2000) in mammals, feathers in birds (Chuong *et al.*, 2000; Wu *et al.*, 2004), or macrochaete sensory bristles in flies (Posakony, 1994). In all of the small organs where these structures are generated, multiple signaling interactions are required to mobilize the appropriate regulatory states, in the appropriate spatial relations to one another, and on the fine scale of individual cells and cell layers. Yet feathers and teeth and peripheral sensory organs are the kinds of structure that are clade specific, and that have evolved in many different forms, so we know *a priori* that evolutionary deployment of signaling cassettes in their development must be a highly flexible and easily changeable process. The same conclusion follows from the basic fact that six signal transduction systems, *viz.* Notch, EGF, FGF, TGFβ, Hedgehog, and Wnt, are utilized to provide spatial inputs in a myriad of diverse subcircuits in the development of any bilaterian, and in different contexts in different bilaterians (for reviews substantiating the diversity of developmental utilization of each of these cassettes see, respectively, Cadigan and Nusse, 1997; McMahon *et al.*, 2003; Kingsley, 1994; Artavanis-Tsakonis *et al.*, 1995; Szebenyi and Fallon, 1999; Shilo, 2003). That is, though in their internal biochemical interactions these signal transduction cassettes are highly conserved across the Bilateria, they are indeed "plugged in" at all levels of gene regulatory networks, in all sorts of regulatory spatial specification processes. This is in contrast to the dedicated functions proposed for kernels. Since signaling cassettes are used in the development of the diversely elaborated body parts of the different animal clades which define the Classes, Orders, and Families of bilaterian phyla, their redeployment has been a major mechanism of evolutionary diversification in body plan at all levels, at least from the Early Cambrian down to the present (Davidson and Erwin, 2006).

A similarly obvious *a priori* argument pertains to evolutionary deployment and redeployment of all the other I/O devices which determine whether in any given context regulatory subcircuits are allowed to operate. Below we consider in this light the evolutionary import of various *hox* gene functions: these genes offer pointed illustrations of I/O switches that in evolution alter body plan, they themselves are not responsible for constructing the spatial pattern of regulatory states that underlie the formation of the body part. In general terms, the linkages that produce the inputs into network subcircuits, and those which lead their outputs to the next subcircuit in the network, can all be considered I/O linkages (e.g., Fig. 5.1). As a class these may be predicted to be evolutionarily labile, as discussed below. Change in them is what used to be referred to as changing the "embryological address" to which a given function is directed.

Linnaean Framework for the Evolutionary Consequences of Alterations in Gene Regulatory Networks

We come then to the conclusion summarized in Fig. 5.2 (Davidson and Erwin, 2006), which relates changes in different kinds of network parts to the different qualities of the resulting evolutionary consequences. Developmental gene regulatory networks are inhomogeneous in structure and discontinuous and modular in organization, and so changes in them will have inhomogeneous and discontinuous effects in evolutionary terms. Changes at the network periphery, such as the continuous changes in differentiation gene batteries and in the immediate upstream linkages which determine the deployment of these batteries, affect the terminal stages of development of each body part: these changes are reflected in Linnaean terms as species and generic differences. Changes in the internal portions of the network, specifically redeployment of plug-ins and making and breaking I/O linkages affect regional regulatory state specification, hence pattern formation, and hence the morphology of body parts. These kinds of change imperfectly reflect the Class, Order, and Family level diversification of animals. The basic stability of phylum-level morphological characters since the advent of bilaterian assemblages may be due to the extreme conservation of network kernels. The most important consequence is that contrary to classical evolution theory, the processes that drive the small changes observed as species diverge cannot be taken as models for evolution of the body plans of animals. These are as apples and oranges, so to speak, and that is why it is necessary to apply new principles that derive from the structure/function relations of gene regulatory networks to approach the mechanisms of body plan evolution.

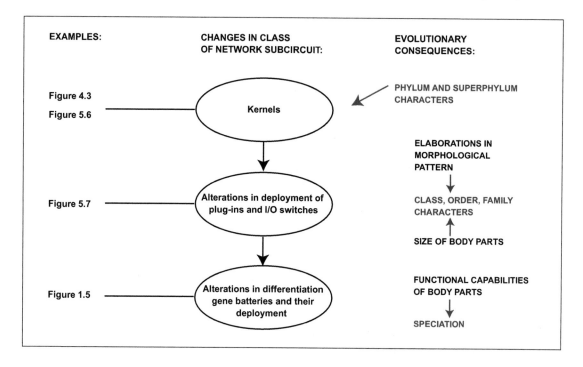

FIGURE 5.2 Changes in the different parts of a gene regulatory network and their qualitatively diverse evolutionary consequences. The center column indicates changes in different network components; the right column the evolutionary consequences expected, which affect different taxonomic levels of diversification (red). The left column refers to examples of the network feature or effects of change therein portrayed elsewhere in this book (from Davidson and Erwin, 2006).

BILATERIAN KERNELS, PREDICTED AND REAL

The kernel concept has two faces: the kernels of developmental gene regulatory networks could in principle be recognized either by their intrinsic properties and functionalities, or by comparisons between the networks of distant organisms that reveal extensive and detailed conservation of their subcircuit architecture. Unfortunately, as the preceding Chapter illustrates, there are yet too few networks solved to permit an evaluation of the generality of the kernel concept by either route. Yet the stakes are high because of the evolutionary implications, and also the power of this idea to organize a vast amount of comparative descriptive data on individual gene use in various animals, which in a frequently loose way has been used to support the orthology of developmental processes across the Bilateria. Here "orthology" has the evolutionary significance of direct descent from a common ancestor; that is, the proposition that given developmental processes of distant animals are fundamentally the same because they share inherited regulatory transactions that the ancestor also used. In what follows we examine several suggestive areas of apparent orthology in gene use. But perceived similarity in time and place of expression does not necessarily indicate orthologous network linkages, and that is ultimately where the argument turns. As we shall see, furthermore, putative phylum and superphylum kernels are more accessible on present evidence than are *trans*-bilaterian kernels; yet there is evidence to be explored at all three levels.

Caveats and a Caution

One instructive example is the essential role of *pax6/eyeless* (*ey*) in the developmental morphogenesis of eyes across the Bilateria. This became a paradigm for evolutionary comparison that might illuminate the phylogenetic origins of developmental processes, but for our present purposes it also serves as a parable and a caution. The story began with the discovery of Halder *et al.* (1995) that *ey*, which is required for eye formation in *Drosophila*, is an ortholog of *pax6*, which is required for eye formation in mammals. This was followed by dramatic experiments of Quiring *et al.* (1995), Halder *et al.* (1998b), and others, in which it was shown that ectopic expression of *ey* in other than the eye imaginal disc could divert these discs to development of amazingly well-formed ectopic eyes. It was soon discovered that other regulatory genes participate in the specification of the eye field in *Drosophila* and are linked to *ey* in a regulatory subcircuit: specifically, another member of the *ey/pax6* gene family, *twin of eyeless* (*toy*), is upstream of *ey*, while downstream of *ey* are three genes, *eyes absent* (*eya*), *sine oculis* (*so*), and *dachshund* (*dac*) the products of which interact with one another. These genes are also recursively wired in feedback relations with one another and with *ey* (for reviews, Gehring and Ikeo, 1999; Donner and Maas, 2004; Silver and Rebay, 2005). All are required for the development of anterior head structures and for the normal development of eye tissue, and to some extent they can all induce ectopic eyes on overexpression (e.g., Bonini *et al.*, 1998). In addition, two other genes, *eyegone* (*eyg*), another member of the *pax6* gene family, and *optix*, like *so* a member of the *six* family of regulatory genes, act in parallel during eye development in *Drosophila*. Strikingly, genes belonging to the same families as *Drosophila eya*, *so*, *optix*, and *dac* as well as *ey* are also expressed in the developing eyes of vertebrates, although in each case there are multiple members of these gene families in vertebrates, and which if any is playing an orthologous role is difficult to determine. Nonetheless, in mice mutations of some members of these families in addition to *pax6* display eye defects (reviewed by Donner and Maas, 2004; Silver and Rebay, 2005). This seemed presumptively a perfect case for the proposition that there survives a *trans*-bilaterian eye development subcircuit, an orthologous architecture descendent from the common bilaterian ancestor, in which *ey/pax6* is the "master" regulator because its expression sets in train the other members of the subcircuit. Our question is whether this is an allowable interpretation, in which case it might provide an example of an evolutionally conserved network kernel for a body part that almost all bilaterians generate.

Despite the enormously impressive peripheral and descriptive evidence, this case does not in fact conform to the definitions required above of a *trans*-bilaterian kernel, though it could well constitute a perfectly excellent and valuable example of a kernel of the arthropod phylum or pan-arthropod superphylum. This would be of no mean importance as the arthropods are a diversified clade that was already well developed in the Early Cambrian (Walossek, 1999; Chen, 2004). In *Drosophila* the *ey* subcircuit has all the internal qualities of a kernel: it is recursively wired, it does operate early in the developmental process in which the body part is formed, and interference with expression of any of its genes is indeed incompatible with eye formation. But there are three prohibitive difficulties with the idea that this is a pan-bilaterian kernel for eye development.

First, the requirement that a common regulatory subcircuit underlies the extremely different developmental processes by which the eyes of insects, cephalopod mollusks, scallops, jellyfish, and mammals are formed (Gehring and Ikeo, 1999) gave pause, and stimulated the search for an alternative explanation for the apparently universal engagement of *ey*/*pax6* or one of its relatives in eye development. And there is indeed a very reasonable alternative: *pax6* (and/or its close relative *pax2*) is in modern organisms a canonical regulator of the visual pigment differentiation gene batteries of the retina, and of eye lens differentiation gene batteries, and this is likely to have been its primitive role rather than developmental pattern formation and specification of eye morphogenetic domains (Scott, 1994; discussed by Gehring and Ikeo, 1999; Davidson, 2001; Erwin and Davidson, 2002; Blanco *et al.*, 2005). Thus since it was already being expressed in different ancestral evolutionary lineages in whatever passed for their light-sensitive structures, it could be further utilized for novel developmental purposes as these lineages evolved their diverse eyes. This puts the onus of the argument on the question of whether the linkages of the genes of the network required for pattern formation in eye development are or are not orthologous to those of *Drosophila* in any given non-arthropod animal.

The second difficulty is that when examined in detail, too few of the regulatory relationships of the subnetwork genes beyond *pax6/ey* itself are in fact the same in vertebrates as in *Drosophila*. For example, the mouse *eya* genes are not downstream of *pax6* as they are in *Drosophila*, and mutations in them do not affect eye formation though they affect other tissues (Donner and Maas, 2004; Silver and Rebay, 2005). Nor are the mouse *dach* genes necessary for eye morphogenesis (or any other morphological attributes of the eye; Davis *et al.*, 2001; Backman *et al.*, 2003). In mouse *pax6* mutants the *dach1* gene is still expressed in the neural retina, though it may be downstream of *pax6* in the lens as in the *Drosophila* eye; however, the feedback linkages by which *dac* maintains expression of *ey* and *so* in *Drosophila* are missing in the mouse (reviewed by Donner and Maas, 2004). Finally, in vertebrates *six3* and *six6* are the most important of the *six* class genes for eye development, but these are orthologs of *Drosophila optix*, which is not part of the *Drosophila ey* subcircuit, rather than orthologs of *so*, which is. In some parts of the eye of some vertebrate species *six3* or *six6* is dependent on *pax6* expression, as is *so* in *Drosophila*, but in others not. Thus some linkages between *ey/pax6* and *so/six3* or *six6* might be similar, but these are paralogous not orthologous to the *six* class gene of the *Drosophila ey* subcircuit.

The third difficulty with the concept that the *ey/pax6* subnetwork is a pan-bilaterian kernel is that a very similar subcircuit is used in vertebrate muscle, involving *pax3* instead of *pax6*, but in which a *six* gene, an *eya* gene, and a *dach* gene are similarly linked and similarly interactive downstream; and another variant of this subcircuit is utilized in otic placodes (reviewed by Silver and Rebay, 2005). Thus these regulators constitute a cassette which is something like a plug-in that can be put to various use, rather than a dedicated apparatus underlying developmental definition of a given body part. So, in summary, there is no pan-bilaterian eye kernel that consists of the same genes linked in the same way in an orthologous gene network subcircuit.

Eyes *per se* aside, the caution that emerges is that evidence that orthologous genes participate in apparently similar developmental processes of diverse animals is only a hunting license in the search for true kernels. The gold standard requirement is that the subcircuit regulatory linkages of these genes in these animals be the same or closely similar. Of course no observation on regulation of given genes in a developmental process apparently shared across a great phylogenetic gulf is totally immune from alternative explanations of convergent evolution, rather than descent from a common evolutionary starting point. But few would seriously suggest that the starfish/sea urchin kernel in Fig. 4.3 is the product of convergent evolution; there are too many detailed, exact architectural features held in common. While such cases remain rare, two recent improvements on conventional comparative "evo-devo" descriptions have made it possible at least to approach the question of kernel generality in a way that provides some intellectual security from the possibility of alternative explanations, even if the prospective kernels are not yet in view or their linkages are only partially explicit. These are the use of genomics methodologies to enable rigorous comparative investigations of expression of whole cohorts of regulatory genes in given developmental processes; and the growing accumulation of *cis*-regulatory and perturbation data that permit a direct assessment of network subcircuit orthology. Here we apply those tools that are at hand: our object is to assess in advance, so far as is possible, the likelihood that there do really exist definitive body parts shared across Bilateria, across bilaterian superphyla, or within phyla, which are likely to be developmentally organized by orthologous network kernels.

Regional Specification in the Bilaterian Nervous System

The brains of vertebrates and insects have a tripartite structure (forebrain/proto-cerebrum, midbrain/deuterocerebrum, hindbrain/tritocerebrum) followed by a nerve cord extending posteriorly. The *hox* genes are expressed in their famous colinear order in the nerve cords of both groups, almost entirely posterior to the brain, such that the most anterior domain of expression, that of *hoxa1* and *hoxb1* in mice, for instance, is located in the hindbrain (rhombomere 4), and similarly their ortholog *labial* in *Drosophila* is expressed in the posterior part of the trito-cerebrum. Several striking similarities in the relative patterns of expression of other transcription factors have been noted in comparing the developing brains of mice and flies. These apparent homologies have led to a strong argument that they were the properties of the "urbilaterian brain" from which ecdysozoan and deuterostome brains alike descend (Arendt and Nübler-Jung, 1999; Reichert and Simeone, 2001; Lichtneckert and Reichert, 2005). Here we consider the possibility that conserved network kernels underlie the regional specification of the central nervous systems of arthropods and vertebrates. The argument starts out on a better tack than its equivalent in the case of bilaterian eyes, which encounters rough sailing immediately because of the lack of homology in the morphological structures of different kinds of eyes. An additional advantage is the wealth of data on relative spatial patterns of expression of regulatory genes, which as we shall see turns out to be remarkably useful.

Figure 5.3A displays diagrammatically the anterior to posterior domains of expression of four key genes in *Drosophila*, mouse, and hemichordate (after Lichtneckert and Reichert, 2005). In the *Drosophila* and mouse diagrams the major morphological subunits of the CNS are aligned to one another, and it can be seen that the anterior *otd/otx* domain, the posterior *unpg/gbx2* domain, the *pax2/5/8* domain at the anterior end of the latter, and the anterior-most *lab/hox1* stripe are all in similar places relative to one another and to the morphological subdivisions of the brain. In terms of the underlying regulatory apparatus this requires that the outputs of the four regional regulatory states that these gene expression patterns represent contribute, respectively, to the development of similar subparts of the CNS. Since the earlier embryonic development of *Drosophila* and vertebrate embryos occur very differently (cf. Chapter 3), the upstream linkages into the CNS regionalization system cannot be imagined to be homologous. The implica-tion is that the genes participating in the regulatory network that executes brain regionalization must interact with one another in a manner that is to some extent similar, to account for the similar relative position of their domains of expression: here we consider only four of these genes, but below the repertoire expands considerably.

In Fig. 5.3B is direct evidence that regulatory interaction between two of these genes, *otd/otx* and *unpg/gbx2*, is indeed responsible for their mutual spatial bound-ary of expression. Most remarkably, the regulatory interactions appear the same in *Drosophila* and in mouse and other vertebrates (Millet *et al.*, 1999; Martinez-Barbera *et al.*, 2001; Kikuta *et al.*, 2003; reviewed by Rhinn and Brand, 2001; Wurst and

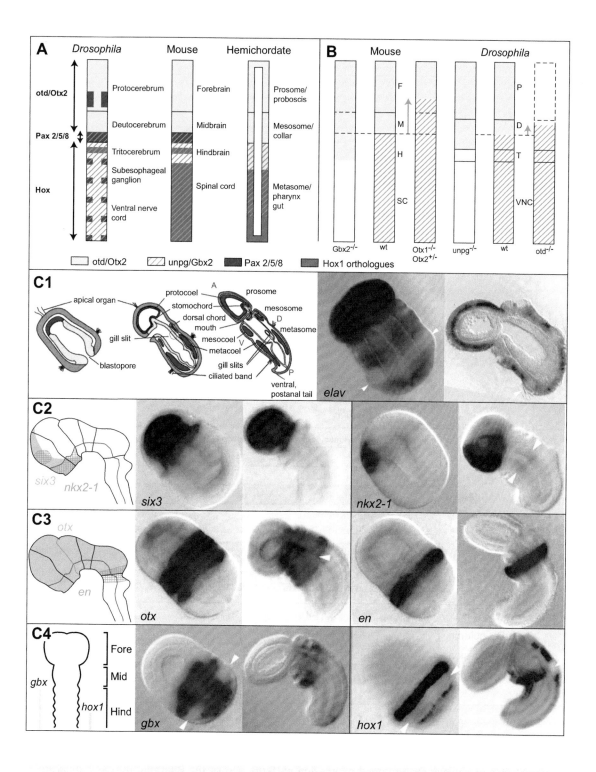

FIGURE 5.3 For legend see pages 202–203.

(Continues)

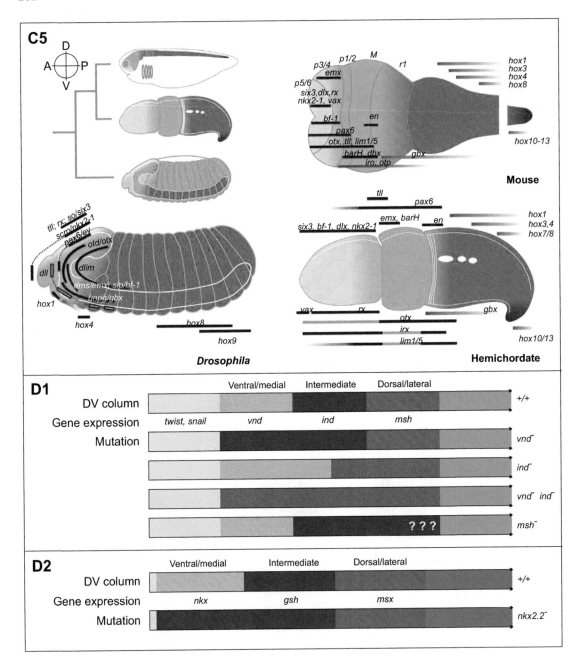

FIGURE 5.3 (continued)

Homologies in nervous system gene expression patterns across the Bilateria. (A) Basic tripartite regional organization of developing nervous systems (from Lichtneckert and Reichert, 2005).

Drosophila has a centralized ventral nervous system, the mouse a centralized dorsal nervous system, the hemichordate a diffuse epidermal nerve net (see text), and yet in terms of gene expression patterns during nervous system development they display the following similarities. At the anterior end (protocerebrum/forebrain in *Drosophila* and mouse, respectively) the *otd/otx2* gene is expressed; at the deuterocerebrum/hindbrain junction in *Drosophila* and mouse is a domain of *pax2/5/8* expression which overlaps the anterior end of the domain of expression of the *unpg/gbx2* gene where it abuts the *otd/otx* domain; in the hemichordate only *gbx2* expression overlaps posterior *otx2* expression here, but in all three animals extends posteriorly into the ventral nerve cord/dorsal spinal cord/metasomal nerve net; and in all three animals the most anterior domain of *hox1* expression is within the domain of *unpg/gbx2* expression in the equivalent of the hindbrain region. Data in this diagram for *Drosophila* are from Hirth *et al.*, 2003; for mouse, from Wurst and Bally-Cuif, 2001; for the hemichordate, from Lowe *et al.*, 2003. (B) Mutually repressive interactions between *otd/otx* and *unpg/gbx2* genes in *Drosophila* and mouse, from Lichtneckert and Reichert, 2005. Diagram is based on gene expression patterns in *Drosophila* mutants (Hirth *et al.*, 2003) and gene knockouts in mouse (Joyner *et al.*, 2000). In both organisms, lack of *otd/otx* expression causes anterior shift of *unpg/gbx2* expression domains and lack of *unpg/gbx2* expression causes posterior shift of *otd/otx* expression domains. (C) Expression of regulatory genes in the diffuse nervous system of the hemichordate *Saccoglossus kowalevskii* (from Lowe *et al.*, 2003). Drawings in leftmost panels of C2–C4 represent developing mouse brain. (C1) Relevant embryonic stages (A, anterior; D, dorsal; P, posterior; V, ventral), and demonstration of the diffuse epidermal nervous system by *in situ* hybridization with a panneuronal marker (Elav) seen in whole mount (left) and section (right). (C2) Expression of *six3* and *nkx2-1*, in late gastrula (left) and in one-gill slit (right) hemichordate embryos, visualized by whole mount *in situ* hybridization; as shown in drawings at left, these genes are expressed in the embryonic mouse embryo forebrain. (C3) Expression of *otx* and *engrailed* genes in hemichordate embryos, as in C2; in mouse expression of *otx* extends back to midbrain/hindbrain boundary, and expression of *engrailed* occurs at this boundary. (C4) Expression of *gbx* and *hox1* in hemichordate embryos, as in C2, 3; anterior boundaries of expression of these genes are in hindbrain in mouse. (C5) Comparative summary of expression of 22 regulatory genes: Upper right, embryonic vertebrate central nervous system (dorsal view); r1, rhombomere 1; M, midbrain; p, prosomeres. Lower right, diffuse nervous system of whole hemichordate body, three-gill slit stage (lateral view); lower left, embryonic *Drosophila* central nervous system, indicated by the white outline (lateral view). The phylogenetic relationship of these animals is indicated at upper left and the relative anterior/posterior positions in the three drawings are matched according to the blue shading. (D) Medial to lateral patterns of expression of *vnd/nkx2.2*, *ind/gsh*, and *msh/msx* genes in *Drosophila* (D1) and mouse (D2) central nervous systems, from Cornell and Von Ohlen (2000). Wild type is at top in both sets of diagrams; below are the consequences on spatial gene expression and downstream neuronal fate designations of the mutations indicated at right. The mutual repression by which the boundaries of the expression domains are maintained is directly supported by gene expression studies in *Drosophila* (see text), but the effect of the *nkx2.2* mutation in the mouse is inferred from the absence of the normal floor plate and appearance of intermediate type cell fates in the medial zone (Briscoe *et al.*, 1999).

Bally-Cuif, 2001; Hirth *et al.*, 2003; Lichtneckert and Reichert, 2005). In both mouse and *Drosophila*, if the *otd/otx* gene is mutated, the anterior brain does not form and the domain of *unpg/gbx2*, and of caudal brain structures, extends ectopically forward; similarly, mutation of *unpg/gbx2* causes posterior expansion of the domain of *otd/otx* expression. In *Xenopus*, furthermore, the repressive effect of the *gbx2* ortholog on *otx2* expression and the repressive effect of *otx2* (at high levels) on *gbx2* are shown to be direct transcriptional functions (Glavic *et al.*, 2002; Tour *et al.*, 2002a,b). Experiments in which *gbx2* expression was made to occur ectopically in the mouse hindbrain show that repression of other regulatory genes is its main patterning function (Li *et al.*, 2005). There may be additional linkages in common between *Drosophila* and vertebrates, though whether they are direct is usually unknown. In *Drosophila*, *otd* mutants lose expression of the *pax2/5/8* gene *pox neuro* (Hirth *et al.*, 2003), which is normally expressed just posterior to the end of the *otd* domain (Fig. 5.3A). Correspondingly, in *Xenopus*, ectopic expression of *otx2* causes activation of *pax2*, a direct effect since it can be obtained in the absence of protein synthesis (Tour *et al.*, 2002a), and in several experiments in which *otx* expression was caused to occur in ectopic locations the expression of *pax2* was shifted concomitantly (e.g., Broccoli *et al.*, 1999; Katahira *et al.*, 2000). In addition, the expression of *lab* is shifted anteriorly in *unpg* mutants (Hirth *et al.*, 2003), and similarly in the mouse normal spatial expression of *hox1* genes in the hindbrain depends on *gbx2* expression and in *gbx2* mutants the anterior boundary of *hox1* expression moves to a more anterior position in the hindbrain (Wassarman *et al.*, 1997; Li *et al.*, 2005; for reviews of the effects on other genes of experimental shift in the *otx/gbx* boundary position, Joyner *et al.*, 2000; Simeone, 2000; Rhinn and Brand, 2001). In summary, then, not only is there in common among these distant bilaterians a remarkable and very particular spatial relationship in the very early development of the brain between the domains of expression of the *otd/otx*, *unpg/gbx*, *pax2/5/8* and *lab/hox1* regulatory genes, but there is some indication that the transcriptional regulatory linkages are similar as well, and a few are already shown to be direct. Considering that these expression patterns are representatives of regional regulatory states that include many other genes as well (see below), this could be the tip of an iceberg, that is, kernels that are used for patterning the anterior end of the nervous system. This possibility is strengthened, in an unexpected way, by the results reproduced in Fig. 5.3C. Here we see examples of transcriptional regulatory gene expression domains in the developing nervous system of a hemichordate, from a recent study on orthologs of no less than 22 different genes known to function during development of the CNS of vertebrates (Lowe *et al.*, 2003). Hemichordates are the sister phylum of echinoderms; the hemichordate-echinoderm clade and the chordates are the main components of the deuterostome superphylum. But neither hemichordates nor echinoderms have centralized nervous systems as do chordates, in which the dorsal CNS is a definitive trait, and as do arthropods, in which the CNS is ventral. Instead the nervous system of hemichordates and echinoderms consists of a dense but diffuse epidermal nerve net. In some anterior regions of hemichordates neuronal cell bodies account for >50% of the epidermal cells (reviewed by Lowe *et al.*, 2003). Figure 5.3C1 shows expression of a pan-neural marker which lights up all neurons in the directly developing embryos of the hemichordate *Saccoglossus*, and the epidermal location of the nervous system is clearly displayed in the section on the right. Many other bilaterians also have diffuse as opposed to centralized nervous systems, including acoel flatworms, which are likely basal to other surviving bilaterians (Ruiz-Trillo *et al.*, 2002, 2004) and cnidarians, an outgroup for Bilateria. Though of course all of these are the modern products of evolutionary process and are not to be considered representative of "primitive" status, there is here a reasonable phylogenetic argument that the ancestral bilaterian nervous system was also diffuse. The chordates and the arthropods would then have centralized their nervous systems in their different ways independently, inserting largely non-neural epidermal domains in their embryonic ectoderm on either ventral or dorsal sides (Lowe *et al.*, 2003). However centralization occurred, the main point for us is that commonalities in the patterning of this very differently constructed bilaterian nervous system with those of chordates or arthropods can have nothing to do with homologies in the morphological embryogenesis of their nervous systems. Instead any such commonalities must reveal directly shared properties of the gene regulatory apparatus that controls spatial expression of these genes.

Examples of the localized expression domains described by specific regulatory genes in the hemichordate nervous system are shown in Figs. 5.3C2–4, and a comparative summary that includes mouse and *Drosophila* is reproduced from Lowe *et al.* (2003) in Fig. 5.3C5. These results reveal an underlying regulatory system controlling these genes that establishes the relative anterior/posterior position of their domains of expression. Astoundingly, this apparatus operates in a similar way in the development of the centralized brain and spinal column of the vertebrates as in the development of the diffuse epidermal nervous system of the hemichordate. Note that "anterior" here means the region of the neural tube where the forebrain will form in the vertebrate CNS, but the epidermis of the whole of the first of the three major body parts (prosome) in the hemichordate; "middle" means the midbrain of the vertebrate CNS but in the hemichordate the epidermal nerve net surrounding the whole middle body segment (mesosome) and the anterior part of the third segment (metasome); and "posterior" means hindbrain and spinal chord in the vertebrate CNS, but the epidermal nerve net of the posterior metasome in the hemichordate. For example, of six genes expressed in the forebrain of the developing mouse all are expressed in the epidermal nerve net of the *Saccoglossus* prosome (*six3, rx, dlx, vax, nkx2-1, bf-1*; examples are shown in Fig. 5.3C2). Ten genes, of which the domains of expression in the vertebrate include the midbrain, are expressed in the mesosome and/or anterior metasome of *Saccoglossus* (*tll, pax6, emx, barH, otp, dbx, lim1/5, irx, otx, en*; for examples, Fig. 5.3C3), that is, more posteriorly than the vertebrate forebrain genes. Furthermore, those genes of which the domains of expression end near the midbrain/hindbrain boundary set by the *otx/gbx* interface in vertebrates are expressed back to the anterior metasome in the hemichordate (Fig. 5.4C4). However, though *gbx* expression extends posteriorly in *Saccoglossus* as in vertebrates, it overlaps rather than abuts *otx* expression as in both vertebrates and *Drosophila* (cf. Fig. 5.3A). We have seen explicitly in Chapters 3 and 4 that regional regulatory states such as those that are indicated by these transcription factor gene expressions are set up through the operation of gene regulatory network subcircuits: it may be that we are here observing the output of highly conserved subcircuits which operate throughout the deuterostome superphylum. Their role would be to establish anterior/posterior regional distinctions in the organismal nervous system, irrespective of differences and distinctions in both the upstream embryological inputs and the downstream development of the nervous system. This amounts to the prediction that there awaits discovery of a deuterostome superphylum (i.e., hemichordate plus vertebrate) kernel, which underlies the patterning of the nervous system and includes many or all of these genes.

The extent to which this patterning system might actually be pan-bilaterian is harder to see. The comparison to *Drosophila* in Fig. 5.3C5 shows overlap in relative location of expression with the deuterostome regulatory organization most clearly in respect to the anterior-most genes, i.e., in addition to those discussed above in respect to Fig. 5.3A, B, *viz. rx, so/six3, scro/nkx2-1, dll/dlx, slp/bf-1*. Other evidence suggests the possibility of conservation of patterning functions in the CNS between *Drosophila* and vertebrates for *tll, ems/emx*, and *barH* genes among others (reviewed by Lowe *et al.*, 2003; Lichtneckert and Reichert, 2005).

Thus far we have focused on what by all considerations is a definitive character of Bilateria, the anterior/posterior organization of the nervous system, particularly on that portion of it anterior of the domain of expression of any of the *hox* genes (midbrain and forebrain in vertebrates and arthropods). But there is still another aspect of regulatory gene expression in the development of vertebrate and arthropod CNS that bespeaks the possible existence of a network patterning kernel. This is the bilateral patterning of the CNS from the ventral midline in *Drosophila*, and from the dorsal midline in vertebrates, by three orthologous transcription factors expressed during embryogenesis in adjacent longitudinal stripes (Weiss *et al.*, 1998; reviewed by Cornell and Von Ohlen, 2000). As shown in the diagrams of Fig. 5.3D, these genes are *vnd/nkx* (multiple *nkx* family members) medially, *ind/gsh* in the intermediate position, and *msh/msx* most laterally. Once again the types of neurons and structures to which the different domains marked out by the regulatory pattern give rise downstream are different in insects and vertebrates: it is the early domain-making pattern of regulatory states which is conserved. So could be the regulatory interactions among these genes. In *Drosophila*, mutations in these genes reveal that their expression is required for the regional identity of the neuronal structures developing in each domain, and also that the boundaries of the stripes devolve from mutual repression. Thus *vnd* represses *ind* in the medial domain, so that in the absence of *vnd* expression, *ind* expression spreads toward the midline; similarly *ind* represses *msh* in the intermediate domain (Weiss *et al.*, 1998; Fig. 5.3D1). Misexpression experiments on *vnd* demonstrate that the factor encoded by this gene can act as a direct transcriptional silencer, and insertion of Vnd target sites in a heterologous enhancer results in repression of the activity of the latter in the endogenous stripe of *vnd* expression (Cowden and Levine, 2003). Furthermore, a *cis*-regulatory module controlling *ind* expression contains multiple Vnd sites (Stathopoulos and Levine, 2005). In vertebrates much less is known, but one key result is reproduced in Fig. 5.3D2: just as in *Drosophila vnd* mutants *ind* expression spreads medially, in *nkx2.2* mutants neuronal cell types normally forming from the intermediate *gsh* domain appear in the ventral column where *nkx2.2* is normally expressed, and those normally forming there are missing (Briscoe *et al.*, 1999).

One possible objection to the idea that this medial-lateral patterning system is the output of an ancient bilaterian subcircuit might be that it would be directly inconsistent with the radial organization of the diffuse epidermal nervous system of the hemichordate in respect to the dorsal-ventral aspect of the body plan, were that representative of ancestral forms. But it is interesting that almost uniquely among the genes studied in *Saccoglossus* by Lowe *et al.* (2003), the *nkx2-1* gene is not expressed radially, but (as in the vertebrate brain) its domain of expression is ventrally confined (Fig. 5.3C2).

As noted above, phylum-specific kernels are easier to identify. Shifting the phylogenetic focus down to the level of the chordates, more direct DNA level evidence for detailed conservation of the regulatory apparatus underlying CNS patterning becomes available. First, in terms of gene expression domains, the basal tripartite organization of the brain pertains across the whole phylum; it is evident from ascidians to vertebrates (Fig. 5.4A; Raible and Brand, 2004; Takahashi and Holland, 2004; Cañestro *et al.*, 2005). There are indeed some gross differences, in that it appears that extant cephalochordates have lost the midbrain-hindbrain boundary region, and the vertebrates have differentiated a midbrain region marked by expression of the gene *dmbx*. Early in its development the hindbrain of vertebrates is transiently organized in segmental subunits, the rhombomeres, each of which gives rise to certain ganglia and to certain populations of neural crest (reviewed by Gavalas *et al.*, 2003). Each rhombomere expresses a given transcriptional regulatory state (Davidson, 2001, for review). Though invertebrate chordates do not have rhombomeres at all, and do not produce the facial and lower head ganglia that vertebrates do, during embryonic development the patterns of gene expression in the equivalent midbrain and hindbrain region of the embryonic ascidian are strikingly similar to those of vertebrates (Fig. 5.4B).

Gene regulatory network subcircuits responsible for specification of rhombomeres 3, 4, 5, and 6 are emerging from extensive data on the effects of gene knockouts in mice on expression of other genes, and from *cis*-regulatory analyses of some of these same genes (Barrow *et al.*, 2000; Davidson, 2001; Manzanares *et al.*, 2002; Gavalas *et al.*, 2003). The rhombomere 3–5 subcircuits are shown

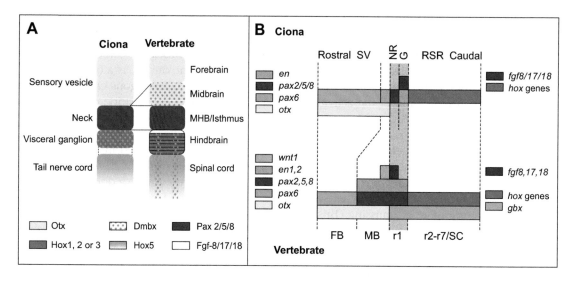

FIGURE 5.4 Chordate hindbrain gene expression patterns and evidence for vertebrate rhombomere specification kernel. (A) Tripartite organization of brain in ascidians and vertebrates, indicated by gene expression domains, from Takahashi and Holland (2004). The forebrain, midbrain/hindbrain boundary region, and hindbrain domains are marked by expression of *otx*, *pax2/5/8*, and anterior *hox* genes as in other bilaterians (cf. Fig. 5.3) despite the relatively very simple structure of the ascidian sensory vesicle compared to the vertebrate brain. The *gbx* gene is not present in the *Ciona* genome (Imai *et al.*, 2004). Only the vertebrates express the *dmbx* gene in the midbrain, though later both vertebrates and ascidians express this gene in cells of the hindbrain near the anterior limit of *hox* gene expression. (B) Another alignment of embryonic ascidian and vertebrate brains with special reference to the gene expression patterns around the midbrain/hindbrain boundary (from Raible and Brand, 2004). Abbreviations: SV, sensory vesicle; NR, neck region; G, visceral ganglion; RSR, rhombrospinal region (anterior end of tail nerve cord); FB, forebrain; MB, midbrain; r1-7, rhombomeres; SC, spinal cord. (C) Genomic control circuitry for rhombomere specification, and conservation across vertebrates. (C1) Rhombomere specification network subcircuits in mouse. Network diagrams, as views from the nuclei of rhombomeres 3, 4, and 5 (cf. Chapter 4), are revised from Davidson (2001), where earlier references are reviewed; additional data incorporated are from *cis*-regulatory studies of Kwan *et al.*, 2001; Scemama *et al.*, 2002; Manzanares *et al.*, 2002; and from mutational analyses of Gavalas *et al.*, 2003. (C2) *hoxa2* cis-regulatory module diagrammatically displaying sites that control expression in the hindbrain, rhombomeres 3 and 5, and in neural crest emanating from rhombomere 4 (from Tümpel *et al.*, 2002). (C3) Conservation of target site sequences in *cis*-regulatory module shown diagrammatically in (C2) in mouse, human, chicken, and shark (from Tümpel *et al.*, 2002). RE2,3,4, the Krox20, and the BoxA sites are all conserved across this phylogenetic distance.

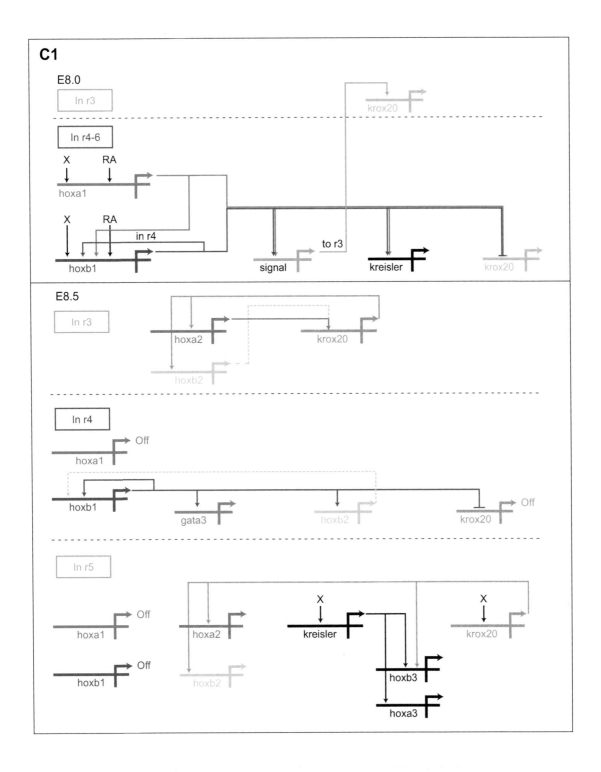

FIGURE 5.4 (continued)

(Continues)

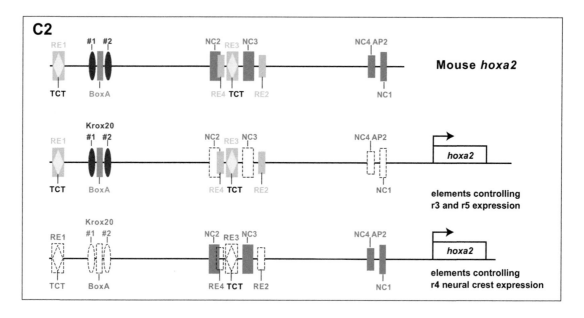

FIGURE 5.4 (continued)

in Fig. 5.4C1. In mouse the main interacting players in these subcircuits, which have rhombomere-specific feedback loops and cross-regulatory linkages suggestive of kernel-like wiring, are the *hoxa1*, *hoxa2*, *hoxb1*, *hoxb2*, *hoxa3*, *hoxb3*, *krox20*, and *kreisler* genes. There is now much direct *cis*-regulatory evidence in support of the inputs displayed in these subcircuits (*op. cit.*, and also Kwan *et al.*, 2001; Scemama *et al.*, 2002; Tümpel *et al.*, 2002). For our present context, the point is that here also can be found direct evidence that the same inputs operate the respective genes of the rhombomere specification subcircuits throughout the vertebrates and even beyond. For example, among the *cis*-regulatory modules at key nodes in these subcircuits are those that direct *hoxa2* expression to rhombomeres 3 and 5 (E8.5, Fig. 5.4C1). As illustrated in Fig. 5.4C2, 3 (Tümpel *et al.*, 2002) the specific target sites of this key node of the rhombomere specification subcircuit are conserved from sharks to chickens to mammals; so also are other sites that direct *hoxa2* expression in neural crest emanating from rhombomere 4. Note that the presence of the Krox20 site in these genes implies that the feedback linkage from the *krox20* gene to the *hoxa2* gene in rhombomeres 3 and 5 is conserved. The *cis*-regulatory module controlling rhombomeric expression of a *hoxb2* gene is also conserved at the target site level from teleost fish to humans

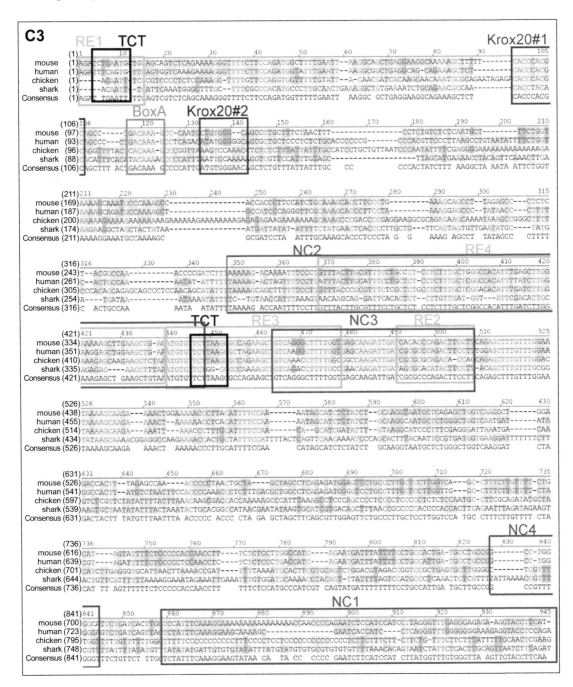

FIGURE 5.4 (continued)

(Scemama *et al.*, 2002), including the sites for the Krox20 inputs required in rhombomeres 3 and 5, and for the Hoxb1 input required in rhombomere 4 (cf. Fig. 5.4C1). A third example is the conservation of the Kreisler sites in the *cis*-regulatory module controlling *hoxa3* expression in rhombomere 5 (Fig. 5.4C1), which are again conserved from shark to man (Manzanares *et al.*, 2001). These *cis*-regulatory modules do not demonstrate the conservation of the complete subcircuit, but the multiple conserved target sites in each directly imply the presence of a significant portion of the circuitry in all these species. The basic circuitry could have survived even in invertebrate chordates which lack rhombomeric segmentation: thus a *cis*-regulatory module from the amphioxus *hox3* gene directs expression of a reporter construct to the posterior rhombomeres, just where mouse *hoxb3* is expressed (Manzanares *et al.*, 2000). Possibly we are seeing here a phylum-specific network kernel, perhaps analogous to the echinoderm endoderm specification kernel shown in Fig. 4.3.

The anterior-posterior organization of the nervous system is one of the great definitive features of Bilateria. The proposition that its conservation in Bilateria is due to their common use of shared developmental gene regulatory kernels will obviously remain an issue until the regulatory linkages are out and have been resolved experimentally at the genomic sequence level. That nervous system patterning kernels exist is a precisely defined and precisely testable proposition, experimentally now within our grasp, and there is clearly sufficient presumptive evidence to justify tests at the gene network level. More generally, whatever the outcome of searches for pan-bilaterian or superphylum kernels in modern animals, there is a good likelihood that extremely conserved, phylum-specific kernels underlying phylum-specific developmental features exist, and can be recovered. Even were this likelihood only upheld as more and more gene regulatory networks are unraveled and compared, and even if true pan-bilaterian kernels remain elusive, it would still fundamentally illuminate the phylum level constraints on diversity of body plan since the Early Cambrian.

The Bilaterian Gut

As remarked above, the bilaterian through gut appears *a priori* to invite the speculation that shared regulatory kernels underlie the presence in all major phyla of homologous gut parts, mouth, pharynx, midgut, hindgut, and anus. There are at present various hints that even across great phylogenetic gulfs the regulatory apparatus that Bilateria utilize in constructing the gut retains common mechanistic features, but demonstration of common linkages is in general so far missing. An example of such a hint is the essential role of genes encoding Gata class transcription factors in development of the gut. In *Drosophila* the *gata* class *serpent* gene (Rehorn *et al.*, 1996) and the *gatae* gene (Okumura *et al.*, 2005) are specifically required for midgut formation; in *C. elegans* a series of *gata* genes is needed for midgut formation, including *end1* and *end3*, which perform specification functions in the gene regulatory network for gut development, and *elt2* and *7*, which are direct operators of gut differentiation genes (for reviews, Davidson, 2001; Maduro and Rothman, 2002); in sea urchins, the *gatae* gene is essential (cf. Fig. 4.2); and in vertebrates, the *gata4,5,6* genes drive gut endoderm specification (discussed below). In addition, other particular families of regulatory genes are commonly associated with gut development, in *C. elegans*, *Drosophila*, and vertebrates, including *foxa/fkh* genes, *sox* family genes, *hnf4* (nuclear hormone receptor family) genes (for reviews, Cremazy *et al.*, 2001; Lengyel and Iwaki, 2002; Maduro and Rothman, 2002; and the following discussion).

Other genes are universally expressed in the bilaterian hindgut and the invaginations by which it arises. Three such genes are: the *brachyury* gene (Holland *et al.*, 1995; Peterson *et al.*, 1999; Technau, 2001), expression of which is required for gastrular invagination and normal gut formation in *Drosophila* (i.e., the *brachyenteron, byn* gene), in sea urchins, and in vertebrates (Conlon and Smith, 1999; Peterson and McClay, 2003; reviewed by Lengyel and Iwaki, 2002); the *caudal* (*cad/cdx*) gene, a parahox gene expressed at the posterior end of the body plan from *Drosophila* to vertebrates, and later in the gut, and which is needed for hindgut invagination and development in *Drosophila* (Wu and Lengyel, 1998); and again the *foxa/hnf3β/fkh* genes. In *Drosophila*, *cad* is directly upstream and required for activation of *fkh* (Wu and Lengyel, 1998). In constructing an argument for a pan-bilaterian regulatory subcircuit for hindgut development including these three genes, Lengyel and Iwaki (2002) reviewed evidence that they are expressed in concert at the site of hindgut invagination in *Drosophila*, mouse, and *Xenopus*. However, while coincident expression indeed adds up to a reasonable hunting license for deeply embedded pan-bilaterian gut subcircuits, examination reveals prohibitive complications in this case, even at the essentially superficial level of gene use. For example, in *Drosophila*, the *byn* gene is activated by a terminal regulatory system that is obviously not general, beginning with Torso signaling which activates the *tailless* (*tll*) gene, which is in turn required for *byn* activation (Wu and Lengyel, 1998); but in deuterostomes *tll* is only expressed in the brain, at the other end of the animal (cf. Fig. 5.3). In sea urchins *bra* is activated by *gatae* at the hindgut/blastopore (Fig. 4.2), but in *Drosophila* the *gata* genes are only expressed in the midgut, never in the hindgut; indeed their role is to repress *byn* in the midgut (Okumura *et al.*, 2005). In *C. elegans* the *caudal* ortholog, *pal1*, is indeed expressed in posterior blastomeres, but it directly controls expression of genes producing ectoderm, muscle, and neurons, and only indirectly, via signaling, does it affect posterior endoderm development (Baugh *et al.*, 2005). These diverse intergenic relations clearly indicate differences across the Bilateria in the wiring of the network subcircuits of which they are a part. One possibility is that as usual in embryogenesis (cf. Chapter 3) the earliest phases of gut formation are very clade specific, and that only at the stage when the regional regulatory states of the gut itself are to be specified, might highly conserved regulatory kernels be called into play. Much of the evidence so far adduced, and many of the difficulties that become apparent on looking under this corner of the rug, concern early embryonic events.

An interesting possible example of an evolutionarily conserved, later gut specification regulatory module found in *Drosophila* is the use of two different, anciently
separated paralogs of the *odd-skipped* family of Zn finger regulators in patterning
the small intestine at the anterior hindgut-midgut junction. One of these paralogs,
drumstick (*drm*), interacts directly with another transcriptional repressor, Lines,
preventing it in turn from blocking the action of the second *odd-skipped* paralog,
bowel (Iwaki *et al.*, 2001; Green *et al.*, 2002; Johansen *et al.*, 2003). The latter is an
activator required for small intestine fate, and its ectopic expression in posterior
hindgut suffices for conversion to small intestine fate (Johansen *et al.*, 2003). The
same genes are apparently required to pattern the junction between the foregut
and hindgut, as is also the *fkh* gene (Hoch and Pankratz, 1996; Johansen *et al.*,
2003). In *C. elegans* the orthologs of *drm* and *bowel* are also required for
intestine development (Buckley *et al.*, 2004). This suggests the retention of this
patterning subcircuit across the ecdysozoan superphylum. Tantalizingly, there is
a human gene of the *odd-skipped* family that is expressed in the colon (Katoh,
2002). Even more tantalizing is the relation between genes encoding a Sox E class
factor, *sox100B* in *Drosophila* (Loh and Russell, 2000) and *sox9* in mammals, and
fkh/foxa1,2,3 in the regional patterning of the gut. These genes are expressed in
overlapping domains in the gut during regionalization stages in both mammals
and flies. A *cis*-regulatory module of the human *sox9* gene which in transgenic
mice drives hindgut expression, includes a target site for a Foxa class factor;
most remarkably, this same site is conserved, along with a cluster of other sites,
in enhancers of the cognate genes of *Drosophila*, *Ciona*, and vertebrates (Bagheri-
Fam *et al.*, 2006). Though the goods are certainly not yet in hand irrespective of
frequent claims that regional gut patterning processes in *Drosophila* and chordates
are fundamentally homologous, these claims could turn out to be correct in the
essential sense of shared regulatory circuitry.

It is again easier to find phylum or subphylum level gut specification kernels.
Within the deuterostomes the patterns of circuit conservation for this process form
a nested, almost Linnaean set. In vertebrates, even across the teleost/tetrapod
divide, the circuitry is preserved from early embryonic specification into gut formation. Fig. 5.5A1 dramatically illustrates in the very early zebrafish embryo the
co-expression in prospective endoderm cells of four regulatory genes that are
linked in the core of the endoderm specification subcircuit, *viz.* the *gata5* gene,
the homeodomain gene *bon/mix*, the *foxa2* gene, and the *sox17* gene (Reiter
et al., 2001). Ectopic expression of the Gata5 factor suffices to cause co-expression
of *sox17* and *bon/mix* (Fig. 5.5A2, 3; Reiter *et al.*, 2001). Though there is yet no
direct *cis*-regulatory evidence, the epistatic relations among these genes, and the
upstream signaling functions that are required for their activation, can be portrayed as in Fig. 5.5B1. These relations are remarkably similar to those which are
obtained in *Xenopus*, as shown in Fig. 5.5B2. The upper part of Fig. 5.5B2, i.e.,
that indicating the activation of the *xnr* genes and the *mix* and *mixer* genes
downstream of the maternal VegT factor we have already seen at the network
level, where there is *cis*-regulatory support (Fig. 4.4B2). There is further evidence

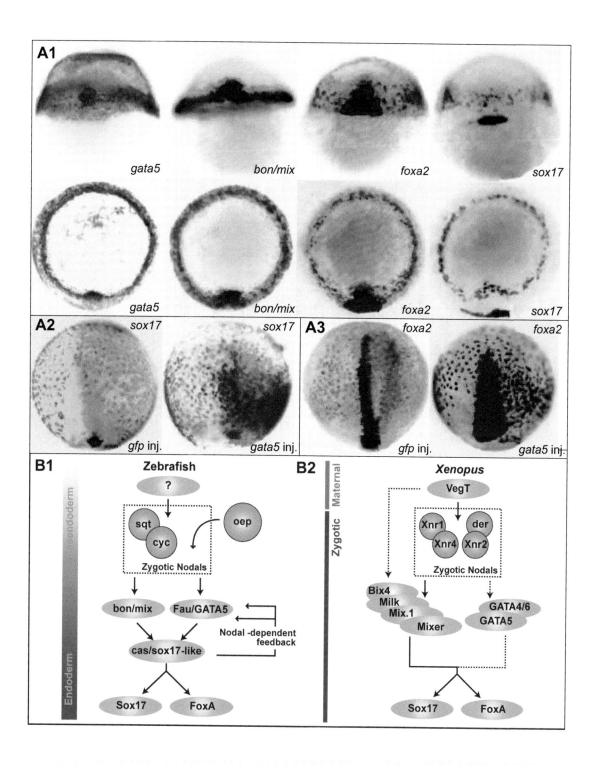

FIGURE 5.5 For legend see pages 216–217.

(Continues)

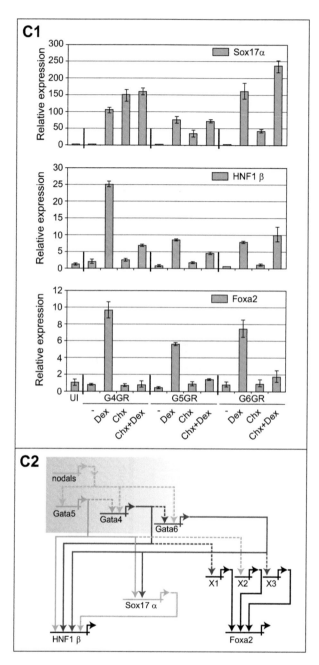

FIGURE 5.5 Endoderm specification in vertebrates. (A) Regulatory gene expression in zebrafish endoderm. (A1) Co-expression of four regulatory genes in marginal cells from which will arise endoderm,

that the interaction between the *Xenopus gata* genes and the *sox17* and *foxa2* genes is direct (Afouda *et al.*, 2005). This comes from experiments in which mRNA encoding the Gata factors was injected into the animal pole of fertilized eggs, the animal pole blastomeres were later removed and cultured, and expression of these downstream genes was demonstrated to occur even in the presence of protein synthesis inhibitors (e.g., see Fig. 5.5C1). In *Xenopus* and in mammals there are several *gata* genes that function during endoderm development (*gata 4,5,* and *6*), and their respective roles are made explicit in the network subcircuit shown in Fig. 5.5C2 (Afouda *et al.*, 2005). Here we see that in response to low intensity signaling by nodal class factors (i.e., in the embryo the Xnr's), *gata6* directly activates *hnf1β* and *sox17*, while *gata4* and *gata5* are induced at higher signaling levels and thereupon also generate direct inputs into these genes; activation of *foxa2*, however, is indirect.

in late blastula stage embryos, 4.3 h postfertilization (from Reiter *et al.*, 2001). Top row displays dorsal views; second row, animal pole views. (A2) Demonstration that Gata5 suffices for ectopic activation of *sox17* transcription; and (A3) that it suffices for *foxa2* transcription (from Reiter *et al.*, 2001). One blastomere of 2-cell zebrafish embryos was injected with either *gfp* mRNA as a control, together with *lacz* mRNA (light blue stain) to serve as an injection marker (left), or with *gata5* mRNA, plus *lacz* mRNA (right). Presence of *sox17* transcript (A2) or *foxa2* transcript (A3) was then monitored by whole mount *in situ* hybridization (dark blue stain). (B) Epistatic relations among signaling and regulatory genes in specification of zebrafish and *Xenopus* embryonic endoderm. Diagrams are slightly modified from Shivdasani (2002); *foxa* has been added as a Gata target gene (see parts A3 and C of this Figure, and text). Dotted line indicates an inferred interaction; solid lines indicate experimentally supported interactions. (B1) Zebrafish; names of genes are indicated both as generally known, and as abbreviations of names derived from zebrafish mutations. (B2) *Xenopus*; there are multiple members of the *mix* gene family in *Xenopus*, but only one so far known in zebrafish (*bon*); for the maternal VegT factor, cf. Chapter 3. (C) Direct and indirect linkages from *gata* genes to other endodermal regulatory genes (from Afouda *et al.*, 2005). (C1) Use of protein synthesis inhibitor (cycloheximide; Chx) to distinguish direct from indirect induction of transcriptional expression of Gata target genes: mRNA encoding a Gata4-glucocorticoid ligand binding domain (G4GR), or corresponding Gata5 (G5GR) or Gata6 (G6GR) fusions was injected into the animal pole of *Xenopus* eggs. Animal pole cells were later removed and cultured with or without dexamethasone (Dex) to permit access of the encoded factor to the nucleus, in presence or absence of Chx. Expression of the indicated target genes was monitored by quantitative PCR. Treatment with Dex in every case shown results in induction of these target genes, but only if induction also occurs in presence of Chx as well as Dex can the response be considered direct. For example, *sox17α* is a direct target of Gata4. (C2) Network subcircuit displaying response of the three *gata* genes to Nodal signaling of different intensity (shading; activin was used experimentally as a surrogate nodal factor), and their direct (solid lines) or indirect (dashed lines) induction of target gene transcription. Here "X" represents an unknown intermediate gene. Note the feed forward relation between *gata5, sox17α,* and *hnf1β*.

When we look outside the vertebrates, however, in *Ciona* or the sea urchin, this particular network cannot pertain, since in these animals there is only one gene of the *gata4,5,6* class. Nor in sea urchins, for example, is endoderm specification initiated by nodal class signaling, but rather in an entirely different way (Fig. 4.2). In sea urchins nodal signaling is instead required for oral ectoderm specification (Duboc *et al.*, 2004), and in *Ciona* nodal signaling is required for patterning the neural plate (Hudson and Yasuo, 2005). There is one essential relationship that lies deep in the core of both sea urchin and vertebrate networks, and this is expression of a *gata* gene upstream of, and required for, expression of a *foxa* gene. But while this is a direct linkage in the echinoderm endoderm kernel of Fig. 4.3, it is indirect in *Xenopus* (Fig. 5.5C2). Furthermore, in the echinoderm kernel *brachyury* is under similar control to *foxa*, while as we saw in Fig. 4.4, in *Xenopus* the *brachyury* gene has entirely different inputs. While to some extent these comparisons suffer from absence of evidence, which, as the adage goes, is no substitute for evidence of absence, we know enough to see that there is little support for a pan-deuterostome endoderm specification kernel, just as there is little support for a pan-bilaterian endoderm specification kernel. But there is very likely to be a vertebrate endoderm specification kernel, just as there is a quite different echinoderm one. And as above, there could be later gut regionalization kernels that do have a *trans*-phyletic distribution, and that across Bilateria underlie the development of the foregut, midgut, hindgut, and their sub parts.

A Possible Pan-Bilaterian Kernel Underlying Specification of the Heart Progenitor Field

Hearts occur throughout the Bilateria, though they are structured very differently in different clades. In its essentials, the heart is a tubular mesodermal structure the function of which is to contract rhythmically, so as to pump body fluids around the circulatory system, open or closed. Heart contractile protein differentiation gene batteries and their regulatory controllers differ from those of the main body musculature (e.g., Fig. 1.5). In *Drosophila* the heart is a linear tubular structure formed from the dorsal mesoderm (Fig. 3.4) which circulates the hemolymph by pumping it from a posterior chamber which contains intake ostia, out through an aorta that extends forward from the contractile chamber. In vertebrates the heart is far more complex in its chambered morphological organization and functional plumbing, and it arises initially from ventral mesoderm. But, it was discovered some years ago that the progenitor fields for the heart in *Drosophila* and in mammals are identified by expression of orthologous regulatory genes, *tinman* in *Drosophila* and *nkx2.5* in vertebrates (see Fig. 3.5B). This now turns out to indicate the existence of more extensive underlying similarities. As Fig. 5.6 illustrates, *cis*-regulatory and genetic data permit the construction of remarkably similar gene network subcircuits for control of the establishment of heart progenitor fields in *Drosophila* and in vertebrates (see legend for sources of linkages; earlier compilations on which this Figure is based are Cripps and Olson, 2002; Zaffran

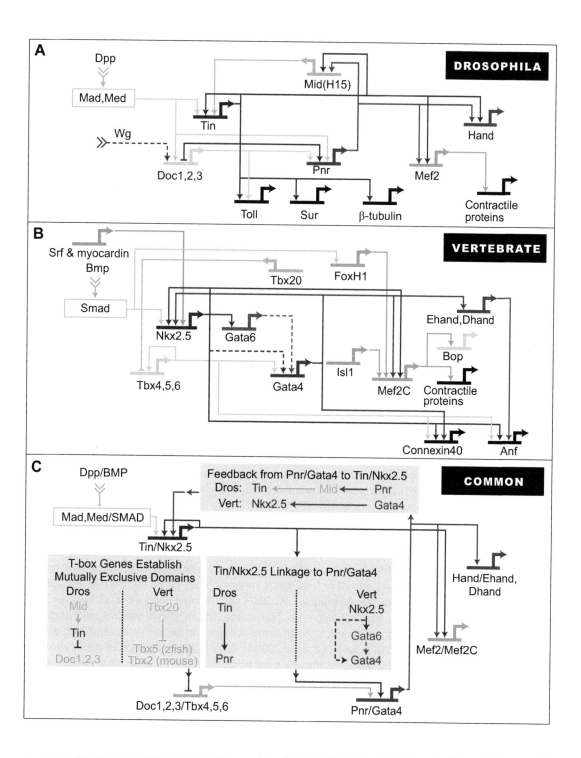

FIGURE 5.6 For legend see page 220.

(Continues)

and Frasch, 2002; Reim *et al.*, 2005). In Fig. 5.6A and B the orthologous genes are similarly colored (except that all downstream differentiation genes are in black). These diagrams show that the equivalent genes of the same families, *tin/nkx2.5*, *doc1,2,3/tbx4,5,6*, *pnr/gata4,6*, *mef2/mef2c*, and *mid/tbx20*, all occupy similar positions in the heart specification subcircuits (Fig. 5.6C). What is in some ways most remarkable is the implication that it is these circuitry functions that are the essential conserved features; as indicated in the gray boxes of Fig. 5.6C three

FIGURE 5.6 (continued)
A putative pan-bilaterian kernel for heart specification. Network subcircuits are shown that include a core set of genes belonging to orthologous families in *Drosophila* and mouse or other vertebrates (from Davidson and Erwin, 2006). Diagrams are formulated using BioTapestry visualization (see Chapter 4). Cognate regulatory genes are indicated in the same colors, downstream genes in black. Absence of a linkage means only that the linkage is not known to exist, not that it is known not to exist. Dashed lines indicate interactions for which there is some evidence but which are not as solidly established as for interactions indicated by solid lines. (A) *Drosophila* heart specification network subcircuit. Sources are as follows (for earlier references see Cripps and Olson, 2002; Zaffran and Frasch, 2002): Linkages of *pnr* to *hand* and *tin* to *hand*, Kölsch and Paululat, 2002; *mid* (and *h15*) to *tin*, Reim *et al.*, 2005; *pnr* to *mid* (and *h15*) and *tin* to *mid* (and *h15*), Reim *et al.*, 2005, Miskolczi-McCallum *et al.*, 2005, and Qian *et al.*, 2005; *tin* repression of *doc*, Reim *et al.*, 2005; *tin* to *sur*, Cripps and Olson, 2002, Reim *et al.*, 2005, and Nasonkin *et al.*, 1999; *doc* to *pnr*: inferred from early co-expression of *pnr* and *doc* and dependence of downstream *pnr* target genes including *mid* on *doc* expression (Reim *et al.*, 2005) in certain later cells of the heart that do not then express *tin*; *tin* to *toll* and *doc* to *toll*, Wang *et al.*, 2001; *dpp* to *pnr*, Kleindienst and Bodmer, 2003. *doc* is considered equivalent to *tbx4/5/6* of vertebrates since there may have been a common ancestor of these vertebrate Tbox genes (Reim *et al.*, 2003). (B) Vertebrate heart specification network subcircuit. Amniotes have primary and secondary heart fields, but this is a derived condition since fish have only the equivalent of the primary field, so linkages of both primary and secondary fields are here included. Sources are as follows (see Cripps and Olson, 2002, for earlier references): For *tbx5* autoregulation, Sun *et al.*, 2004; for linkages from *gata4* and *isl1* to *mef2c* (in anterior heart field, not primary heart field), Dodou *et al.*, 2004; *foxh1* and *nkx2.5* to *mef2c* (primary heart field), von Both *et al.*, 2004; for possible *gata6* to *gata4* linkage, Rossant *et al.*, 2004, and Koutsourakis *et al.*, 1999; *tbx5* to *gata4*, Bruneau *et al.*, 2001 and Heicklen-Klein and Evans, 2004; *gata4*, *nkx2.5*, and *tbx5* to *connexin40*, Linhares *et al.*, 2004; *dhand* to *atrial natriuretic peptide* (Anp), Dai *et al.*, 2002; *smad* to *nkx2.5*, Cripps and Olson, 2002, and Lee *et al.*, 2004; for possible repression of *tbx5* by *tbx20* (in zebrafish), Szeto *et al.*, 2002 (however, this does not occur in *Xenopus*; Brown *et al.*, 2005), and repression of *tbx2* by *tbx20* (in mouse), Singh *et al.*, 2005; for possible linkage *nkx2.5* to *gata4*, presence of putative *nkx2.5* sites in *gata4* heart *cis*-regulatory expression module (Linhares *et al.*, 2004); *mef2c* to *bop*, Phan *et al.*, 2005. (C) Subcircuit linkages that occur in both *Drosophila* and vertebrates, i.e., the shared heart specification kernel. The gray boxes portray different ways in which the same two nodes of the subcircuits are linked in *Drosophila* and vertebrates.

different functions of these kinds are carried out in slightly different ways in these subcircuits but always to the same end. Thus in *Drosophila* there is a linkage from *tin* to *pnr* and in vertebrates there is a linkage from *nkx2.5* to *gata6* and thence probably to *gata4*; in *Drosophila* there is a feedback from *pnr* to *tin* via the Tbox gene *mid*, and in vertebrates there is a feedback from *gata4* to *nkx2.5*; in both *Drosophila* and vertebrates exclusive domains of Tbox gene expression are established by repressive interactions. Note that these are among the key aspects that endow these subcircuits with their definitive property of recursive wiring. Other such wiring features are the feed forward relation in both organisms among the *tin/nkx2.5, pnr/gata4,6* genes and the *mef2* genes; in both, the inputs from *doc1,2,3/tbx4,5,6* into *pnr/gata4*; and finally the dual inputs into *tin/nkx2.5* from a Smad/Tgfβ class signal transduction factor and from *tin/nkx2.5* autoregulation. Though it is undoubtedly incomplete, and is based partly on recently emergent information which will probably change on further study, the detailed similarities in circuit design and in the specific roles of the participant regulatory genes shown in Fig. 5.6 are clearly much too impressive to ignore. And so, to step back and consider the larger significance of the diagrams that confront us in Fig. 5.6, what this exercise demonstrates is that pan-bilaterian kernels exist, for in all ways what is shown here conforms to the definition of exactly that.

Distribution of Kernels

If one pan-bilaterian kernel exists so must others, likely among them those for which the arguments were discussed in the preceding sections. But there are many other possible cases as well. For example, there could be kernels that provide the regulatory basis of the cellular immune systems that Bilateria utilize, despite their diverse effector genes (Hoffman *et al.*, 1999; Rothenberg and Davidson, 2003); kernels that underlie the patterning of the mesoderm (Jagla *et al.*, 2001), and kernels that underlie the patterning of neuronal as opposed to ectodermal cells (Rebeiz *et al.*, 2005), all basic bilaterian developmental functions. Many shared bilaterian characters have not been discussed at all here, characters that require spatial patterning functions in the development of given body parts, among them the external ectodermal integument, external sensory apparatus, the nephric secretory apparatus, the mouth and foregut, and so forth, each of which provides a hunting ground for shared regulatory circuitry. However, note that the body parts that are considered here, the nervous system with its anterior-posterior organization, the midgut and hindgut, and the heart, constitute a significant portion of the definitive bilaterian body plan. Another fundamental bilaterian feature is metamerism, though it very possibly arose independently in the various bilaterian superphyla. It is not possible at present to know what aspects of bilaterian developmental gene regulatory circuitry exist in cnidarians, since as of this writing no *cis*-regulatory experiment has ever been done in a cnidarian embryo. But these animals are not as different from bilaterians as once universally believed.

Not only did we and they diverge from one another relatively shortly in geologic time before there occurred the diversification of the Bilateria (Peterson *et al.*, 2004), but the organization of their early regulatory spatial organization is bilaterian-like. Thus their larvae express regulatory genes asymmetrically in the dorso-ventral as well as anterior-posterior axes just as do bilaterians (Finnerty *et al.*, 2004), and though they are supposed to be diploblastic, mesodermal and myogenic regulatory genes are expressed in certain of their life cycle stages (reviewed by Seipel and Schmid, 2005). The point here is not that bilaterian kernels arose only with the Bilateria, though this must in part be true. Rather it is that whenever they were assembled, the existence of these highly conserved, indispensable subcircuits may explain the persistence of the definitive characteristics of the bilaterian superclade ever since its appearance in the fossil record (Davidson and Erwin, 2006).

A frequent confounding observation, in considering conservation of developmental regulatory circuitry, is that even famously persistent regulatory apparatus is sometimes just not there. In the context of the examples of this discussion, for instance, in *Ciona* the *gbx* gene cannot play the key role it does in patterning the midbrain/hindbrain junction in vertebrates and *Drosophila* because this gene is not present in the *Ciona* genome (Imai *et al.*, 2004). However, it is present in the sea urchin genome, and from the phylogeny, its absence in *Ciona* is obviously due to a loss in the ascidian lineage. We need not speculate on whether its dispensability is related to the extreme paucity of the centralized ascidian "brain" relative to that of *Drosophila* or vertebrates (perhaps this gene is expressed in the posterior part of the developing epidermal nerve net of the sea urchin, as in *Saccoglossus*). Similarly, *C. elegans* does not have a body part that performs the function of the heart in other ecdysozoans or in chordates. Instead the *nkx2* family gene *ceh22* is expressed in the pharynx, which like a heart pulsates rhythmically, but for the different purpose of pumping bacteria the animal has ingested into its gut. The *ceh22* gene is required for pharyngeal muscle function, as are *nkx2.5* and *tinman* required for expression of contractile protein genes in mice and *Drosophila*. However, the *cis*-regulatory module driving pharyngeal expression of *ceh22* responds to a pharyngeal Foxa factor, which is not the case for *nkx2.5* or *tinman*, and no aspects of the heart specification circuitry of Fig. 5.6 upstream of *nkx2.5* or *tinman* have been detected in the *ceh22 cis*-regulatory control system (Vilimas *et al.*, 2004). At least that part of the heart kernel is absent in *C. elegans*. In other words, losing a subcircuit, like pulling the regulatory plug on it and turning it off (see below), is easily done if in some clade its output is dispensable, i.e., the body part is not needed or has been deemphasized. Or put more generally, asymmetry in the evolutionary occurrence of loss, as opposed to the occurrence of assembly, is another definitive property of kernels.

In summary, a main prediction that emerges from the foregoing is that there will be found many kernels that are phylum-specific (or perhaps subphylum-specific, e.g., vertebrate-specific), and that these kernels will explain the conservation of major phyletic characters ever since the Cambrian (Davidson and Erwin, 2006).

A good example is probably the anterior-posterior specification subcircuit of the chordate (or perhaps vertebrate) hindbrain (Fig. 5.5C). Cephalochordates and vertebrates are represented in the Early Cambrian assemblages (Chen *et al.*, 1999; Shu *et al.*, 1999), and so it is not a reach to imagine that the *hox* gene cross-regulations of Fig. 5.5C1 were operating in the development of the brains of these animals 520 million years ago, since these groups retain today at least some (perhaps most) detailed aspects of the same regulatory apparatus. Another such case is probably to be found in the regulatory instructions for the emergence and function of neural crest in vertebrates (Fig. 5.5C2; Meulemans and Bronner-Fraser, 2004). Similarly we may suspect the existence of conserved network kernels that provide the regulatory logic underlying the annelid segmentation generator; the arthropod appendage patterning system, the development of the molluscan shell gland; of the radial echinoderm water vascular system; and so forth.

SUBPHYLETIC EVOLUTION OF BILATERIAN BODY PLANS: *hox* GENES AS SOURCES OF LABILE NETWORK INPUT/OUTPUT SWITCHES

Diverse Functional Consequences of *hox* Gene Expression

hox genes do many things and affect developmental processes and thus morphology in many different ways. The only way to know what any regulatory gene actually does is to know the direct *cis*-regulatory targets of its gene products, and fortunately there have now been identified over 50 targets of various Hox factors (for review, Pearson *et al.*, 2005). Direct targets include such diverse downstream effectors of morphological process as an apoptosis gene (Lohmann *et al.*, 2002), genes encoding cell adhesion effector proteins (e.g., Chen and Ruley, 1998; Bruhl *et al.*, 2004), and genes controlling cell cycle activity (Bromleigh and Freeman, 2000). Numerous additional genes of a great variety of function have been identified as Hox targets in microarray studies, including genes encoding various enzymes, cell motility genes, intercellular signaling genes, and more cell adhesion genes (Knosp *et al.*, 2004; Cobb and Duboule, 2005; Lei *et al.*, 2005; Williams *et al.*, 2005). But our interest here is in the role of *hox* genes in major evolutionary alterations of body part morphology. Such alterations require reorganization of the spatial distribution of regulatory states during development, that is, alteration of the operation of gene regulatory networks. The *hox* gene targets which are directly relevant are therefore the *cis*-regulatory systems of genes encoding other transcription factors, the products of which are among the outputs of spatial patterning subcircuits.

hox genes affect the development of many body parts which differ in morphology among the diverse clades within a phylum, and this in turn illuminates aspects of gene regulatory network organization that are evolutionarily labile. A particularly revealing mode of operation of *hox* genes is regional interference, by direct repression, with the operation of a spatial patterning system required for construction of some morphological feature of the body part. Regional repression of pattern institution is clearly to be distinguished from the function of building the pattern. Thus loss of the *hox* gene repressor function results in extension of the same pattern to the spatial domain where the repression was executed in the normal situation. This kind of mechanism is a classic case of a function which operates by controlling the output of a developmental regulatory system, rather than generating this output. That is, in terms of network structure, it is to be thought of as an external Boolean switch, or as described above, an I/O control on the regulatory state produced by a patterning subcircuit. An example was considered in detail in Chapter 2, where the *cis*-regulatory mechanism by which the Ubx and Antp *hox* gene products shut down the *distalless* gene in the abdomen of *Drosophila* was reviewed (Fig. 2.5D). Expression of *dll* is essential to patterning the appendages, and this regional *cis*-regulatory repression function is responsible for the lack of legs on the abdominal segments of adult insects (shown initially by Vachon *et al.*, 1992). Conversely, just as often, *hox* genes are seen to act as if they are switching on a patterning subcircuit in a given spatial domain of the organism. The positive and negative *hox* switches of course can act only within the spatial domain of expression of the *hox* gene, and as we take up below, variation in the spatial domains of *hox* gene expression is a major source of evolutionary change: these domains determine where the switches are to operate.

Multiple Examples: Regional Repression and Installation of Patterning Output, and the Evolutionary Diversification of Body Parts

In Fig. 5.7 this theme is illustrated in a variety of taxonomic and morphological contexts. As an introduction, in Fig. 5.7A1 is the skeleton of an embryonic python (Cohn and Tickle, 1999). Snakes evolved from lizard-like ancestors which had both forelimbs and hindlimbs. The python belongs to a clade of snakes which during development still form hindlimb buds (though these fail to grow in the embryo) but which form no forelimb buds. The lack of the forelimb bud is correlated with the extended domains of *hoxc8* and *hoxc6* expression in the python embryo: these genes are expressed in most tetrapods only up to the level in the lateral plate mesoderm of the flank where the forelimb buds form, but in the python they are expressed all the way up to the head (Fig. 5.7A2). Limb bud formation requires a complex set of transcriptional and signaling gene expressions, and the extension of these flank mesoderm *hox* expression domains across the position where the buds would form appears to have canceled the operation of bud specification. There is no molecular mechanism here. However, the example is typical of many correlations between an evolutionary alteration in a spatial domain of *hox* gene expression, and the cancellation in that domain of a patterning event which occurs elsewhere in the body plan, here the organization of the hindlimb bud, just beyond the posterior limit of *hoxc6* and *hoxc8* expression (Fig. 5.7A2).

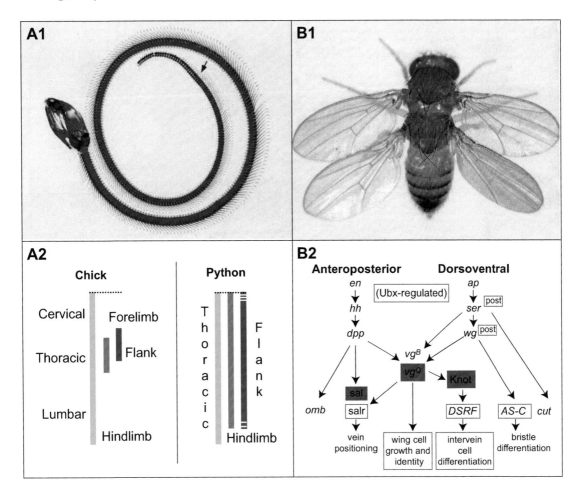

FIGURE 5.7 *hox* genes as regional off or on switches, applied to developmental patterning functions.
(A) Correlation of middle *hox* gene expression domains with thoracic body plan in python and chicken (from Cohn and Tickle, 1999). (A1) Complete skeleton of late python embryo. Position of hindlimb bud (which has been removed to enable visualization of axial morphology) is indicated by arrow, and vertebrae are homologous in structure from this point forward to head; the forelimb bud does not form. (A2) Domains of *hoxb5* (green), *hoxc6* (red), and *hoxc8* (blue) expression in axial, paraxial, and lateral plate mesoderm in chicken and python embryos. (B) Direct *Ubx* repression in the *Drosophila* haltere of transcriptional control circuitry for wing patterning. (B1) Four-winged fly, resulting from total loss of *Ubx* function (image from E.B. Lewis; Bender *et al.*, 1983). The haltere imaginal discs of the third thoracic segment have executed the same wing patterning functions in the development of this animal as have

(Continues)

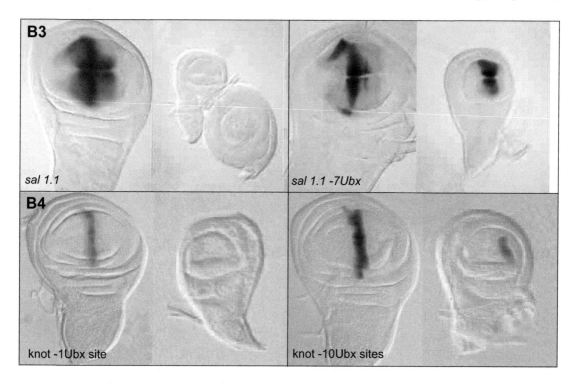

FIGURE 5.7 (continued)

the wing imaginal discs of the second thoracic segment. (B2) Summary of known repressive effects of Ubx factor on genes of the wing patterning network, from Weatherbee *et al.* (1998), modified by inclusion of data from Galant *et al.* (2002). These repressive interactions normally preclude development of wing structures in the haltere. Open red boxes indicate genes that are repressed in the wing disc in clones ectopically expressing *Ubx*, and that are activated in the haltere in clones lacking *Ubx* function; solid red boxes indicate additional evidence for direct repressive Ubx effects. VgQ denotes an expression construct that includes a *cis*-regulatory module from the *vestigial* gene, which is required for wing outgrowth, and is inactive in the haltere disc; this construct is derepressed in a haltere clone lacking *Ubx* function (Weatherbee et al., 1998). (B3) Direct *cis*-regulatory repression of the *spalt* (*sal*) gene by Ubx in the haltere disc (from Galant et al., 2002). An expression construct in which the *spalt sal* (*sal*) wing *cis*-regulatory module drives a *lacz* reporter (*sal1.1*) is expressed in wing (left) but not haltere (right). However, when all seven Ubx monomer target sites present in this construct are mutated (*sal1.1-7Ubx*), expression is derepressed in the haltere though not affected in the wing. (B4) Direct repression of *knot cis*-regulatory module by Ubx (from Hersh and Carroll, 2005). In the wing disc *knot* is expressed along the A/P axis, but is silent in the haltere disc. A *knot cis*-regulatory construct driving a *lacz* reporter that mediates correct expression in the wing disc remains silent in the haltere disc after mutation of one of the ten Ubx sites it contains (left); however, mutation of all 10 Ubx sites causes derepression of this construct in the haltere (right). (C) Lumbar repression of

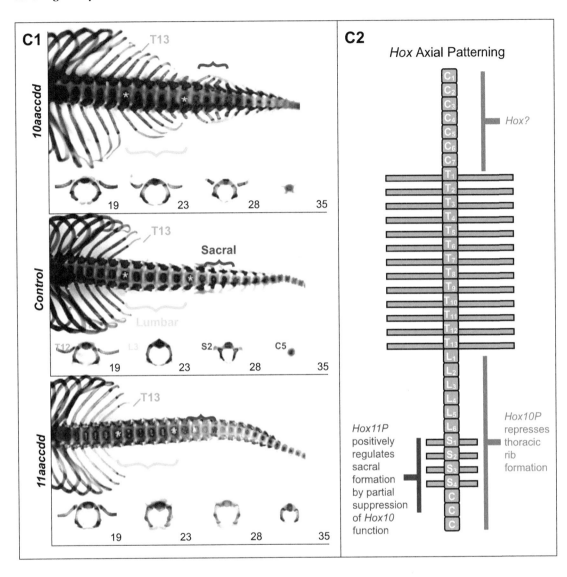

thoracic and sacral vertebral patterning by *hox10* and *hox11* genes in the mouse (from Wellik and Capecchi, 2003). (C1) Axial skeletal preparations from mice of genotypes indicated at left: 10aaccdd, homozygous triple knockout of *hox10a*, *hox10c*, and *hox10d* paralogs; 11aaccdd, same for *hox11* paralogs. Yellow asterisks and brackets mark lumbar, and red asterisks and brackets mark sacral vertebrae. At the bottom of each panel is shown a cross section of the indicated thoracic, lumbar, sacral,

(Continues)

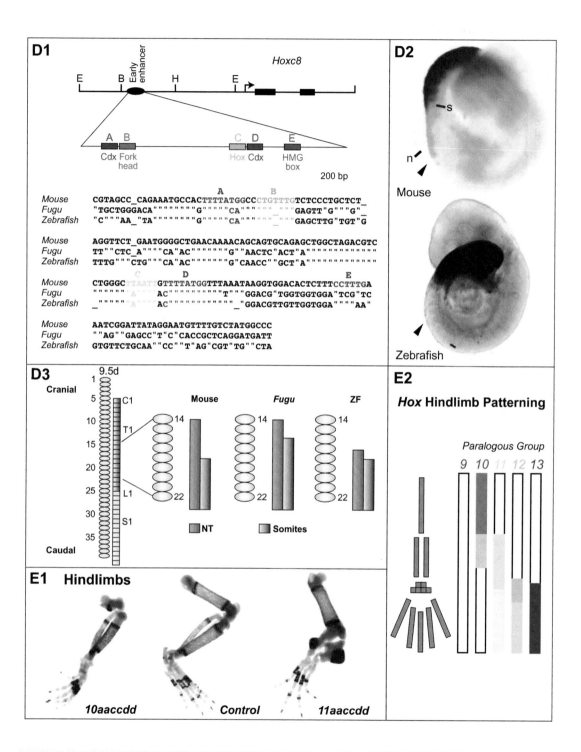

FIGURE 5.7 (continued)

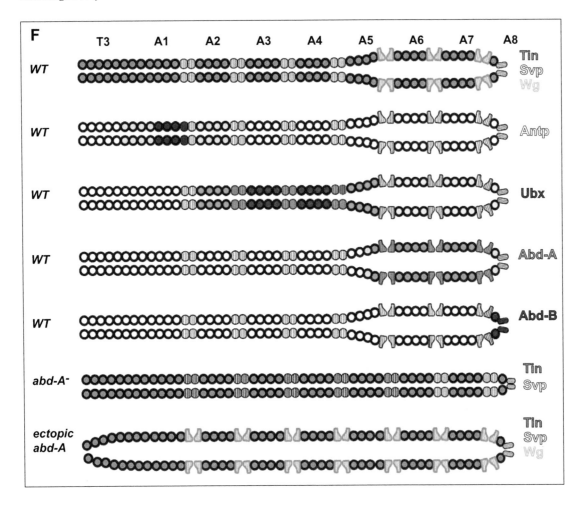

and caudal vertebrae. (C2) Summary of effects of triple *hox* mutations on axial patterning. The normal structure is shown in the diagram, and the functions executed by the *hox10* and *hox11* genes, as inferred from the triple mutant phenotypes, are indicated at the sides; P, paralog group. (D) *cis*-Regulatory control of *hoxc8* expression domain and examples of its evolutionary variation. (D1) Diagram of genomic region, and sequences of *cis*-regulatory modules controlling early axial expression of *hoxc8* gene, from mouse, *Fugu*, and zebrafish (Anand *et al.*, 2003). These two teleosts diverged about 150 million years ago. Note that among the color-coded target sites is a Hox factor site, indicative of *hox* gene cross-regulation. (D2) Expression in transgenic mice of *cis*-regulatory constructs under control of the early enhancers shown in D1 from mouse and from zebrafish. The arrowhead indicates somite 14. (D3) Summary of anterior boundaries of expression domains of the *hoxc8* early enhancer reporter constructs. Differences in enhancer target site sequences account for distinctions in spatial expression (D1–D3 from Anand *et al.*, 2003). (E) Positive inputs of *hox10* and *hox11* genes into

(Continues)

What this means is that the patterning subcircuit exists separate from the switch *per se*: hence the view that these *hox* gene functions are, in terms of network logic, I/O devices.

The example in Fig. 5.5B, where the mechanism of a similar kind of phenomenon is illustrated at the *cis*-regulatory level, may be imagined as a paradigm for how *hox* gene products switch off patterning subcircuits, preventing them from establishing the regulatory states needed for developmental progression. The *Ubx* gene of *Drosophila* (of the same class as the *hox6-hox8* genes of vertebrates) is required for the distinction between the wing, which arises from an imaginal disc on the second thoracic segment, and the haltere, which arises from the homologous disc on the third thoracic segment. The haltere is a small balancing organ which lacks many of the distinctive features of the wing, features that are the end result of the operation of developmental gene regulatory networks of the wing imaginal disc that set up the vein patterns, the intervein wing blade membrane, determine the location of the major sensory bristles, the size of the wing, and so forth. At the relevant stage the *Ubx* gene expression domain includes the haltere but not the wing imaginal disc. In the first direct demonstration of this major mode of *hox* gene function, Bender *et al.* (1983) showed that loss of *Ubx* activity results in transformation of the haltere into a perfectly formed wing, producing a four-winged fly (illustration from E. B. Lewis is reproduced in Fig. 5.7B1). That is, the whole wing patterning network now operates in the haltere disc as well as in the wing disc, and since the loss of function of *Ubx* results in the expression of this network in the haltere, the normal role of *Ubx* is to turn off its operation there. How the *Ubx* gene executes this switch function was determined in experiments in which clones of cells lacking *Ubx* function were produced in the haltere disc and clones of cells ectopically expressing *Ubx* were produced in the wing disc (Weatherbee *et al.*, 1998; Galant *et al.*, 2002). Within these clones expression of various genes participating in the wing patterning network was monitored.

FIGURE 5.7 (continued)

hindlimb patterning (from Wellik and Capecchi, 2003). (E1) Effects on hindlimb structure of same triple mutants as in (C). (E2) Diagrammatic summary: the *hox10* genes are required for formation of upper leg structures and the *hox11* genes for the lower leg structures, excepting the foot. (F) Regional requirement for *Ubx* function in development of the *Drosophila* heart, from Lo *et al.* (2002). The dorsal vessel is shown in a late embryo; T, thoracic segment; A, abdominal segment; expression of *tinman* (orange), *wingless* (*wg*, green), and the nuclear receptor gene *svp* (yellow) are indicated; anterior is left. The heart is the enlarged posterior chamber attached to the abdominal wall of segments A6–8 and the aorta extends forward into the thorax. The pairs of *svp* plus *wg* expressing cardioblasts in the heart are at the locations of the intake valves or ostia. Below are shown individually the adjacent domains of expression of *antp*, *Ubx*, *abdA*, and *abdB* in the dorsal vessel. The final two diagrams (color coded as in the wild type diagram at top) display the consequences for morphology and gene expression of absence of abdA function, and of ectopic expression of abdA in the aortal region.

Ubx expression does not affect the initial phase of transcriptional regionalization, in which the anterior/posterior and dorsal/ventral regulatory domains are established similarly in both wing and haltere imaginal discs (Williams *et al.*, 1994; Gómez-Skarmeta *et al.*, 1996; reviewed in Davidson, 2001). But in the haltere disc, *Ubx* expression results in the repression of multiple genes in the subsequent regulatory network, blocking multiple aspects of wing patterning. The summary of these results is reproduced in Fig. 5.7B2 (Weatherbee *et al.*, 1998; Galant *et al.*, 2002). For at least three of the affected genes the mechanism is direct *cis*-regulatory repression. Thus a *cis*-regulatory module of the *vestigial* gene, which encodes a key wing transcriptional co-activator, is prevented by *Ubx* expression from driving transcription of the reporter in the haltere (Weatherbee *et al.*, 1998; Galant *et al.*, 2002). A second example is the *spalt* regulatory gene, normally also expressed in the wing but not the haltere. The *cis*-regulatory module accounting for *spalt* expression in the wing includes seven Ubx target sites, and when all these are mutated expression of a reporter construct now extends to the haltere (Galant *et al.*, 2002; Fig. 5.7B3). The third example, the *knot* gene, is normally also expressed in the wing but not the haltere (Hersh and Carroll, 2005). Expression of this regulatory gene is required for activation of the *srf* gene (cf. Fig. 5.7B2), a major controller of downstream genes that generate the intervein wing membrane, and loss of *knot* function eliminates an intervein region of the wing. *cis*-Regulatory analysis again revealed multiple Ubx target sites as well as a positive site required for activation of *knot* in the wing, and mutation of the Ubx sites causes derepression of a *knot* reporter construct in the haltere (Fig. 5.7B4). To summarize, in the haltere imaginal disc, the Ubx gene product binds in the *cis*-regulatory modules of genes that are required for advanced stages of wing patterning, and (probably together with cofactors) in these *cis*-regulatory modules it acts as a dominant repressor. It thus precludes operation of the wing patterning network, which otherwise would operate in the haltere, accounting for the difference in the body parts that derive from these otherwise similarly competent imaginal discs.

The four-winged fly was once a bit mysterious, but is no more. With this analysis under our belt as a model case, we can imagine similar mechanisms for many other startling *hox* gene effects on the body plan, particularly those that have figured in subphyletic evolution. Of course the prerequisite for solving such mechanisms is knowledge of the gene regulatory network underlying the regional specification process; the corollary is that there will be no solution to these cases until these networks are at least partially known. In Fig. 5.7C is an example a little like the four-winged fly, in which the phenotype of the loss of *hox* gene function is loss of morphological diversity along the anterior/posterior axis of a major body part, so that we can see that the major role of the *hox* genes in question is to repress the operation of a patterning system in a given spatial domain. Figure 5.7C1 shows the skeletal structure of the posterior portion of the mouse vertebral column and ribs, in a wild type control; in the absence of three paralogous *hox10* genes; and in the absence of three paralogous *hox11* genes (Wellik and Capecchi, 2003). As summarized in Fig. 5.7C2, the role of the *hox10* genes is essentially to repress the rib-forming developmental system in the lumbar region of the spine, where these *hox* genes are expressed. It is not that the *hox10* genes "make" the lumbar spine; rather they act by pulling the plug on the operation of the developmental patterning system underlying rib formation, which otherwise would and could operate in their domain of expression.

Since this aspect of body plan varies among different groups of vertebrates, it is clearly malleable, and one aspect is the malleability of the *cis*-regulatory control of the location of *hox* gene expression domains. This is illustrated for the axial *cis*-regulatory control module of the *hoxc8* gene in Fig. 5.7D (Anand *et al.*, 2003). The sequences of this module are largely conserved between mouse, chicken, and fish (Belting *et al.*, 1998; Anand *et al.*, 2003; Fig. 5.7D1). When inserted in an expression construct, the fish modules direct expression to the mouse axial structures (Fig. 5.7D2). However, due in part to differences in the element "E" region of the *cis*-regulatory sequence, the anterior limit of the domain of expression mediated by these expression constructs is different when driven by a zebrafish fish *hoxc8* enhancer than by the cognate mouse enhancer. As summarized in Fig. 5.7D3, the *hoxc8* enhancers even of two fish differ in the anterior boundaries of the expression domains they generate in transgenic mice. These observations are of direct evolutionary interest because the different domains of *hoxc8* expression correlate with the extent of axial thoracic structures in the body plans of different vertebrate clades (Fig. 5.7A; Belting *et al.*, 1998).

Hox transcription factors function in different ways to activate or repress their many target genes: they bind the sites they recognize as monomers, or as multimers, or often in complexes with a canonical repertoire of other DNA-binding cofactors (for review, see Pearson *et al.*, 2005). Thus by the same logic that allows us to see that in cases where they act as repressors they "de-install" patterning subcircuits (e.g., Figs. 5.7A–D), when behaving as (required) activators they may act to install such subcircuits. The remainder of Fig. 5.7 is devoted to what would appear to be likely examples of positive *hox* gene switch function. Fig. 5.7E (Wellik and Capecchi, 2003) shows that the same triple *hox10* and *hox11* mutants discussed above lack, respectively, the upper or lower limb skeletal structures; the prediction is that these *hox* genes are providing necessary "AND" logic inputs to regulatory genes participating in the gene regulatory subcircuits that set up the chondrogenic spatial domains from which the respective skeletal structures derive. In Fig. 5.7F (Lo *et al.*, 2002) we return to the *Drosophila* heart, but at a later stage than that to which the specification subcircuit shown in Fig 5.6A pertains. At the top the dorsal vessel can be seen to consist of a posterior contractile chamber (the heart proper), and an aorta. The heart is located in abdominal segments A6–A8, and is equipped with three bilateral pairs of specialized cells at the sites of intake valves or ostia. The aorta extends anteriorly into the thorax. The contiguous domains of expression in the dorsal vessel of four *hox* genes are shown below; only *abdA* is expressed throughout the contractile heart chamber. In the absence of this transcription factor the whole dorsal vessel, as marked by the expression of *tinman* (see Figs. 2.5A and Fig. 5.6), turns into an aorta. If AbdA protein is expressed ectopically in the aorta region the whole dorsal tube becomes a heart, and there is no aorta (Lo *et al.*, 2002; Ryan *et al.*, 2005). The AbdA factor is required for the operation of the regulatory system underlying the creation of the contractile chamber, including the cells that produce the ostia, which express some different genes than do the adjacent cardial cells (see legend); thus, as Fig. 5.7F shows, if the *abdA* gene is expressed in the domain where normally the aorta forms, this region also produces ostial cells.

Many of the gross regional effects of *hox* gene gain or loss of function mutations on morphology that have been logged over the last several decades are reminiscent of evolutionary diversification. The initial example was the four-winged fly; more basal insects indeed normally produce four identical wings. These kinds of differences occur at Class, Order, and Family levels, as in Fig. 5.2, as well as at lower levels. The general import of Fig. 5.7 is that *hox* genes (among their other roles) act regionally to permit or preclude operation of developmental patterning subcircuits, or as put here, as I/O switches. In terms of network logic these switches can be thought of as external to the subcircuits themselves, even though the operation of these switches is mediated by positive or negative Boolean *cis*-regulatory functions executed at nodes of the network patterning subcircuits. A large number of other examples could have been included, ranging from relatively major morphological effects such as considered in Fig. 5.7, to minute morphological effects such as the control by *Ubx* of macrochaete bristle location by *Ubx* in the third as opposed to the second leg of *Drosophila* (Rozowski and Akam, 2002), a species-specific character. In beetles, *Ubx* function is required for the innate wing patterning network to operate (in the hindwing), because without it expression of three different wing patterning regulatory genes fails (Tomoyasu *et al.*, 2005). Clearly, the patterning subcircuits change in evolution much more slowly than do the positive and negative external switches determining their spatial deployment: for example, wings and macrochaete bristles are formed similarly in development across a range of insects in which they are located differently, and the same for the development of limbs and ribs in the diverse body plans of vertebrates. Given the existence of I/O switches for patterning subcircuits, metamerism becomes particularly useful, since in many of the above examples the metameric unit is the domain of subcircuit function. Thus change in the spatial location of these switches leads to metameric diversification. A reasonable question, then, is what mechanisms account for the relatively rapid evolutionary change in the deployment of *hox* I/O switches.

hox Gene Subnetworks and Spatial Evolution

Several different aspects of *hox* gene function are evolutionarily labile, all relevant to this conundrum. For instance, a change in the coding sequence of the Ubx protein itself, which accounts for its ability to repress leg formation, occurred in the evolution of insects from arthropod ancestors (Ronshaugen *et al.*, 2002; Galant and Carroll, 2002). Since *hox* genes are expressed so variously in related organisms, it is obvious that the *cis*-regulatory modules which control their exact temporal and spatial domains of expression must be a major source of evolutionary variation (Fig. 5.7). This is of course true of any regulatory gene involved in spatial patterning at any network level. A spectacular case is the rapid morphological change caused by alteration of the spatial *cis*-regulatory control apparatus of the *pitx1* gene in the morphogenesis of pelvic spines in stickleback fish (Shapiro *et al.*, 2004). Populations of these fish that diverged only 10,000 years ago differ in the presence or absence of these spines, and the difference depends on the spatial regulation of the *pitx1* gene in the pelvic region of the larva, which is evidently required for their development. Here the loss of spines is due to loss of regulatory function, so that the gene does not get expressed appropriately. It is easy to understand how loss of a particular phase of a necessary gene function due to destructive mutation of a particular *cis*-regulatory module could occur rapidly and frequently.

A key to the evolutionary role of *hox* genes lies in the richness of their own spatial regulatory mechanisms. The design of the exquisitely specific *cis*-regulatory modules that control their spatial expression (e.g., Fig. 5.4C) appears no different qualitatively than that of many other exquisitely specific developmentally active *cis*-regulatory modules, as reviewed in Chapter 2. But *hox* genes have additional properties. For one thing, *hox* genes function as spatially vectorial expression systems with respect to one another, in that they generate adjacent or nested domains of expression within a body part, usually along the anterior-posterior axis (though the same vectorial patterning system has also been co-opted for function in an orthogonal direction in tetrapod appendages; for review, Davidson, 2001). Furthermore, the spatial domains of *hox* gene expression may often depend on their own cross-regulation. The rhombomere specification subnetwork illustrated in Fig. 5.4C1 provides an excellent illustration of *hox* gene cross-regulation in its architectural context. Similarly, in the *Drosophila* heart discussed above, the boundary of *Antp* expression is set by Ubx repression (Ryan *et al.*, 2005). The extensive cross-regulation of *hox* genes implies that when in evolution one spatial domain of *hox* gene expression changes, so may others with reference to it. For example, in comparing spiders, crustaceans, insects, and centipedes, the spatial domains of expression of all anterior *hox* genes except *labial* differ markedly, with respect to the boundaries of the homologous head segments (Hughes and Kaufman, 2002). If there are cross-regulatory subcircuits that control these expression boundaries, for example, by mutual repression, then a potentially powerful mechanism for evolutionary change in spatial *hox* gene expression would be predicted, for each *cis*-regulatory change which affects the expression of any one *hox* gene will have indirect effects on expression of those other *hox* genes with which it interacts as well. This would provide a regulatory amplification of the diversity of the spatial effects of given *cis*-regulatory mutations.

In summary, gene regulatory network structure explains some of the prominent effects of *hox* genes on developmental deployment of patterning subcircuits, just as this in turn explains their prominent role as modulators of the body plan in subphyletic evolution. These genes, and other transcription factors that also act as network I/O switches, are scarcely the only sources of evolutionary lability at the body plan level. Another major mechanism of change is obviously redeployment of signal system "plug-ins," for which the arguments do not need to be repeated here. Nor are these mechanisms independent; for example, among the direct target of *hox* genes in *Drosophila* is the gene encoding the signal ligand Dpp (Capovilla and Botas, 1998; Marty *et al.*, 2001).

METAZOAN ORIGINS AND GENE NETWORKS BEFORE KERNELS

"Deconstructing" Bilaterian Gene Regulatory Networks in Time

For much of this book we have been "constructing" the process of development in the terms of the genomic regulatory apparatus by which it is controlled. In this Chapter we have been exploring the evolutionary consequences of the functional heterogeneity of the different kinds of parts of which gene regulatory networks are built. But we must remember that the diverse body plans of bilaterian animals since the Cambrian were and are all products of gene regulatory networks of more or less equivalent complexity, composed of all the different kinds of parts that these gene networks have. To imagine, even at the least detailed level of resolution, the stages of evolution of Metazoa before there were extant bilaterian level gene regulatory networks, is to imagine much simpler kinds of developmental biology, such as could be programmed by the pieces and parts of gene regulatory networks that must have first come into existence. Following is a brief sketch of a pathway "pre-kernel" network evolution might have followed, arrived at by the process of deconstructing modern network architecture into its different parts and considering their relative antiquity. The intent is not to generate a story or a scenario. Rather it is to find a few stepping stones leading toward the ultimately interesting evolutionary issue of defining the key genomic regulatory property that potentiated the appearance of developmental programs similar to those of modern bilaterians.

One thing that is clear from many genome projects, as discussed in Chapter 1, is that the Bilateria share a gene toolkit that includes all the main classes of regulatory genes and signaling "plug-in" apparatus, which therefore was already present in their last common ancestor (reviewed by Erwin and Davidson, 2002). New evidence extends this argument: not only are almost all such genes that anyone looks for found as well in cnidarians, which may be too close to bilaterians, but they are found further afield as well. A rapidly growing literature on gene complements of sponges and even choanoflagellates, a protozoan outgroup that is close to Metazoa (King and Carroll, 2001), reveals many classes of genes that are also in the bilaterian toolkit. Among them are genes encoding various kinds of transcription factors, including homeodomain, T-box, and forkhead factors

(Manuel and Le Parco, 2000; Manuel *et al.*, 2004; reviewed by Coutinho *et al.*, 2003; Adell and Müller, 2004); and many cell signaling and cell adhesion proteins (reviewed by Müller and Müller, 2003; King *et al.*, 2003; King, 2004; Adell and Müller, 2004). Sponges do not generate the complex and functionally specialized body parts that bilaterians (and at least to some extent cnidarians) do, but it is increasingly unlikely that this is due to lack of the kinds of genes used to direct the developmental process. Sponges are metazoans and share common ancestors with the bilaterians. At some point in the evolution of bilaterian lineages regulatory genes were assembled in different ways into the kinds of regulatory networks that underlie bilaterian body plans. It is the stages of this assembly process that we need to think about, beginning with the oldest components of gene regulatory networks.

The oldest components are differentiation gene batteries. Though they have continued to evolve and change, and thus are also in some sense always the newest components of developmental gene regulatory networks, in form they are the most venerable of the different kinds of subcircuits of which networks are composed. This can be inferred from the fact that the most distant from us of extant metazoans, sponges, still have many kinds of differentiated, highly specialized cell types. Among these are cells that line the external and internal surfaces, some of which secrete a collagen-polysaccharide complex, others of which produce flagellae, others of which form pores; cells that secrete the collagenous skeleton, and others that produce the siliceous or calcareous spicules; contractile myocytes; phagocytic cells that produce digestive enzymes; and so forth (reviewed by Brusca and Brusca, 2003). These cell types must each deploy particular transcription regulators which target the *cis*-regulatory control elements of the respective sets of structural genes active in it, i.e., differentiation gene batteries in the sense of the diagrams of Figs. 1.5 and 4.1. So, in considering the evolution of metazoan gene regulatory networks, the initial stage (ancestral to both Bilateria and sponges) should have been one in which the developmental program consisted entirely of extremely shallow subcircuits defining given cell types, in which the battery regulators respond directly to whatever embryological spatial or signaling cues define the prospective locations of the different cell types (Fig. 5.8, "Stage 1"). What the outcome of their development was like is indeterminate. They need not have consisted largely of extracellular materials as do modern sponges, and could have been largely flat or simple microscopic parenchymal forms, but composed of various cell types in which a not so simple array of diverse differentiation genes was expressed regionally (similar arguments are made by Seipel and Schmid, 2005). It is an interesting thought that vertebrate stem cells in culture have some of these same qualities: they give rise to differentiated cell types which express one or another differentiation gene battery, but they perform no space-dividing operations and do no pattern formation, they generate no morphology, and they build no body parts.

The body parts of modern animals are developmentally organized following spatial assignment of diverse regulatory states, as discussed at length in Chapter 3, and what is missing in the stage 1 regulatory apparatus imagined here is the network architecture which performs this function. Both logic and many current observations suggest the mechanism by which arose this kind of architecture. In this second stage of network evolution the division of embryological space into different regulatory domains, which can then be used for different purposes including morphological structure building, is separated from the act of deploying differentiation gene batteries. These ancestral networks would thus achieve the structural character that we have seen repeatedly in the real networks above, an internal regulatory state specification core and a periphery composed of differentiation gene batteries and their immediate regulators. A likely pathway by which this happened is as follows. Since there are already diverse differentiation gene battery regulators deployed in different regions (embryonic "addresses"), there is already the rudiment of a spatially diversified *trans*-regulatory state. Therefore additional subcircuits in which the nodes contain *cis*-regulatory modules that respond to these same *trans*-regulators could run in the same address. Were such *cis*-regulatory modules to arise either in these regulatory genes or in different ones, the result would be that these same transcriptional regulators as are also used to control the differentiation gene batteries could now be used for higher level interactions. The requisite *cis*-regulatory modules could have "escaped" from the differentiation genes, or could have formed *de novo* and been selected for their ability to respond to the transcription factors that are regionally ambient, a process that has been demonstrated in recent evolutionary time (Gompel *et al.,* 2005).

The tell-tale quality of this mechanism is that given regulatory genes are used at different hierarchical levels of the network, in both core and periphery, both for patterning and to activate differentiation genes (Fig. 5.8, "Stage 2"). This feature is still very much with us. A great example is the case of *pax6* in eyes. While there is no good argument for a *trans*-bilaterian eye kernel including *pax6*, as mentioned at the beginning of this Chapter there is another quite good reason for the *trans*-bilaterian inclusion of *pax6* in eye development. This is that *pax6* is an activator of downstream differentiation genes in eyes, including the visual pigment genes that all eyes need. Perhaps this was its original role (Scott, 1994) and because it was expressed at the "visual spot" address it was eventually incorporated, independently, in the different morphogenetic programs for eye development. The use of upper level patterning regulators also for operation of far downstream differentiation genes is seen in many modern contexts, e.g., for *tinman/nkx2.5* in heart, *cad/cdx* in intestine, and *otd/otx* in brain (reviewed by Davidson, 2001). The multilevel use of regulatory genes in the development of given body parts is not an index of the atavistic survival of specific regulatory linkages, for the circuitry has changed as body part morphology has diversified. But the mechanism, evolutionary intercalation into the network of new circuitry using the same regulators, is atavistic as a process, a process that has never stopped.

Conditions permitting, there is inherent in this mechanism an explosive power of diversification in morphology and function of "proto-body parts" in our imagined Stage 2, quite unlike the relatively staid stability that has obtained since

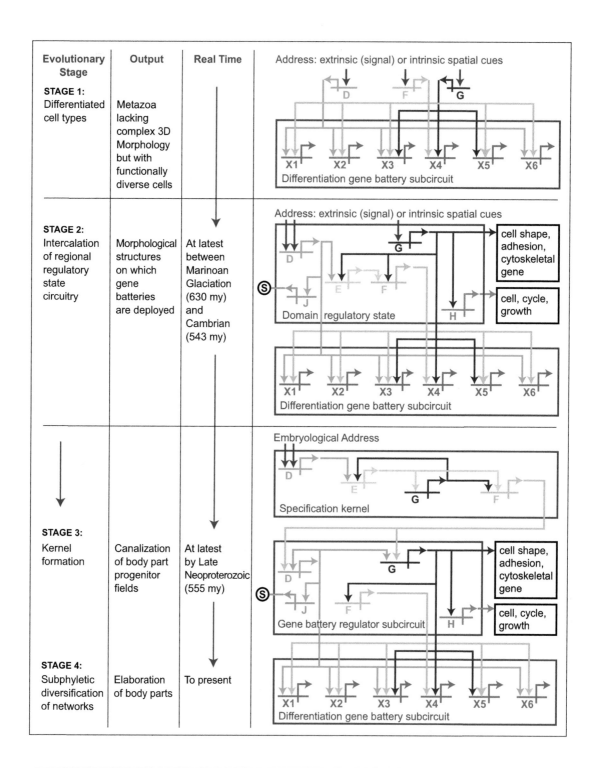

FIGURE 5.8 For legend see page 239.

the Cambrian. The kernel concept can explain the canalization of gene networks and of developmental process since the triumph of the bilaterian versions of animal body plans. In terms of network evolution, the secret of the bilaterian lockdown of potentialities lies in the advent of kernel circuits that canalized the spatial subdivision process underlying development of given body parts, because of their recursive wiring, so that the penalty of circuit change is disaster. According to molecular phylogeny done on proteins shared across the Bilateria, this is likely to have occurred sometime before the Cambrian and after the end of the last of the great world wide glaciations (Fig. 5.8, "Stage 3"), for that is the interval within which the bilaterian clades diversified (Aris-Brosou and Yang, 2002; Peterson *et al.*, 2004; Douzery *et al.*, 2004). This is also the era when the fossil record first indicates the presence of complex animals (Zhang *et al.*, 1998; Erwin and Davidson, 2002; Chen *et al.*, 2004; Peterson and Butterfield, 2005; Bottjer, 2005) including obvious macroscopic bilaterian forms (e.g., Fedonkin and Waggoner, 1997). In the sense of these stages of network evolution, this was the third discrete kind of process by which the networks of modern animals arose (Fig. 5.8, "Stage 3). The subphyletic processes of body part diversification and speciation, post-kernel installation, represent the current phase (Fig. 5.8, "Stage 4").

Concluding Reflection: The Principle of Animal Development and Evolution

There is a functional requirement at the genome level which is implicit but not explicit in the foregoing, for all but perhaps the initial stage of network evolution. This is the capacity of the *cis*-regulatory nodes of the network to process and integrate biological information. We began in Chapter 1 of this book with the concept of regulatory information processing, as illustrated in the diagram of Fig. 1.2, where cartoon *cis*-regulatory modules are portrayed as recipients of information carried to them regarding cell lineage, signaling from adjacent cells, and spatial and temporal situation. Bilaterian development would be impossible absent this fundamental property of genomic *cis*-regulatory systems. The operational features of *cis*-regulatory information processing were explored in Chapter 2, where it is seen to underlie the amazing range of *cis*-regulatory designs that are utilized to set up regulatory states in embryonic space. As shown in Chapter 3, installation of spatial regulatory states is the common basis of all forms of bilaterian embryogenesis. However, the network apparatus discussed in Chapter 4, which provides the mechanism for heritably programming the developmental

A speculation: stages of gene regulatory network evolution. The network diagrams are derived from those in Fig. 4.1, where the properties of the circuitry portrayed are discussed. Initial stage: network complexity is that of a differentiation gene battery. Second stage: intercalation of circuitry for regional specification of regulatory state. Third stage: installation of recursively wired kernel that canalizes initial developmental organization of the body part. Fourth stage: diversification of network architecture and of body part morphology at the subphyletic level.

installation of regulatory states, generates an additional level of information processing, that which emerges from the operation of the subcircuits of the network. We considered in this connection the image of a delocalized computational device, the individual elements of which are the *cis*-regulatory modules of the regulatory genes at its nodes, and the function of which depends on these modules and on the network wiring architecture.

Genomically encoded information processing at the primary level of the *cis*-regulatory module, and information processing at the derived level of the network subcircuit, must be thought of as the principle that makes the bilaterian style of development possible. Therefore it is also the principle that made the evolution of bilaterian grade animals possible. Given the components, the assembly of endless varieties of the computer would have followed, and until kernel canalization, the great bilaterian lockdown of possibilities, we can permit ourselves to imagine a strange juxtaposition: environmental selection operating on biological computer design. Current molecular phylogenies (Peterson *et al.*, 2004; Peterson and Butterfield, 2005) indicate relatively shallow dates only about 50 million years prior to the divergence of Bilateria for the divergence of Metazoa, measured as the branch point where sponge and bilaterian ancestors diverge. Perhaps this is a distant reflection of the time in geological history when information processing *cis*-regulatory modules arose.

The time is almost upon us when we will be able to build *cis*-regulatory modules and network subcircuits in the laboratory and test their developmental operation in living systems. Considering evolution of body plans in terms of network circuitry, as a history of assembly of grades of network organization, is to transform this vexed subject into a prospectus for laboratory research. From such research will ultimately come by experimental demonstration the revelation that evolution and development merge in the Regulatory Genome.

REFERENCES

Ackers, G. K., Johnson, A. D., and Shea, M. A. (1982). Quantitative model for gene regulation by λ phage repressor. *Proc. Natl. Acad. Sci. USA* **79**, 1129–1133.

Adams, M. D. *et al.* (2000). The genome sequence of *Drosophila melanogaster*. *Science* **287**, 2185–2195.

Adell, T. and Müller, W. E. G. (2004). Isolation and characterization of five Fox (Forkhead) genes from the sponge *Suberites domuncula*. *Gene* **334**, 35–46.

Afouda, B. A., Cia-Uitz, A., and Patient, R. (2005). GATA4, 5 and 6 mediate TGFβ maintenance of endodermal gene expression in *Xenopus* embryos. *Development* **132**, 763–774.

Albright, S. R. and Tjian, R. (2000). TAFs revisited: More data reveal new twists and confirm old ideas. *Gene* **242**, 1–13.

Allen, D. L., Sartorius, C. A., Sycuro, L. K., and Leinwand, L. A. (2001). Different pathways regulate expression of the skeletal myosin heavy chain genes. *J. Biol. Chem.* **276**, 43524–43533.

Amore, G. *et al.* (2003). *Spdeadringer*, a sea urchin embryo gene required separately in skeletogenic and oral ectoderm gene regulatory networks. *Dev. Biol.* **261**, 55–81.

Anand, S. *et al.* (2003). Divergence of *Hoxc8* early enhancer parallels diverged axial morphologies between mammals and fishes. *Proc. Natl. Acad. Sci. USA* **100**, 15666–15669.

Anderson, K. V., Jurgens, G., and Nüsslein-Volhard, C. (1985). Establishment of dorsal-ventral polarity in the *Drosophila* embryo: Genetic studies on the *toll* gene product. *Cell* **42**, 779–789.

Andrioli, L. P. M., Vasisht, V., Theodosopoulou, E., Oberstein, A., and Small, S. (2002). Anterior repression of a *Drosophila* stripe enhancer requires three position-specific mechanisms. *Development* **129**, 4931–4940.

Andrioli, L. P. M., Oberstein, A., Corado, M., Yu, D., and Small, S. (2004). Groucho-dependent repression by Sloppy-paired 1 differentially positions anterior pair-rule stripes in the *Drosophila* embryo. *Dev. Biol.* **276**, 541–551.

Ansel, K. M., Lee, D. U., and Rao, A. (2003). An epigenetic view of helper T-cell differentiation. *Nature Immunol.* **4**, 616–623.

Aparicio, S. *et al.* (2002). Whole-genome shotgun assembly and analysis of the genome of *Fugu rubripes*. *Science* **297**, 1301–310.

Apitz, H. *et al.* (2004). Identification of regulatory modules mediating specific expression of the *roughest* gene in *Drosophila melanogaster*. *Dev. Genes Evol.* **214**, 453–459.

Arbeitman, M. N. *et al.* (2002). Gene expression during the life cycle of *Drosophila melanogaster*. *Science* **297**, 2270–2275.

Arendt, D. and Nübler-Jung, K. (1999). Comparison of early nerve cord development in insects and vertebrates. *Development* **126**, 2309–2325.

Aris-Brosou, S. and Yang, Z. (2002). Effects of models of rate evolution on estimation of divergence dates with special reference to the metazoan 18S ribosomal RNA phylogeny. *Syst. Biol.* **51**, 703–714.

Arnone, M. and Davidson, E. H. (1997). The hardwiring of development: Organization and function of genomic regulatory systems. *Development* **124**, 1851–1864.

Arnone, M. I. *et al.* (1997). Green fluorescent protein in the sea urchin: New experimental approaches to transcriptional regulatory analysis in embryos and larvae. *Development* **124**, 4649–4659.

Arnosti, D. N., Barolo, S., Levine, M., and Small, S. (1996). The *eve* stripe 3 enhancer employs multiple modes of transcriptional synergy. *Development* **122**, 205–214.

Artavanis-Tsakonas, S., Matsuno, K., and Fortini, M. E. (1995). Notch signaling. *Science* **268**, 225–232.

Ashburner, M. (1989). "*Drosophila:* A Laboratory Handbook." Cold Spring Harbor Press, Cold Spring Harbor, NY.

Azpiazu, N. and Frasch, M. (1993). *tinman* and *bagpipe:* Two homeobox genes that determine cell fates in the dorsal mesoderm of *Drosophila. Genes Dev.* **7**, 1325–1340.

Azpiazu, N. Lawrence, P. A., Vincent, J.-P., and Frasch, M. (1996). Segmentation and specification of the *Drosophila* mesoderm. *Genes Dev.* **10**, 3183–3194.

Backman, M., Machon, O., van den Bout, C. J., and Krauss, S. (2003). Targeted disruption of mouse Dach1 results in postnatal lethality. *Dev. Dyn.* **226**, 139–144.

Bae, E., Calhoun, V. C., Levine, M., Lewis, E. B., and Drewell, R. A. (2002). Characterization of the intergenic RNA profile at *Abdominal-A* and *Abdominal-B* in the *Drosophila* bithorax complex. *Proc. Natl. Acad. Sci. USA* **99**, 16847–16852.

Bagheri-Fam, S. *et al.* (2006). Long range upstream and downstream enhancers control distinct subsets of the complex spatiotemporal *Sox9* expression pattern. *Dev. Biol.* **291**, 382–397.

Balázsi, G., Barabási, A.-L., and Oltvai, Z. N. (2005). Topological units of environmental signal processing in the transcriptional regulatory network of *Escherichia coli. Proc. Natl. Acad. Sci. USA* **102**, 7841–7846.

Barolo, S. and Posakony, J. W. (2002). Three habits of highly effective signaling pathways: Principles of transcriptional control by developmental cell signaling. *Genes Dev.* **16**, 1167–1181.

Barrow, J. R., Stadler, H. S., and Capecchi, M. R. (2000). Roles of *Hoxa1* and *Hoxa2* in patterning the early hindbrain of the mouse. *Development* **127**, 933–944.

Baugh, L. R., Hill, A. A., Slonim, D. K., Brown, E. L., and Hunter, C. P. (2003). Composition and dynamics of the *Caenorhabditis elegans* early embryonic transcriptome. *Development* **130**, 889–900.

Baugh, L. R. *et al.* (2005). The homeodomain protein PAL-1 specifies a lineage-specific regulatory network in the *C. elegans* embryo. *Development* **132**, 1843–1854.

Baumeister, R., Liu, Y., and Ruvkun, G. (1996). Lineage-specific regulators couple cell lineage asymmetry to the transcription of the *Caenorhabditis elegans* POU gene *unc-86* during neurogenesis. *Genes Dev.* **10**, 1395–1410.

Baylies, M. K., Bate, M., and Ruiz Gomez, M. (1998). Myogenesis: A view from *Drosophila. Cell* **93**, 921–927.

Beck, F., Tata, F., and Chawengsaksophak, K. (2000). Homeobox genes and gut development. *BioEssays* **22**, 431–441.

Bell, A. C., West, A. G., and Felsenfeld, G. (2001). Insulators and boundaries: Versatile regulatory elements in the eukaryotic genome. *Science* **291**, 447–450.

Belting, H.-G., Shashikant, C. S., and Ruddle, F. H. (1998). Modification of expression and *cis*-regulation of *Hoxc8* in the evolution of diverged axial morphology. *Proc. Natl. Acad. Sci. USA* **95**, 2355–2360.

Bender, W. *et al.* (1983). Molecular genetics of the Bithorax complex in *Drosophila melanogaster. Science* **221**, 23–29.

Berezikov, E. *et al.* (2005). Phylogenetic shadowing and computational identification of human microRNA genes. *Cell* **120**, 21–24.

Berg, O. G. and von Hippel, P. H. (1987). Selection of DNA binding sites by regulatory proteins. Statistical-mechanical theory and application to operators and promoters. *J. Mol. Biol.* **193**, 723–750.

Berman, B. P. *et al.* (2002). Exploiting transcription factor binding site clustering to identify *cis*-regulatory modules involved in pattern formation in the *Drosophila* genome. *Proc. Natl. Acad. Sci. USA* **99**, 757–762.

Blackman, R. K., Sanicola, M., Raftery, L. A., Gillevet, T., and Gelbart, W. M. (1991). An extensive 3′ *cis*-regulatory region directs the imaginal disk expression of *decapentaplegic*, a member of the TGF-β family in *Drosophila. Development* **111**, 657–665.

Blanco, J., Girard, F., Kamachi, Y., Kondoh, H., and Gehring, W. J. (2005). Functional analysis of the chicken δ*1-crystallin* enhancer activity in *Drosophila* reveals remarkable evolutionary conservation between chicken and fly. *Development* **132**, 1895–1905.

Bodmer, R. (1993). The gene *tinman* is required for specification of the heart and visceral muscles in *Drosophila. Development* **118**, 719–729.

Bogarad, L. D., Arnone, M. I., Chang, C., and Davidson, E. H. (1998). Interference with gene regulation in living sea urchin embryos: Transcription factor Knock Out (TKO), a genetically controlled vector for blockade of specific transcription factors. *Proc. Natl. Acad. Sci. USA* **95**, 14827–14832.

Bolouri, H. and Davidson, E. H. (2002). Modeling DNA sequence-based *cis*-regulatory gene networks. *Dev. Biol.* **246**, 2–13.

Bolouri, H. and Davidson, E. H. (2003). Transcriptional regulatory cascades in development: Initial rates, not steady state, determine network kinetics. *Proc. Natl. Acad. Sci. USA* **100**, 9371–9376.

Bonini, N. M., Leiserson, W. M., and Benzer, S. (1998). Multiple roles of the *eyes absent* gene in *Drosophila. Dev. Biol.* **196**, 42–57.

Bottjer, D. J. (2005). The early evolution of animals. *Sci. Am.* **293**, 42–47.

Bourque, G., Pevzner, P. A., and Tesler, G. (2004). Reconstructing the genomic architecture of ancestral mammals: Lessons from human, mouse, and rat genomes. *Genome Res.* **14**, 507–516.

Bowerman, B. (1998). Maternal control of pattern formation in early *Caenorhabditis elegans* embryos. *Curr. Topics Dev. Biol.* **39**, 73–117.

Bowerman, B., Draper, B. W., Mello, C. C., and Preiss, J. R. (1993). The maternal gene *skn-1* encodes a protein that is distributed unequally in early *C. elegans* embryos. *Cell* **74**, 443–452.

Brink, C. (2003). Promoter elements in endocrine pancreas development and hormone regulation. *Cell. Mol. Life Sci.* **60**, 1033–1048.

Briscoe, J. *et al.* (1999). Homeobox gene *Nkx2.2* and specification of neuronal identity by graded Sonic hedgehog signaling. *Nature* **398**, 622–627.

Britten, R. J. (1984). Mobile elements and DNA repeats. *Carlsberg Res. Commun.* **49**, 169–178.

Britten, R. J. (1996). DNA sequence insertion and evolutionary variation in gene regulation. *Proc. Natl. Acad. Sci. USA* **93**, 9374–9377.

Britten, R. J. and Davidson, E. H. (1969). Gene regulation for higher cells: A theory. *Science* **165**, 349–358.

Britten, R. J. and Davidson, E. H. (1971). Repetitive and non-repetitive DNA sequences and a speculation on the origins of evolutionary novelty. *Quart. Rev. Biol.* **46**, 111–138.

Broccoli, V., Boncinelli, E., and Wurst, W. (1999). The caudal limit of Otx2 expression positions the isthmic organizer. *Nature* **401**, 164–168.

Broitman-Maduro, G., Maduro, M. F., and Rothman, J. H. (2005). The noncanonical binding site of the MED-1 GATA factor defines differently regulated target genes in the *C. elegans* mesendoderm. *Dev. Cell* **8**, 427–433.

Bromleigh, V. C. and Freeman, L. P. (2000). p21 is a transcriptional target of HOXA10 in differentiating myelomonocytic cells. *Genes Dev* **14**, 2581–2586.

Brown, C. T. *et al.* (2002). New computational approaches for analysis of *cis*-regulatory networks. *Dev. Biol.* **246**, 86–102.

Brown, D. D. *et al.* (2005). Tbx5 and Tbx20 act synergistically to control vertebrate heart morphogenesis. *Development* **132**, 553–563.

Bruhl, T. *et al.* (2004). Homeobox A9 transcriptionally regulates the EphB4 receptor to modulate endothelial cell migration and tube formation. *Circ. Res.* **94**, 743–751.

Bruneau, B. G. *et al.* (2001). A murine model of Holt-Oram syndrome defines roles of the T-box transcription factor tbx5 in cardiogenesis and disease. *Cell* **106**, 709–721.

Brusca, R. C. and Brusca, G. J. (2003). "Invertebrates," 2nd Ed. Sinauer Associates, Sunderland, MA.

Buchberger, A., Nomokonova, N., and Arnold, H.-H. (2003). Myf5 expression in somites and limb buds of mouse embryos is controlled by two distinct distal enhancer activities. *Development* **130**, 3297–3307.

Buckley, M. S., Chau, J., Hoppe, P. E., and Coulter, D. E. (2004). *odd-skipped* homologs function during gut development in *C. elegans. Dev. Genes Evol.* **214**, 10–18.

Butler, J. E. F. and Kadonaga, J. T. (2001). Enhancer-promoter specificity mediated by DPE or TATA core promoter motifs. *Genes Dev.* **15**, 2515–2519.

C. elegans Sequencing Consortium (1998). Genome sequence of the nematode *C. elegans:* A platform for investigating biology. *Science* **282**, 2012–2018.

Cabrera, C. V., Lee, J. J., Ellison, J. W., Britten, R. J., and Davidson, E. H. (1984). Regulation of cytoplasmic mRNA prevalence in sea urchin embryos: Rates of appearance and turnover for specific sequences. *J. Mol. Biol.* **174**, 85–111.

Cadigan, K. M. and Nusse, R. (1997). Wnt signaling: A common theme in animal development. *Genes Dev.* **11**, 3286–3305.

Cai, C.-L. *et al.* (2005). T-box genes coordinate regional rates of proliferation and regional specification during cardiogenesis. *Development* **132**, 2475–2487.

Cai, H. N., Arnosti, D. N., and Levine, M. (1996). Long-range repression in the *Drosophila* embryo. *Proc. Natl. Acad. Sci. USA* **93**, 9309–9314.

Calestani, C., Rast, J. P., and Davidson, E. H. (2003). Isolation of mesoderm specific genes in the sea urchin embryo by differential macroarray screening. *Development* **130**, 4587–4596.

Calhoun, V. C. and Levine, M. (2003). Long-range enhancer-promoter interactions in the *Scr-Antp* interval of the *Drosophila* Antennapedia complex. *Proc. Natl. Acad. Sci. USA* **100**, 9878–9883.

Calhoun, V. C., Stathopoulos, A., and Levine, M. (2002). Promoter-proximal tethering elements regulate enhancer-promoter specificity in the *Drosophila Antennapedia* complex. *Proc. Natl. Acad. Sci. USA* **99**, 9243–9247.

Callebaut, M. (2005). Origin, fate, and function of the components of the avian germ disc region and early blastoderm: Role of ooplasmic determinants. *Dev. Dyn.* **233**, 1194–1216.

Calvo, D. *et al.* (2001). A POP-1 repressor complex restricts inappropriate cell type-specific gene transcription during *Caenorhabditis elegans* embryogenesis. *EMBO J.* **20**, 7197–7208.

Cameron, R. A. *et al.* (2000). A sea urchin genome project: Sequence scan, virtual map, and additional resources. *Proc. Natl. Acad. Sci. USA* **97**, 9514–9518.

Cameron, R. A., Oliveri, P., Wyllie, J., and Davidson, E. H. (2004). *cis*-Regulatory activity or randomly chosen genomic fragments from the sea urchin. *Gene Exp. Patt.* **4**, 205–213.

Cameron, R. A. *et al.* (2005). An evolutionary constraint: Strongly disfavored class of change in DNA sequence during divergence of *cis*-regulatory modules. *Proc. Natl. Acad. Sci. USA* **102**, 11769–11774.

Cañestro, C., Bassham, S., and Postlethwait, J. (2005). Development of the central nervous system in the larvacean *Oikopleura dioica* and the evolution of the chordate brain. *Dev. Biol.* **285**, 298–315.

Capovilla, M. and Botas, J. (1998). Functional dominance among Hox genes: Repression dominates activation in the regulation of *dpp*. *Development* **125**, 4949–4957.

Carvajal, J. J., Cox, D., Summerbell, D., and Rigby, P. W. J. (2001). A BAC transgenic analysis of the *Mrf4/Myf5* locus reveals interdigitated elements that control activation and maintenance of gene expression during muscle development. *Development* **128**, 1857–1868.

Casey, E. S., O'Reilly, M.-A. J., Conlon, F. L., and Smith, J. C. (1998). The T-box transcription factor Brachyury regulates expression of *eFGF* through binding to a non-palindromic response element. *Development* **125**, 3887–3894.

Chang, J., Kim, I.-O., Ahn, J.-S., and Kim S.-H. (2001). The CNS midline cells control the *spitz* class and *Egfr* signaling genes to establish the proper cell fate of the *Drosophila* ventral neuroectoderm. *Int. J. Dev. Biol.* **45**, 715–724.

Chang, J., Jeon, S.-H., and Kim, S. H. (2003). The hierarchical relationship among the *spitz/Egfr* signaling genes in cell fate determination in the *Drosophila* ventral neuroectoderm. *Mol. Cells* **15**, 186–193.

Chang, S., Johnston, R. J., Jr., and Hobert, O. (2003). A transcriptional regulatory cascade that controls left-right asymmetry in chemosensory neurons of *C. elegans*. *Genes Dev.* **17**, 2123–2137.

Chang, S., Johnston, R. J., Jr., Frøkjær-Jensen, C., Lockery, S., and Hobert, O. (2004). MicroRNAs act sequentially and asymmetrically to control chemosensory laterality in the nematode. *Nature* **430**, 785–789.

Chapman, M. A. *et al.* (2003). Comparative and functional analyses of *LYL1* loci establish marsupial sequences as a model for phylogenetic footprinting. *Genomics* **81**, 249–259.

Chen, G. and Courey, A. J. (2000). Groucho/TLE family proteins and transcriptional repression. *Gene* **249**, 1–16.

Chen, J. and Ruley, H. E. (1998). An enhancer element in the EphA2 (Eck) gene sufficient for rhombomere-specific expression is activated by HOXA1 and HOXB1 homeobox proteins. *J. Biol. Chem.* **273**, 24670–24675.

Chen, J.-Y. (2004). "The Dawn of Animal World." Jiangsu Science & Technology Publishing House, Nanjing, China.

Chen, J.-Y., Huang, D.-Y., and Li, C.-W. (1999). An early Cambrian craniate-like chordate. *Nature* **402**, 518–522.

Chen, J.-Y., Bottjer, D. J., Oliveri, P., Dornbos, S. Q., Gao, F., Ruffins, S., Chi, H., Li, C.-W. and Davidson, E. H. (2004). Small bilaterian fossils from 40 to 55 million years before the Cambrian. *Science* **305**, 218–222.

Cheng, G., Hagen, T. P., Dawson, M. L., Barnes, K. V., and Menick, D. R. (1999). The role of GATA, CArG, E-box, and a novel element in the regulation of cardiac expression of the Na$^+$-Ca^{2+} exchanger gene. *J. Biol. Chem.* **274**, 12819–12826.

Christiaen, L., Bourrat, F., and Joly, J.-S. (2005). A modular *cis*-regulatory system controls isoform-specific *pitx* expression in ascidian stomodaeum. *Dev. Biol.* **277**, 557–566.

Chuang, C., Wikramanayake, A. H., Mao, C., Li, X., and Klein, W. (1996). Transient appearance of *Strongylocentrotus purpuratus* Otx in micromere nuclei: Cytoplasmic retention of SpOtx possibly mediated through an α-actinin intersection. *Dev. Genet.* **19**, 231–237.

Chuong, C.-M., Chodankar, R., Widelitz, R. B., and Jiang, T.-X. (2000). *Evo-Devo* of feathers and scales: Building complex epithelial appendages. *Curr. Opin. Genet. Dev.* **10**, 449–456.

Clements, D. and Woodland, H. R. (2003). VegT induces endoderm by a self-limiting mechanism and by changing the competence of cells to respond to TGF-β signals. *Dev. Biol.* **258**, 454–463.

Clyde, D. E. *et al.* (2003). A self-organizing system of repressor gradients establishes segmental complexity in *Drosophila*. *Nature* **426**, 849–853.

Cobb, J. and Duboule, D. (2005). Comparative analysis of genes downstream of the Hoxd cluster in developing digits and external genitalia. *Development* **132**, 3055–3067.

Coffman, J. A. and Davidson, E. H. (2001). Oral-aboral axis specification in the sea urchin embryo. I. Axis entrainment by respiratory asymmetry. *Dev. Biol.* **230**, 18–28.

Coffman, J. A., Kirchhamer, C. V., Harrington, M. G., and Davidson, E. H. (1996). SpRunt-1, a new member of the Runt-domain family of transcription factors, is a positive regulator of the aboral ectoderm-specific *CyIIIa* gene in the sea urchin embryos. *Dev. Biol.* **174**, 43–54.

Coffman, J. A., Kirchhamer, C. V., Harrington, M. G., and Davidson, E. H. (1997). SpMyb functions as an intramodular repressor to regulate spatial expression of *CyIIIa* in sea urchin embryos. *Development* **124**, 4717–4727.

Coffman, J. A., McCarthy, J. J., Dickey-Sims, C., and Robertson, A. J. (2004). Oral-aboral axis specification in the sea urchin embryo. II. Mitochondrial distribution and redox state contribute to establishing polarity in *Strongylocentrotus purpuratus*. *Dev. Biol.* **273**, 160–171.

Cohen, B., Simcox, A. A., and Cohen, S. M. (1993). Allocation of the thoracic imaginal primordia in the *Drosophila* embryo. *Development* **117**, 597–608.

Cohn, M. J. and Tickle, C. (1999). Developmental basis of limblessness and axial patterning in snakes. *Nature* **399**, 474–479.

Cohn, M. J. *et al.* (1997). *Hox9* genes and vertebrate limb specification. *Nature* **387**, 97–101.

Conklin, E. G. (1905). The organization and cell lineage of the ascidian egg. *J. Acad. Nat. Sci. (Philadelphia)* **13**, 1–119.

Conlon, F. L. and Smith, J. C. (1999). Interference with *brachyury* function inhibits convergent extension, causes apoptosis, and reveals separate requirements in the FGF and activin signaling pathways. *Dev. Biol.* **213**, 85–100.

Cornell, R. A. and Von Ohlen, T. (2000). *Vnd/nkx, ind/gsh*, and *msh/msx*: Conserved regulators of dorsoventral neural patterning? *Curr. Opin. Neurobiol.* **10**, 63–71.

Courey, A. J. and Jia, S. (2001). Transcriptional repression: The long and the short of it. *Genes Dev.* **15**, 2786–2796.

Coutinho, C. C., Fonseca, R. N., Mansure, J. J. C., and Borojevic, R. (2003). Early steps in the evolution of multicellularity: Deep structural and functional homologies among homeobox genes in sponges and higher metazoans. *Mech. Dev.* **120**, 429–440.

Cowden, J. and Levine, M. (2002). The snail repressor positions notch signaling in the *Drosophila* embryo. *Development* **129**, 1785–1793.

Cowden, J. and Levine, M. (2003). Ventral dominance governs sequential patterns of gene expression across the dorsal-ventral axis of the neuroectoderm in the *Drosophila* embryo. *Dev. Biol.* **262**, 335–349.

Crémazy, F., Berta, P., and Girard, F. (2001). Genome-wide analysis of Sox genes in *Drosophila melanogaster*. *Mech. Dev.* **109**, 371–375.

Crews, S. T. (1998). Control of cell lineage-specific development and transcription by bHLH-PAS proteins. *Genes Dev.* **12**, 607–620.

Cripps, R. M. and Olson, E. N. (2002). Control of cardiac development by an evolutionarily conserved transcription network. *Dev. Biol.* **246**, 14–28.

Crozatier, M., Glise, B., and Vincent, A. (2004). Patterns in evolution: Veins of the *Drosophila* wing. *Trends Genet.* **20**, 498–505.

Cvekl, A. and Piatigorsky, J. (1996). Lens development and crystalline gene expression: Many roles for Pax-6. *BioEssays* **18**, 621–630.

Dai, Y. S., Cserjesi, P., Markham, B. E., and Molkentin, J. D. (2002). The transcription factors GATA4 and dHAND physically interact to synergistically activate cardiac gene expression through a p300-dependent mechanism. *J. Biol. Chem.* **277**, 24390–24398.

Damle, S. and Davidson, E. H. (2006). Cell fate transformation by regulatory intervention in the sea urchin embryo. Submitted.

Dasen, J. S. *et al.* (1999). Reciprocal interactions of Pit1 and GATA2 mediate signaling gradient-induced determination of pituitary cell types. *Cell* **97**, 587–598.

Davidson, B., Shi, W., and Levine, M. (2005). Uncoupling heart cell specification and migration in the simple chordate *Ciona intestinalis*. *Development* **132**, 4811–4818.

Davidson, E. H. (1968). "Gene Activity in Early Development." Academic Press, New York.

Davidson, E. H. (1986). "Gene Activity in Early Development." Academic Press, Orlando, Florida.

Davidson, E. H. (1989). Linage-specific gene expression and the regulative capacities of the sea urchin embryo: A proposed mechanism. *Development* **105**, 421–445.

Davidson, E. H. (1990). How embryos work: A comparative view of diverse modes of cell fate specification. *Development* **108**, 365–389.

Davidson, E. H. (1991). Spatial mechanisms of gene regulation in metazoan embryos. *Development* **113**, 1–26.

Davidson, E. H. (1993). Later embryogenesis: Regulatory circuitry in morphogenetic fields. *Development* **118**, 665–690.

Davidson, E. H. (2001). "Genomic Regulatory Systems: Development and Evolution." Academic Press, San Diego.

Davidson, E. H. and Erwin, D. H. (2006). Gene regulatory networks and the evolution of animal body plans. *Science* **311**, 796–800.

Davidson, E. H. *et al.* (1974). Arrangement and characterization of repetitive sequence elements in animal DNAs. *Cold Spring Harbor Symp. Quant. Biol.* **38**, 295–301.

Davidson, E. H., Cameron, R. A., and Ransick, A. (1998). Specification of cell fate in the sea urchin embryo: Summary and some proposed mechanisms. *Development* **125**, 3269–3290.

Davidson, E. H. *et al.* (2002a). A provisional regulatory gene network for specification of endomesoderm in the sea urchin embryo. *Dev. Biol.* **246**, 162–190.

Davidson, E. H. *et al.* (2002b). A genomic regulatory network for development. *Science* **295**, 1669–1678.

Davis, D. L. *et al.* (2001). A GATA-6 gene heart-region-specific enhancer provides a novel means to mark and probe a discrete component of the mouse cardiac conduction system. *Mech. Dev.* **108**, 105–119.

Davis, R. J. *et al.* (2001). *Dach1* mutant mice bear no gross abnormalities in eye, limb, and brain development and exhibit postnatal lethality. *Mol. Cell. Biol.* **21**, 1484–1490.

de Celis, J. F. (2003). Pattern formation in the *Drosophila* wing: The development of the veins. *BioEssays* **25**, 443–451.

Dehal, P. *et al.* (2002). The draft genome of *Ciona intestinalis:* Insights into chordate and vertebrate origins. *Science* **298**, 2157–2167.

Deininger, P. L. and Batzer, M. A. (2002). Mammalian retroelements. *Genome Res.* **12**, 1333–1344.

Di Gregorio, A. and Levine, M. (1999). Regulation of *Ci-tropomyosin-like*, a Brachyury target gene in the ascidian, *Ciona intestinalis. Development* **126**, 5599–5609.

Dodou, E., Verzi, M. P., Anderson, J. P., Xu, S.-M., and Black, B. L. (2004). Mef2c is a direct transcriptional target of ISL1 and GATA factors in the anterior heart field during mouse embryonic development. *Development* **131**, 3931–3942.

Donner, A. L. and Maas, R. L. (2004). Conservation and non-conservation of genetic pathways in eye specification. *Int. J. Dev. Biol.* **48**, 743–753.

Donoviel, D. B. *et al.* (1996). Analysis of muscle creatine kinase gene regulatory elements in skeletal and cardiac muscles of transgenic mice. *Mol. Cell. Biol.* **16**, 1649–1658.

Douzery, E. J. P., Snell, E. A., Bapteste, E., Delsuc, F., and Philippe, H. (2004). The timing of eukaryotic evolution: Does a relaxed molecular clock reconcile proteins and fossils? *Proc. Natl. Acad. Sci. USA* **101**, 15386–15391.

Drissen, R. *et al.* (2004). The active spatial organization of the β-*globin* locus requires the transcription factor EKLF. *Genes Dev.* **18**, 2485–2490.

Duboc, V., Röttinger, E., Besnardeau, L., and Lepage, T. (2004). Nodal and BMP2/4 signaling organizes the oral-aboral of the sea urchin embryo. *Dev. Cell* **6**, 397–410.

Edgar, L. G. and McGhee, J. D. (1986). Embryonic expression of a gut-specific esterase in *Caenorhabditis elegans. Dev. Biol.* **114**, 109–118.

Emerson, B. M., Lewis, C. D., and Felsenfeld, G. (1985). Interaction of specific nuclear factors with the nuclease-hypersensitive region of the chicken adult β-globin gene: Nature of the binding domain. *Cell* **41**, 21–30.

Emily-Fenouil, F., Ghiglione, C., Lhomond, G., Lepage, T., and Gache, C. (1998). GSK3B/ shaggy mediates patterning along the animal-vegetal axis of the sea urchin embryo. *Development* **125**, 2489–2498.

Erwin, D. H. and Davidson, E. H. (2002). The last common bilaterian ancestor. *Development* **129**, 3021–3032.

Fabre-Suver, C. and Hauschka, S. D. (1996). A novel site in the muscle creatine kinase enhancer is required for expression in skeletal but not cardiac muscle. *J. Biol. Chem.* **271**, 4646–4652.

FANTOM Consortium and the RIKEN Genome Exploration Research Group Phase I & II Team. (2002). Analysis of the mouse transcriptome based on functional annotation of 60,770 full-length cDNAs. *Nature* **420**, 563–573.

Fedonkin, M. A. and Waggoner, B. M. (1997). The Late Precambrian fossil *Kimberella* is a mollusc-like bilaterian organism. *Nature* **388**, 868–871.

Felsenfeld, G. and Groudine, M. (2003). Controlling the double helix. *Nature* **421**, 448–453.

Finnerty, J. R., Pang, K., Burton, P., Paulson, D., and Martindale, M. Q. (2004). Origins of bilateral symmetry: *Hox* and *Dpp* expression in a sea anemone. *Science* **304**, 1335–1337.

Flint, J. *et al.* (2001). Comparative genome analysis delimits a chromosomal domain and identifies key regulatory elements in the α globin cluster. *Human Mol. Genet.* **10**, 371–382.

Fomin, M., Nomokonova, N., and Arnold, H.-H. (2004). Identification of a critical control element directing expression of the muscle-specific transcription factor MRF4 in the mouse embryo. *Dev. Biol.* **272**, 498–509.

Francis, N. J., Kingston, R. E., and Woodcock, C. L. (2004). Chromatin compaction by a polycomb group protein complex. *Science* **306**, 1574–1577.

Fraser, S. E. and Stern, C. D. (2004). Early rostrocaudal patterning of the mesoderm and neural plate. *In* "Gastrulation. From Cells to Embryo" (C. D. Stern, Ed.), pp. 389–403. Cold Spring Harbor Laboratory Press, Cold Spring Harbor, NY.

Freiman, R. *et al.* (2001). Requirement of tissue-selective TBP-associated factor TAF$_{II}$105 in ovarian development. *Science* **293**, 2084–2087.

Fujioka, M., Emi-Sarker, Y., Yusibova, G. L., Goto, T., and Jaynes, J. B. (1999). Analysis of an *even-skipped* rescue transgene reveals both composite and discrete neuronal and early blastoderm enhancers, and multi-stripe positioning by gap gene repressor gradients. *Development* **126**, 2527–2538.

Fujiwara, S. *et al.* (2002). Gene expression profiles in *Ciona intestinalis* cleavage-stage embryos. *Mech. Dev.* **112**, 115–127.

Galant, R. and Carroll, S. B. (2002). Evolution of a transcriptional repression domain in an insect Hox protein. *Nature* **415**, 910–913.

Galant, R., Walsh, C. M., and Carroll, S. B. (2002). Hox repression of a target gene: Extradenticle-independent, additive action through multiple monomer binding sites. *Development* **129**, 3115–3126.

Galau, G. A., Lipson, E. D., Britten, R. J., and Davidson, E. H. (1977). Synthesis and turnover of polysomal mRNAs in sea urchin embryos. *Cell* **10**, 415–432.

Gavalas, A., Ruhrberg, C., Livet, J., Henderson, C. E., and Krumlauf, R. (2003). Neuronal defects in the hindbrain of *Hoxa1*, *Hoxb1* and *Hoxb2* mutants reflect regulatory interactions among these Hox genes. *Development* **130**, 5663–5679.

Gavis, E. R. and Lehman, R. (1992). Localization of *nanos* RNA controls embryonic polarity. *Cell* **71**, 301–313.

Gebelein, B., Culi, J., Ryoo, H. D., Zhang, W., and Mann, R. S. (2002). Specificity of *Distalless* repression and limb primordia development by abdominal Hox proteins. *Dev. Cell* **3**, 487–498.

Gebelein, B., McKay, D. J., and Mann, R. S. (2004). Direct integration of *Hox* and segmentation gene inputs during *Drosophila* development. *Nature* **431**, 653–659.

Gehring, W. J. and Ikeo, K. (1999). Pax6—Mastering eye morphogenesis and eye evolution. *Trends Genet.* **15**, 371–377.

Gerhart, J., Black, S., and Scharf, S. (1983). Cellular and pancellular organization of the amphibian embryo. *In* "Modern Cell Biology," Vol. 2, "Spatial Organization of Eukaryotic Cells," (R. MacIntosh, Ed.), pp. 483–507, Alan R. Liss, New York.

Gerrish, K., Van Velkinburgh, J. C., and Stein R. (2004). Conserved transcriptional regulatory domains of the *pdx-1* gene. *Mol. Endocrinol.* **18**, 533–548.

Gindhart, J. G., Jr. and Kaufman, T. C. (1995). Identification of *Polycomb* and *trithorax* group responsive elements in the regulatory region of the *Drosophila* homeotic gene *Sex combs reduced*. *Genetics* **139**, 797–814.

Glavic, A., Gómez-Skarmeta, J. L., and Mayor, R. (2002). The homeoprotein *Xiro1* is required for midbrain-hindbrain boundary formation. *Development* **129**, 1609–1621.

Godin, R. E., Urry, L. A., and Ernst, S. G. (1996). Alternative splicing of the *Endo16* transcript produces differentially expressed mRNAs during sea urchin gastrulation. *Dev. Biol.* **179**, 148–159.

Goldstein, B. (1995). An analysis of the response to gut induction in the *C. elegans* embryo. *Development* **121**, 1227–1236.

Golembo, M., Raz, E., and Shilo, B.-Z. (1996). The *Drosophila* embryonic midline is the site of Spitz processing, and induces activation of the EGF receptor in the ventral ectoderm. *Development* **122**, 3363–3370.

Golembo, M., Yarnitzky, T., Volk, T., and Shilo, B.-Z. (1999). Vein expression is induced by the EGF receptor pathway to provide a positive feedback loop in patterning the *Drosophila* embryonic ventral ectoderm. *Genes Dev.* **13**, 158–162.

Gómez-Skarmeta, J. L. and Modolell, J. (1996). *araucan* and *caupolican* provide a link between compartment subdivisions and patterning of sensory organs and veins in the *Drosophila* wing. *Genes Dev.* **10**, 2935–2945.

Gómez-Skarmeta, J. L. *et al.* (1995). *cis*-Regulation of *achaete* and *scute:* Shared enhancer-like elements drive their coexpression in proneural clusters of the imaginal discs. *Genes Dev.* **9**, 1869–1882.

Gompel, N., Prud'homme, B., Wittkopp, P. J., Kassner, V. A., and Carroll, S. B. (2005). Chance caught on the wing: *cis*-Regulatory evolution and the origin of pigment patterns in *Drosophila*. *Nature* **433**, 481–487.

Göttgens, B. *et al.* (2000). Analysis of vertebrate *SCL* loci identifies conserved enhancers. *Nature Biotech.* **18**, 181–186.

Göttgens, B. *et al.* (2001). Long-range comparison of human and mouse *SCL* loci: Localized regions of sensitivity to restriction endonucleases correspond precisely with peaks of conserved noncoding sequences. *Genome Res.* **11**, 87–97.

Grapin-Botton, A. and Melton, D. A. (2000). Endoderm development—from patterning to organogenesis. *Trends Genet.* **16**, 124–130.

Grapin-Botton, A., Majithia, A. R., and Melton, D. A. (2001). Key events of pancreas formation are triggered in gut endoderm by ectopic expression of pancreatic regulatory genes. *Genes Dev.* **15**, 444–454.

Gray, P. A. *et al.* (2004). Mouse brain organization revealed through direct genome-scale TF expression analysis. *Science* **306**, 2255–2257.

Gray, S. and Levine, M. (1996). Short-range transcriptional repressors mediate both quenching and direct repression within complex loci in *Drosophila*. *Genes Dev.* **10**, 700–710.

Gray, S., Szymanski, P., and Levine, M. (1994). Short-range repression permits multiple enhancers to function autonomously within a complex promoter. *Genes Dev.* **8**, 1829–1838.

Grayson, J., Bassel-Duby, R., and Williams, R. S. (1998). Collaborative interactions between MEF-2 and Sp1 in muscle-specific gene regulation. *J. Cell. Biochem.* **70**, 366–375.

Green, R. B., Hatini, V., Johansen, K. A., Liu, X.-J., and Lengyel, J. A. (2002). Drumstick is a zinc finger protein that antagonizes *lines* to control patterning and morphogenesis of the *Drosophila* hindgut. *Development* **129**, 3645–3656.

Griffin, C., Kleinjan, D. A., Doe, B., and van Heyningen, V. (2002). New 3′ elements control *Pax6* expression in the developing pretectum, neural retina and olfactory region. *Mech. Dev.* **112**, 89–100.

Gupta, B. P. and Sternberg, P. W. (2002). Tissue-specific regulation of the LIM homeobox gene *lin-11* during development of the *Caenorhabditis elegans* egg-laying system. *Dev. Biol.* **247**, 102–115.

Gurdon, J. B. (1988). A community effect in animal development. *Nature* **336**, 772–774.

Guss, K. A., Nelson, C. E., Hudson, A., Kraus, M. E., and Carroll, S. B. (2001). Control of a genetic regulatory network by a selector gene. *Science* **292**, 1164–1167.

Habener, J. F., Kemp, D. M., and Thomas, M. K. (2005). Minireview: Transcriptional regulation in pancreatic development. *Endocrinol.* **146**, 1025–1034.

Hadchouel, J. *et al.* (2003). Analysis of a key regulatory region upstream of the *Myf5* gene reveals multiple phases of myogenesis, orchestrated at each site by a combination of elements dispersed throughout the locus. *Development* **130**, 3415–3426.

Halder, G., Callaerts, P., and Gehring, W. J. (1995). Induction of ectopic eyes by targeted expression of the *eyeless* gene in *Drosophila*. *Science* **267**, 1788–1792.

Halder, G. *et al.* (1998a). The Vestigial and Scalloped proteins act together to directly regulate wing-specific gene expression in *Drosophila*. *Genes Dev.* **12**, 3900–3909.

Halder, G., Callaerts, P., Flister, S., Walldorf, U., Kloter, U., and Gehring, W. J. (1998b). Eyeless initiates the expression of both *sine oculis* and *eyes absent* during *Drosophila* compound eye development. *Development* **125**, 2181–2191.

Halfon, M. S. *et al.* (2000). Ras pathway specificity is determined by the integration of multiple signal-activated and tissue-restricted transcription factors. *Cell* **103**, 63–74.

Halfon, M. S., Grad, Y., Church, G. M., and Michelson, A. M. (2002). Computation-based discovery of related transcriptional regulatory modules and motifs using an experimentally validated combinatorial model. *Genome Res.* **12**, 1019–1028.

Han, Z. *et al.* (2002). Transcriptional integration of competence modulated by mutual repression generates cell-type specificity within the cardiogenic mesoderm. *Dev. Biol.* **252**, 225–240.

Hans, S. and Campos-Ortega, J. A. (2002). On the organization of the regulatory region of the zebrafish deltaD gene. *Development* **129**, 4773–4784.

Hardison, R. *et al.* (1997). Locus control regions of mammalian β-globin gene clusters: Combining phylogenetic analyses and experimental results to gain functional insights. *Gene* **205**, 73–94.

Hardison, R. C. (2000). Conserved noncoding sequences are reliable guides to regulatory elements. *Trends Genet.* **16**, 369–372.

Harfe, B. D., McManus, M. T., Mansfield, J. H., Hornstein, E., and Tabin, C. J. (2005). The RNaseIII enzyme *Dicer* is required for morphogenesis but not patterning of the vertebrate limb. *Proc. Natl. Acad. Sci. USA* **102**, 10898–10903.

Harland, R. (2004). Dorsoventral patterning of the mesoderm. *In* "Gastrulation. From Cells to Embryo" (C. D. Stern, Ed.), pp. 373–388. Cold Spring Harbor Laboratory Press, Cold Spring Harbor, NY.

Hartenstein, V. (1993). "Atlas of *Drosophila* Development." Cold Spring Harbor Laboratory Press, Cold Spring Harbor, NY.

Harvey, R. P. (1996). *NK-2* homeobox genes and heart development. *Dev. Biol.* **178**, 203–216.

Haumaitre, C. *et al.* (2005). Lack of TCF2/vHNF1 in mice leads to pancreas agenesis. *Proc. Natl. Acad. Sci. USA* **102**, 1490–1495.

He, L. and Hannon, G. J. (2004). MicroRNAs: Small RNAs with a big role in gene regulation. *Nature Rev.* **5**, 522–531.

Heasman, J. *et al.* (1994). Overexpression of cadherins and underexpression of β-catenin inhibit dorsal mesoderm induction in early *Xenopus* embryos. *Cell* **79**, 791–803.

Heasman, J., Kofron, M., and Wylie, C. (2000). β-catenin signaling activity dissected in the early *Xenopus* embryo: A novel antisense approach. *Dev. Biol.* **222**, 124–134.

Heasman, J., Wessely, O., Langland, R., Craig, E. J., and Kessler, D. S. (2001). Vegetal localization of maternal mRNAs is disrupted by VegT depletion. *Dev. Biol.* **240**, 377–386.

Heicklen-Klein, A. and Evans, T. (2004). T-box binding sites are required for activity of a cardiac GATA-4 enhancer. *Dev. Biol.* **267**, 490–504.

Hersh, B. M. and Carroll, S. B. (2005). Direct regulation of *knot* gene expression by Ultrabithorax and the evolution of *cis*-regulatory elements in *Drosophila*. *Development* **132**, 1567–1577.

Hiller, M. A., Lin, T. Y., Wood, C., and Fuller, M. T. (2001). Developmental regulation of transcription by a tissue-specific TAF homolog. *Genes Dev.* **15**, 1021–1030.

Hiller, M. *et al.* (2004). Testis-specific TAF homologs collaborate to control a tissue-specific transcription program. *Development* **131**, 5297–5308.

Hilton, E., Rex, M., and Old, R. (2003). VegT activation of the early zygotic gene *Xnr5* requires lifting of Tcf-mediated repression in the *Xenopus* blastula. *Mech. Dev.* **120**, 1127–1138.

Hinegardner, R. (1974). Cellular DNA content of the Echinodermata. *Comp. Biochem. Physiol.* **49B**, 219–226.

Hinman, V. F., Nguyen, A., Cameron, R. A., and Davidson, E. H. (2003). Developmental gene regulatory network architecture across 500 million years of echinoderm evolution. *Proc. Natl. Acad. Sci. USA* **100**, 13356–13361.

Hirano, T. and Nishida, H. (1997). Developmental fates of larval tissues after metamorphosis in ascidian *Halocynthia roretzi*. I. Origin of mesodermal tissues of the juvenile. *Dev. Biol.* **192**, 199–210.

Hirano, T. and Nishida, H. (2000). Developmental fates of larval tissues after metamorphosis in the ascidian, *Halocynthia roretzi*. II. Origin of endodermal tissues of the juvenile. *Dev. Genes Evol.* **210**, 55–63.

Hirth, F. *et al.* (2003). An urbilaterian origin of the tripartite brain: Developmental genetic insights from *Drosophila*. *Development* **130**, 2365–2373.

Hobert, O. (2004). Common logic of transcription factor and microRNA action. *Trends Biochem. Sci.* **29**, 462–468.

Hoch, M. and Pankratz, M. J. (1996). Control of gut development by *fork head* and cell signaling molecules in *Drosophila*. *Mech. Dev.* **58**, 3–14.

Hochheimer, A. and Tjian, R. (2003). Diversified transcription initiation complexes expand promoter selectivity and tissue-specific gene expression. *Genes Dev.* **17**, 1309–1320.

Hoffman, J. A., Kafatos, F. C., Janeway, C. A., Jr., and Ezekowitz, R. A. B. (1999). Phylogenetic perspectives in innate immunity. *Science* **284**, 1313–1318.

Holland, P. W. H., Koschorz, B., Holland, L. Z., and Herrmann, B. G. (1995). Conservation of *brachyury* (T) genes in amphioxus and vertebrates: Developmental and evolutionary implications. *Development* **121**, 4283–4291.

Holt, R. A. *et al.* (2002). The genome sequence of the malaria mosquito *Anopheles gambiae*. *Science* **298**, 129–149.

Hornstein, E. *et al.* (2005). The microRNA miR-196 acts upstream of Hoxb8 and Shh in limb development. *Nature* **438**, 671–674.

Hotta, K., Takahashi, H., Erives, A., Levine, M., and Satoh, N. (1999). Temporal expression patterns of 39 *Brachyury*-downstream genes associated with notochord formation in the *Ciona intestinalis* embryo. *Develop. Growth Differ.* **41**, 657–664.

Houston, D. W. and Wylie, C. (2004). The role of Wnts in gastrulation. *In* "Gastrulation. From Cells to Embryo" (C. D. Stern, Ed.), pp. 521–538. Cold Spring Harbor Laboratory Press, Cold Spring Harbor, NY.

Howard-Ashby, M., Materna, S. C., Brown, C. T., Chen, L., Cameron, R. A., and Davidson E. H. (2006a). The identification and characterization of homeodomain genes in early *Strongylocentrotus purpuratus* development. Submitted.

Howard-Ashby, M., Materna, S. C., Brown, C. T., Chen, L., Cameron, R. A., and Davidson E. H. (2006b). The identification and characterization of transcription factor families in early *Strongylocentrotus purpuratus* development. Submitted.

Huang, A. M., Rusch, J., and Levine, M. (1997). An anteroposterior Dorsal gradient in the *Drosophila* embryo. *Genes Dev.* **11**, 1963–1973.

Hudson, C. and Yasuo, H. (2005). Patterning across the ascidian neural plate by lateral Nodal signaling sources. *Development* **132**, 1199–1210.

Hughes, C. L. and Kaufman, T. C. (2002). Hox genes and the evolution of the arthropod body plan. *Evo. Dev.* **4**, 459–499.

Hyman, L. H. (1955). "The Invertebrates: Echinodermata. The Coelomate Bilateria," Vol. IV. McGraw-Hill Book Company, Inc., New York, Toronto, London.

Ilagan, J. G., Cvekl, A., Kantorow, M., Piatigorsky, J., and Sax, C. M. (1999). Regulation of α A-crystallin gene expression. *J. Biol. Chem.* **274**, 19973–19978.

Imai, K., Takada, N., Satoh, N., and Satou, Y. (2000). β-catenin mediates the specification of endoderm cells in ascidian embryos. *Development* **127**, 3009–3020.

Imai, K. S., Satoh, N., and Satou, Y. (2002). Early embryonic expression of *FGF4/6/9* gene and its role in the induction of mesenchyme and notochord in *Ciona savignyi* embryos. *Development* **129**, 1729–1738.

Imai, K. S., Hino, K., Yagi, K., Satoh, N., and Satou, Y. (2004). Gene expression profiles of transcription factors and signaling molecules in the ascidian embryo: Towards a comprehensive understanding of gene networks. *Development* **131**, 4047–4058.

Inami, M. *et al.* (2004). CD28 costimulation controls histone hyperacetylation of the interleukin 5 gene locus in developing Th2 cells. *J. Biol. Chem.* **279**, 23123–23133.

International Human Genome Sequencing Consortium (2001). Initial sequencing and analysis of the human genome. *Nature* **409**, 860–921.

Ip, Y. T., Park, R. E., Kosman, D., Bier, E., and Levine, M. (1992). The *dorsal* gradient morphogen regulates stripes of *rhomboid* expression in the presumptive neuroectoderm of the *Drosophila* embryo. *Genes Dev.* **6**, 1728–1739.

Isaac, A. *et al.* (2000). FGF and genes encoding transcription factors in early limb specification. *Mech. Dev.* **93**, 41–48.

Ishida, A. *et al.* (1993). Cloning and chromosome mapping of the human Mel-18 gene which encodes a DNA-binding protein with a new ring-finger motif. *Gene* **129**, 249–255.

Isshiki, T., Pearson, B., Holbrook, S., and Doe, C. Q. (2001). *Drosophila* neuroblasts sequentially express transcription factors which specify the temporal identity of their neuronal progeny. *Cell* **106**, 511–521.

Istrail, S. and Davidson, E. H. (2005). Logic functions of the genomic *cis*-regulatory code. *Proc. Natl. Acad. Sci. USA* **102**, 4954–4959.

Itkin-Ansari, P. *et al.* (2005). NeuroD1 in the endocrine pancreas: Localization and dual function as an activator and repressor. *Dev. Dyn.* **233**, 946–953.

Iwahori, A., Fraidenraich, D., and Basilico, C. (2004). A conserved enhancer element that drives FGF4 gene expression in the embryonic myotomes is synergistically activated by GATA and bHLH proteins. *Dev. Biol.* **270**, 525–537.

Iwaki, D. D., Johansen, K. A., Singer, J. B., and Lengyel, J. A. (2001). *drumstick, bowl,* and *lines* are required for patterning and cell rearrangement in the *Drosophila* embryonic hindgut. *Dev. Biol.* **240**, 611–626.

Iype, T. *et al.* (2004). The transcriptional repressor Nkx6.1 also functions as a deoxyribonucleic acid context-dependent transcriptional activator during pancreatic β-cell differentiation: Evidence for feedback activation of the *nkx6.1* gene by Nkx6.1. *Mol. Endocrinol.* **18**, 1363–1375.

Jackson, P. D. and Hoffmann, F. M. (1994). Embryonic expression patterns of the *Drosophila decapentaplegic* gene: Separate regulatory elements control blastoderm expression and lateral ectodermal expression. *Dev. Dyn.* **199**, 28–44.

Jacquemin, P., LeMaigre, F. P., and Rousseau, G. G. (2003). The Onecut transcription factor HNF-6 (OC-1) is required for timely specification of the pancreas and acts upstream of Pdx-1 in the specification cascade. *Dev. Biol.* **258**, 105–116.

Jaeger, J. *et al.* (2004). Dynamic control of positional information in the early *Drosophila* embryo. *Nature* **430**, 368–371.

Jaenish, R. and Bird, A. (2003). Epigenetic regulation of gene expression: How the genome integrates intrinsic and environmental signals. *Nature Genet. Suppl.* **33**, 245–254.

Jagla, K., Bellard, M., and Frasch, M. (2001). A cluster of *Drosophila* homeobox genes involved in mesoderm differentiation programs. *BioEssays* **23**, 125–133.

Jagla, T., Bidet, Y., Da Ponte, J. P., Dastugue, B., and Jagla, K. (2002). Cross-repressive interactions of identity genes are essential for proper specification of cardiac and muscular fates in *Drosophila*. *Development* **129**, 1037–1047.

Jaynes, J. B. and O'Farrell, P. H. (1991). Active repression of transcription by the Engrailed homeodomain protein. *EMBO J.* **10**, 1427–1433.

Jensen, J. (2004). Gene regulatory factors in pancreatic development. *Dev. Dyn.* **229**, 176–200.

Jenuwein, T. and Allis, C. D. (2001). Translating the histone code. *Science* **293**, 1074–1080.

Jernvall, J. and Thesleff, I. (2000). Reiterative signaling and patterning during mammalian tooth morphogenesis. *Mech. Dev.* **92**, 19–29.

Jiang, J. and Levine, M. (1993). Binding affinities and cooperative interactions with bHLH activators delimit threshold responses to the dorsal gradient morphogen. *Cell* **72**, 741–752.

Jiang, J., Cai, H., Zhou, Q., and Levine, M. (1993). Conversion of a *dorsal*-dependent silencer into an enhancer: Evidence for *dorsal* corepressors. *EMBO J.* **12**, 3201–3209.

Johansen, K. A., Green, R. B., Iwaki, D. D., Hernandez, J. B., and Lengyel, J. A. (2003). The Drm-Bowl-Lin relief-of-repression hierarchy controls fore- and hindgut patterning and morphogenesis. *Mech. Dev.* **20**, 1139–1151.

Johnson, J. M. *et al.* (2003). Genome-wide survey of human alternative pre-mRNA splicing with exon junction microarrays. *Science* **302**, 2141–2144.

Johnston, R. J., Jr., Chang, S., Etchberger, J. F., Ortiz, C. O., and Hobert, O. (2005). MicroRNAs acting in a double-negative feedback loop to control a neuronal cell fate decision. *Proc. Natl. Acad. Sci. USA* **102**, 12449–12454.

Jones, B. W., Abeysekera, M., Galinska, J., and Jolicoeur, E. M. (2004). Transcriptional control of glial and blood cell development in *Drosophila*: *cis*-Regulatory elements of *glial cells missing*. *Dev. Biol.* **266**, 374–387.

Joyner, A. L., Liu, A., and Millet, S. (2000). *Otx2, Gbx2* and *Fgf8* interact to position and maintain a mid-hindbrain organizer. *Curr. Opin. Cell Biol.* **12**, 736–741.

Kaiser. K. and Meisterernst, M. (1996). The human general co-factors. *Trends Biochem. Sci.* **21**, 342–345.

Kammandel, B. *et al.* (1999). Distinct *cis*-essential modules direct the time-space pattern of the *Pax6* gene activity. *Dev. Biol.* **205**, 79–97.

Katahira, T. *et al.* (2000). Interaction between Otx2 and Gbx2 defines the organizing center for the optic tectum. *Mech. Dev.* **91**, 43–52.

Katoh, M. (2002). Molecular cloning and characterization of *OSR1* on human chromosome 2p24. *Int. J. Mol. Med.* **10**, 221–225.

Kenny, A. P., Kozlowski, D., Oleksyn, D. W., Angerer, L. M., and Angerer, R. C. (1999). SpSoxB1, a maternally encoded transcription factor asymmetrically distributed among early sea urchin blastomeres. *Development* **126**, 5473–5483.

Kenny, A. P., Oleksyn, D. W., Newman, L. A., Angerer, R. C., and Angerer, L. M. (2003). Tight regulation of SpSoxB factors is required for patterning and morphogenesis in sea urchin embryos. *Dev. Biol.* **261**, 412–425.

Kikuta, H., Kanai, M., Ito, Y., and Yamasu, K. (2003). *gbx2* homeobox gene is required for the maintenance of the isthmic region in the zebrafish embryonic brain. *Dev. Dyn.* **228**, 433–450.

Kim, J. *et al.* (1996). Integration of positional signals and regulation of wing formation and identity by *Drosophila vestigial* gene. *Nature* **382**, 133–138.

Kim, J., Johnson, K., Chen, H. J., Carroll, S., and Laughon, A. (1997). *Drosophila* Mad binds to DNA and directly mediates activation of *vestigial* by Decapentaplegic. *Nature* **388**, 304–308.

Kimelman, D. and Bjornson, C. (2004). Vertebrate mesoderm induction: From frogs to mice. *In* "Gastrulation. From Cells to Embryo" (C. D. Stern, Ed.), pp. 363–372. Cold Spring Harbor Laboratory Press, Cold Spring Harbor, NY.

Kimura-Yoshida, C. *et al.* (2004). Characterization of the pufferfish *Otx2 cis*-regulators reveals evolutionarily conserved genetic mechanisms for vertebrate head specification. *Development* **131**, 57–71.

King, N. (2004). The unicellular ancestry of animal development. *Dev. Cell* **7**, 313–325.

King, N. and Carroll, S. B. (2001). A receptor tyrosine kinase from choanoflagellates: Molecular insights into early animal evolution. *Proc. Natl. Acad. Sci. USA* **98**, 15032–15037.

King, N., Hittinger, C. T., and Carroll, S. B. (2003). Evolution of key cell signaling and adhesion protein families predates animal origins. *Science* **301**, 361–363.

Kingsley, D. M. (1994). The TGFβ superfamily: New members, new receptors, and new genetic tests of function in different organisms. *Genes Dev.* **8**, 133–146.

Kirchhamer, C. V. and Davidson, E. H. (1996). Spatial and temporal information processing in the sea urchin embryo: Modular and intramodular organization of the *CyIIIa* gene *cis*-regulatory system. *Development* **122**, 333–348.

Kirchhamer, C. V., Yuh, C.-H., and Davidson, E. H. (1996). Modular *cis*-regulatory organization of developmentally expressed genes: Two genes transcribed territorially in the sea urchin embryo, and additional examples. *Proc. Natl. Acad. Sci. USA* **93**, 9322–9328.

Kirouac, M. and Sternberg, P. W. (2003). *cis*-Regulatory control of three cell fate-specific genes in vulval organogenesis of *Caenorhabditis elegans* and *C. briggsae*. *Dev. Biol.* **257**, 85–103.

Kleinjan, D. A., Seawright, A., Childs, A. J., and van Heyningen, V. (2004). Conserved elements in *Pax 6* intron 7 involved in (auto)regulation and alternative transcription. *Dev. Biol.* **265**, 462–477.

Klinedinst, S. L. and Bodmer, R. (2003). Gata factor Pannier is required to establish competence for heart progenitor formation. *Development* **130**, 3027–3038.

Knirr, S. and Frasch, M. (2001). Molecular integration of inductive and mesoderm-intrinsic inputs governs *even-skipped* enhancer activity in a subset of pericardial and dorsal muscle progenitors. *Dev. Biol.* **238**, 13–26.

Knosp, W. M., Scott, V., Bächinger, H. P., and Stadler, H. S. (2004). HOXA13 regulates the expression of bone morphogenetic proteins 2 and 7 to control distal limb morphogenesis. *Development* **131**, 4581–4592.

Kofron, M. *et al.* (1999). Mesoderm induction in *Xenopus* is a zygotic event regulated by maternal VegT via TGFb growth factors. *Development* **126**, 5759–5770.

Kofron, M. *et al.* (2004). New roles for FoxH1 in patterning the early embryo. *Development* **131**, 5065–5078.

Koide, T., Hayata, T., and Cho, K. W. Y. (2005). *Xenopus* as a model system to study transcriptional regulatory networks. *Proc. Natl. Acad. Sci. USA* **102**, 4943–4948.

Kölsch, V. and Paululat, A. (2002). The highly conserved cardiogenic bHLH factor hand is specifically expressed in circular visceral muscle progenitor cells and in all cell types of the dorsal vessel during *Drosophila* embryogenesis. *Dev. Genes Evol.* **212**, 473–485.

Kölzer, S., Fuss, B., Hoch, M., and Klein, T. (2003). *Defective proventriculus* is required for pattern formation along the proximodistal axis, cell proliferation and formation of veins in the *Drosophila* wing. *Development* **130**, 4135–4147.

Kondoh, H., Uchikawa, M., and Kamachi, Y. (2004). Interplay of Pax6 and SOX2 in lens development as a paradigm of genetic switch mechanisms for cell differentiation. *Int. J. Dev. Biol.* **48**, 819–827.

Koutsourakis, M., Langeveld, A., Patient, R., Beddington, R., and Grosveld, F. (1999). The transcription factor GATA6 is essential for early extraembryonic development. *Development* **126**, 723–732.

Krause, M., Harrison, S. W., Xu, S.-Q., Chen, L., and Fire, A. (1994). Elements regulating cell- and stage-specific expression of the *C. elegans* MyoD family homolog *hlh-1*. *Dev. Biol.* **166**, 133–148.

Kurokawa, D. *et al.* (2004). Regulation of *Otx2* expression and its functions in mouse epiblast and anterior neuroectoderm. *Development* **131**, 3307–3317.

Kuwabara, T., Hsieh, J., Nakashima, K., Taira, K., and Gage, F. H. (2004). A small modulatory dsRNA specifies the fate of adult neural stem cells. *Cell* **116**, 779–793.

Kwan, C. T., Tsang, S. L., Krumlauf, R., and Sham, M. H. (2001). Regulatory analysis of the mouse *Hoxb3* gene: Multiple elements work in concert to direct temporal and spatial patterns of expression. *Dev. Biol.* **232**, 176–190.

Landmann, F., Quintin, S., and Labouesse, M. (2004). Multiple regulatory elements with spatially and temporally distinct activities control the expression of the epithelial differentiation gene *lin-26* in *C. elegans*. *Dev. Biol.* **265**, 478–490.

Latinkic′, B. V. *et al.* (1997). The *Xenopus Brachyury* promoter is activated by FGF and low concentrations of activin and suppressed by high concentrations of activin and by paired-type homeodomain proteins. *Genes Dev.* **11**, 3265–3276.

Laurent, M. N., Blitz, I. L., Hashimoto, C., Rothbächer, U., and Cho. K. W.-Y. (1997). The *Xenopus* homeobox gene *Twin* mediates Wnt induction of *Goosecoid* in establishment of Spemann's organizer. *Development* **124**, 4905–4916.

Leaman, D. *et al.* (2005). Antisense-mediated depletion reveals essential and specific functions of microRNAs in *Drosophila* development. *Cell* **121**, 1097–1108.

Lecuit, T. and Cohen, S. M. (1997). Proximal-distal axis formation in the *Drosophila* leg. *Nature* **388**, 139–145.

Lee, C. M., Yu, D. S., Crews, S. T., and Kim, S. H. (1999). The CNS midline cells and *spitz* class genes are required for proper patterning of *Drosophila* ventral neuroectoderm. *Int. J. Dev. Biol.* **43**, 305–315.

Lee, K.-H., Evans, S., Ruan, T. Y., and Lassar, A. B. (2004). SMAD-mediated modulation of YY1 activity regulates the MBP response and cardiac-specific expression of a GATA4/5/6-dependent chick Nkx2.5 enhancer. *Development* **131**, 4709–4723.

Lee, T. I. *et al.* (2002). Transcriptional regulatory networks in *Saccharomyces cerevisiae*. *Science* **298**, 799–804.

Lehman, D. A. *et al.* (1999). *cis*-Regulatory elements of the mitotic regulator, *string/Cdc25*. *Development* **126**, 1793–1803.

Lei, H., Wang, H., Juan, A. H., and Ruddle, F. H. (2005). The identification of *Hoxc8* target genes. *Proc. Natl. Acad. Sci. USA* **102**, 2420–2424.

Lemaire, P., Garrett, N., and Gurdon, J. B. (1995). Expression cloning of siamois, a *Xenopus* homeobox gene expressed in dorsal-vegetal cells of blastulae and able to induce a complete secondary axis. *Cell* **81**, 85–94.

Lengyel, J. A. and Iwaki, D. D. (2002). It takes guts: The *Drosophila* hindgut as a model system for organogenesis. *Dev. Biol.* **243**, 1–19.

Letting, D. L., Chen. Y.-Y., Rakowski, C., Reedy, S., and Blobel, G. A. (2004). Context-dependent regulation of GATA-1 by friend of GATA-1. *Proc. Natl. Acad. Sci. USA* **101**, 467–481.

Leung, B., Hermann, G. J., and Preiss, J. R. (1999). Organogenesis of the *Caenorhabditis elegans* intestine. *Dev. Biol.* **216**, 114–134.

Levine, M. and Davidson, E. H. (2005). Gene regulatory networks for development. *Proc. Natl. Acad. Sci. USA* **102**, 4936–4942.

Lewis, A. L., Xia, Y., Datta, S. K., McMillin, J., and Kellems, R. E. (1999). Combinatorial interactions regulate cardiac expression of the murine adenylosuccinate synthetase 1 gene. *J. Biol. Chem.* **274**, 14188–14197.

Li, J. Y. H., Lao, Z., and Joyner, A. L. (2005). New regulatory interactions and cellular responses in the isthmic organizer region revealed by altering *Gbx2* expression. *Development* **132**, 1971–1981.

Li, X., Chuang, C. K., Mao, C. A., Angerer, L. M., and Klein, W. H. (1997). Two Otx proteins generated from multiple transcripts of a single gene in *Strongylocentrotus purpuratus*. *Dev. Biol.* **187**, 253–266.

Li, X., Wikramanayake, A. H., and Klein, W. H. (1999). Requirement of SpOtx in cell fate decisions in the sea urchin embryo and possible role as a mediator of β-catenin signaling. *Dev. Biol.* **212**, 425–439.

Liberatore, C. M., Searcy-Schrick, R. D., Vincent, E. B., and Yutzey, K. E. (2002). *Nkx-2.5* gene induction in mice is mediated by a Smad consensus regulatory region. *Dev. Biol.* **244**, 243–256.

Lichtneckert, R. and Reichert, H. (2005). Insights into the urbilaterian brain: Conserved genetic patterning mechanisms in insect and vertebrate brain development. *Heredity* **94**, 465–477.

Lien, C.-L., McAnally, J., Richardson, J. A., and Olson, E. N. (2002). Cardiac-specific activity of an *Nkx2–5* enhancer requires an evolutionarily conserved Smad binding site. *Dev. Biol.* **244**, 257–266.

Lin, G. F. *et al.* (2003). T-box binding site mediates the dorsal activation of *myf-5* in *Xenopus* gastrula embryos. *Dev. Dyn.* **226**, 51–58.

Linhares, V. L. F. *et al.* (2004). Transcriptional regulation of the murine Connexin40 promoter by cardiac factors Nkx2–5, GATA4 and Tbx5. *Cardiovas. Res.* **64**, 402–411.

Liu, T., Wu., J., and He, F. (2000). Evolution of *cis*-acting elements in 5′ flanking regions of vertebrate actin genes. *J. Mol. Evol.* **50**, 22–30.

Livi, C. B. and Davidson, E. H. (2006). Expression and function of *Spblimp1/krox:* An alternatively spliced transcription factor with a central role in sea urchin endomesoderm specification. *Dev. Biol.*, in press.

Lo, P. C. H., Skeath, J. B., Gajewski, K., Schulz, R. A., and Frasch, M. (2002). Homeotic genes autonomously specify the anteroposterior subdivision of the *Drosophila* dorsal vessel into aorta and heart. *Dev. Biol.* **251**, 307–319.

Logan, C. Y. and McClay, D. R. (1997). The allocation of early blastomeres to the ectoderm and endoderm is variable in the sea urchin embryo. *Development* **124**, 2213–2223.

Logan, C. Y., Miller, J. R., Ferkowicz, M. J., and McClay, D. R. (1999). Nuclear β-catenin is required to specify vegetal cell fates in the sea urchin embryo. *Development* **126**, 345–357.

Logan, M., Simon, H. G., and Tabin, C. (1998). Differential regulation of T-box and homeobox transcription factors suggests roles in controlling chick limb-type identity. *Development* **125**, 2825–2835.

Loh, S. H. Y. and Russell, S. (2000). A *Drosophila* group E *Sox* gene is dynamically expressed in the embryonic alimentary canal. *Mech. Dev.* **93**, 185–188.

Lohmann, I., McGinnis, N., Bodmer, M., and McGinnis, W. (2002). The *Drosophila* Hox Gene Deformed sculpts head morphology via direct regulation of the apoptosis activator reaper. *Cell* **110**, 457–466.

Longabaugh, W. J. R., Davidson, E. H., and Bolouri, H. (2005). Computational representation of developmental genetic regulatory networks. *Dev. Biol.* **283**, 1–16.

Lowe, C. J. *et al.* (2003). Anteroposterior patterning in hemichordates and the origins of the chordate nervous system. *Cell* **113**, 853–865.

Lumsden, A. and Krumlauf, R. (1996). Patterning the vertebrate neuraxis. *Science* **274**, 1109–1115.

Lunde, K. *et al.* (2003). Activation of the *knirps* locus links patterning to morphogenesis of the second wing vein in *Drosophila*. *Development* **130**, 235–248.

MacLellan, W. R., Lee, T.-C., Schwartz, R. J., and Schneider, M. D. (1994). Transforming growth factor-β response elements of the skeletal α-actin gene. *J. Biol. Chem.* **269**, 16754–16760.

Maduro, M. F. and Rothman, J. H. (2002). Making worm guts: The gene regulatory network of the *Caenorhabditis elegans* endoderm. *Dev. Biol.* **246**, 68–85.

Maduro, M. F., Lin, R., and Rothman, J. H. (2002). Dynamics of a developmental switch: Recursive intracellular and intranuclear redistribution of *Caenorhabditis elegans* POP-1 parallels Wnt-inhibited transcriptional repression. *Dev. Biol.* **248**, 128–142.

Maduro, M. F., Kasmir, J. J., Zhu, J., and Rothman, J. H. (2005). The Wnt effector POP-1 and the PAL-1/caudal homeoprotein collaborate with SKN-1 to activate *C. elegans* endoderm development. *Dev. Biol.* **285**, 510–523.

Makabe, K. W., Kirchhamer, C. V., Britten, R. J., and Davidson, E. H. (1995). *cis*-Regulatory control of the *SM50* gene, an early marker of skeletogenic lineage specification in the sea urchin embryo. *Development* **121**, 1957–1970.

Malik, S. and Roeder, R. G. (2000). Transcriptional regulation through Mediator-like coactivators in yeast and metazoan cells. *Trends Biochem. Sci.* **25**, 227–283.

Manabe, I. and Owens, G. K. (2001). The smooth muscle myosin heavy chain gene exhibits smooth muscle subtype-selective modular regulation *in vivo*. *J. Biol. Chem.* **276**, 39076–39087.

Manuel, M. and Le Parco, Y. (2000). Homeobox gene diversification in the calcareous sponge *Sycon raphanus*. *Mol. Phylogenet. Evol.* **17**, 97–107.

Manuel, M., Le Parco, Y., and Borchiellini, C. (2004). Comparative analysis of Brachyury T-domains, with the characterization of two new sponge sequences, from a hexactinellid and a calcisponge. *Gene* **340**, 291–301

Manzanares, M. *et al.* (2000). Conservation and elaboration of *Hox* gene regulation during evolution of the vertebrate head. *Nature* **408**, 854–857.

Manzanares, M. *et al.* (2001). Independent regulation of initiation and maintenance phases of *Hoxa3* expression in the vertebrate hindbrain involve auto- and cross-regulatory mechanisms. *Development* **128**, 3595–3607.

Manzanares, M. *et al.* (2002). Krox20 and kreisler co-operate in the transcriptional control of segmental expression of *Hoxb3* in the developing brain. *EMBO J.* **21**, 365–376.

Mao, C.-A. *et al.* (1996). Altering cell fates in sea urchin embryos by overexpressing SpOtx, an Orthodenticle-related protein. *Development* **122**, 1489–1498.

Margulies, E. H., Green, E. D., and NISC Comparative Sequencing Program. (2003). Detecting highly conserved regions of the human genome by multispecies sequence comparisons. *Cold Spring Harbor Symp. Quant. Biol.* **68**, 255–263.

Marín, F. and Charnay, P. (2000). Positional regulation of *Krox-20* and *mafB/kr* expression in the developing hindbrain: Potentialities of prospective rhombomeres. *Dev. Biol.* **218**, 220–234.

Markstein, M., Markstein, P., Markstein, V., and Levine, M. S. (2002). Genome-wide analysis of clustered Dorsal binding sites identifies putative target genes in the *Drosophila* embryo. *Proc. Natl. Acad. Sci. USA* **99**, 763–768.

Markstein, M. *et al.* (2004). A regulatory code for neurogenic gene expression in the *Drosophila* embryo. *Development* **131**, 2387–2394.

Martinez-Barbera, J. P., Signore, M., Boyl, P. P., Puelles, E., Acampora, D., Gogoi, R., Schubert, F., Lumsden, A., and Simeone, A. (2001). Regionalisation of anterior neuroectoderm and its competence in responding to forebrain and midbrain inducing activities depend on mutual antagonism between OTX2 and GBX2. *Development* **128**, 4789–4800.

Marty, T. *et al.* (2001). A HOX complex, a repressor element and a 50 bp sequence confer regional specificity to a DPP-responsive enhancer. *Development* **128**, 2833–2845.

Mastrangelo, I. A., Courey, A. J., Wall, J. S., Jackson, S. P., and Hough, P. V. C. (1991). DNA looping and Sp1 multimer links – A mechanism for transcriptional synergism and enhancement. *Proc. Natl. Acad. Sci. USA* **88**, 5670–5674.

Materna, S. C., Howard-Ashby, M., Gray, R. F., and Davidson, E. H. (2006). The C_2H_2 zinc fingers of the *Strongylocentrotus purpuratus* genome. Submitted.

McDonald, J. A., Fujioka, M., Odden, J. P., Jaynes, J. B., and Doe, C. Q. (2003). Specification of motoneuron fate in *Drosophila*: Integration of positive and negative transcription factor inputs by a minimal *eve* enhancer. *J. Neurobiol.* **57**, 193–203.

McGhee, J. D. and von Hippel, P. H. (1974). Theoretical aspects of DNA-protein interactions: Co-operative and non-co-operative binding of large ligands to a one-dimensional homogeneous lattice. *J. Mol. Biol.* **86**, 469–489.

McMahon, A. P., Ingham, P. W., and Tabin, C. J. (2003). Developmental roles and clinical significance of hedgehog signaling. *Curr. Topics Dev. Biol.* **53**, 1–114.

Mericskay, M. *et al.* (2000). An overlapping CArG/octamer element is required for regulation of *desmin* gene transcription in arterial smooth muscle cells. *Dev. Biol.* **226**, 192–208.

Meulemans, D. and Bronner-Fraser, M. (2004). Gene-regulatory interactions in neural crest evolution and development. *Dev. Cell* **7**, 291–299.

Michaut, L. *et al.* (2003). Analysis of the eye developmental pathway in *Drosophila* using DNA microarrays. *Proc. Natl. Acad. Sci. USA* **100**, 4024–4029.

Mickey, K., Mello, C. C., Montgomery, M., Fire, A., and Preiss, J. (1996). An inductive interaction in 4-cell stage *C. elegans* embryos involves APX-1 expression in the signaling cell. *Development* **122**, 1791–1798.

Miller, J. R. *et al.* (1999). Establishment of Disheveled that is dependent on cortical rotation. *J. Cell Biol.* **146**, 427–437.

Millet, S., Campbell, K., Epstein, D. J., Losos, K., Harris, E., and Joyner, A. L. (1999). A role for *Gbx2* in repression of *Otx2* and positioning the mid/hindbrain organizer. *Nature* **401**, 161–164,

Milo, R. *et al.* (2002). Network motifs: Simple building blocks of complex networks. *Science* **298**, 824–827.

Minokawa, T., Wikramanayake, A. H., and Davidson, E. H. (2005). *cis*-Regulatory inputs of the *wnt8* gene in the sea urchin endomesoderm network. *Dev. Biol.* **288**, 545–558.

Miskolczi-McCallum, C. M., Scavetta, R. J., Svendsen, P. C., Soanes, K. H., and Brook, W. J. (2005). The *Drosophila melanogaster* T-box genes midline and H15 are conserved regulators of heart development. *Dev. Biol.* **278**, 459–472.

Miya, T. and Nishida, H. (2003). An Ets transcription factor, HrEts, is target of FGF signaling and involved in induction of notochord, mesenchyme, and brain in ascidian embryos. *Dev. Biol.* **261**, 25–38.

Moore, G. P., Scheller, R. H., Davidson, E. H., and Britten, R. J. (1978). Evolutionary change in the repetition frequency of sea urchin DNA sequences. *Cell* **15**, 649–660.

Morel, V. and Schweisguth, F. (2000). Repression by Suppressor of Hairless and activation by Notch are required to define a single row of *single-minded* expressing cells in the *Drosophila* embryo. *Genes Dev.* **14**, 377–388.

Morgan, T. H. (1934). "Embryology and Genetics." Columbia University Press, New York.

Morrisey, E. E. *et al.* (1998). GATA6 regulates HNF4 and is required for differentiation of visceral endoderm in the mouse embryo. *Genes Dev.* **12**, 3579–3590.

Mouse Genome Sequencing Consortium. (2002). Initial sequencing and comparative analysis of the mouse genome. *Nature* **420**, 520–562.

Müller, F. and Tora, L. (2004). The multicoloured world of promoter recognition complexes. *EMBO J.* **23**, 2–8.

Müller, J. and Bienz, M. (1991). Long-range repression conferring boundaries of *Ultrabithorax* expression in the *Drosophila* embryo. *EMBO J.* **10**, 3147–3155.

Müller, W. E. G., and Müller, I. M. (2003). Analysis of the sponge [Porifera] gene repertoire: Implications for the evolution of the metazoan body plan. *Prog. Mol. Subcell. Biol.* **37**, 1–33.

Mutskov, V. and Felsenfeld, G. (2004). Silencing of transgene transcription precedes methylation of promoter DNA and histone H3 lysine 9. *EMBO J.* **23**, 138–149.

Nakada, Y., Parab, P., Simmons, A., Omer-Abdalla, A., and Johnson, J. E. (2004). Separable enhancer sequences regulate the expression of the neural bHLH transcription factor neurogenin 1. *Dev. Biol.* **271**, 479–487.

Nakahara, K. and Carthew, R. W. (2004). Expanding roles for miRNAs and siRNAs in cell regulation. *Curr. Opin. Cell Biol.* **16**, 127–133.

Nakamura, Y., Makabe, K. W., and Nishida, H. (2003). Localization and expression pattern of type I postplasmic mRNAs in embryos of the ascidian *Halocynthia roretzi*. *Gene Exp. Patterns* **3**, 71–75.

Nakayama, T. *et al.* (2003). CD8 T cell-specific downregulation of histone hyperacetylation and gene activation of the IL-4 gene locus by ROG, repressor of GATA. *Immunity* **19**, 281–294.

Nambu, J. R., Lewis, J. O., Wharton, K. A., Jr., and Crews, S. T. (1991). The *Drosophila single-minded* gene encodes a helix-loop-helix protein that acts as a master regulator of CNS midline development. *Cell* **67**, 1157–1167.

Nasevicius, A. and Ekker, S. C. (2000). Effective targeted gene "knockdown" in zebrafish. *Nature Genet.* **26**, 218–220.

Nasiadka, A., Dietrich, B. H., and Krause, H. M. (2002). Anterior-posterior patterning in the *Drosophila* embryo. *In* "Advances in Developmental Biology and Biochemistry" (M. DePamphilis, Ed.), Vol. **12**, pp. 155–204. Elsevier Sciences, B. V., Amsterdam, New York.

Nasonkin, I. *et al.* (1999). A novel sulfonylurea receptor family member expressed in the embryonic *Drosophila* dorsal vessel and tracheal system. *J. Biol. Chem.* **274**, 29420–29425.

Nelson, C. E. *et al.* (1996). Analysis of *Hox* gene expression in the chick limb bud. *Development* **122**, 1449–1466.

Newman-Smith, E. H. and Rothman, J. H. (1998). The maternal-to-zygotic transition in embryonic patterning of *Caenorhabditis elegans*. *Curr. Opin. Genet. Dev.* **8**, 472–480.

Nguyen, H. T. and Xu, X. (1998). *Drosophila mef2* expression during mesoderm development is controlled by a complex array of *cis*-acting regulatory modules. *Dev. Biol.* **204**, 550–566.

Nibu, Y., Zhang, H., and Levine, M. (1998). Interaction of short-range repressors with *Drosophila* CtBP in the embryo. *Science* **280**, 101–104.

Nishida, H. (1987). Cell lineage analysis in ascidian embryos by intracellular injection of a tracer enzyme. *Dev. Biol.* **121**, 526–541.

Nishida, H. (1997). Cell fate specification by localized cytoplasmic determinants and cell interactions in ascidian embryos. *Int. Rev. Cytol.* **176**, 245–306.

Nishida, H. (2002a). Patterning the marginal zone of early ascidian embryos: Localized maternal mRNA and inductive interactions. *BioEssays* **24**, 613–624.

Nishida, H. (2002b). Specification of developmental fates in ascidian embryos: Molecular approach to maternal determinants and signaling molecules. *Int. Rev. Cytol.* **217**, 227–276.

Nishida, H. (2005). Specification of embryonic axis and mosaic development in ascidians. *Dev. Dyn.* **233**, 1177–1193.

Nishida, H. and Sawada, K. (2001). *macho-1* encodes a localized mRNA in ascidian eggs that specifies muscle fate during embryogenesis. *Nature* **409**, 724–729.

Nocente-McGrath, C., Brenner, C. A., and Ernst, S. G. (1989). *Endo16*, a lineage-specific protein of the sea urchin embryo, is first expressed just prior to gastrulation. *Dev. Biol.* **136**, 264–272.

Novina, C. D. and Sharp, P. A. (2004). The RNAi revolution. *Nature* **430**, 161–164.

Ochoa-Espinosa, A. *et al.* (2005). The role of binding site cluster strength in Bicoid-dependent patterning in *Drosophila*. *Proc. Natl. Acad. Sci. USA* **102**, 4960–4965.

Odom, D. T. *et al.* (2004). Control of pancreas and liver gene expression by HNF transcription factors. *Science* **303**, 1378–1382.

Okumura, T., Matsumoto, A., Tanimura, T., and Murakami, R. (2005). An endoderm-specific GATA factor gene, *dGATAe*, is required for the terminal differentiation of the *Drosophila* endoderm. *Dev. Biol.* **278**, 576–586.

Oliveri, P. and Davidson, E. H. (2004). Gene regulatory network controlling embryonic specification in the sea urchin. *Curr. Opin. Genet. Dev.* **14**, 351–360.

Oliveri, P. and Davidson, E. H. (2006). Gene regulatory network architecture and exclusion of alternative cell fates during embryonic specification. Submitted.

Oliveri, P., Carrick, D. M., and Davidson, E. H. (2002). A regulatory gene network that directs micromere specification in the sea urchin embryo. *Dev. Biol.* **246**, 209–228.

Oliveri, P., McClay, D. R., and Davidson, E. H. (2003). Activation of *pmar1* controls specification of micromeres in the sea urchin embryo. *Dev. Biol.* **258**, 32–43.

Oliveri, P., Tu, Q., and Davidson, E. H. (2006a). Architectural features of the skeletogenic gene regulatory network in the sea urchin embryo. Submittted.

Oliveri, P., Davidson, E. H., and McClay, D. R. (2006b). Early role of the *foxa* gene in endoderm specification in the sea urchin embryo. Submittted.

Olson, E. N. (1990). MyoD family: A paradigm for development. *Genes Dev.* **4**, 1454–1461.

Omori, M. *et al.* (2003). CD8 T cell-specific downregulation of histone hyperacetylation and gene activation of the IL-4 gene locus by ROG, repressor of GATA. *Immunity* **19**, 281–294.

Otim, O., Amore, G., Minokawa, T., McClay, D. R., and Davidson, E. H. (2004). *SpHnf6*, a transcription factor that executes multiple functions in sea urchin embryogenesis. *Dev. Biol.* **273**, 226–243.

Otte, A. P. and Kwaks, T. H. J. (2003). Gene repression by Polycomb group protein complexes: A distinct complex for every occasion? *Curr. Opin. Genet. Dev.* **13**, 448–454.

Panganiban, G. *et al.* (1997). The origin and evolution of animal appendages. *Proc. Natl. Acad. Sci. USA* **94**, 5162–5166.

Papatsenko, D. and Levine, M. (2005). Quantitative analysis of binding motifs mediating diverse spatial readouts of the Dorsal gradient in the *Drosophila* embryo. *Proc. Natl. Acad. Sci. USA* **102**, 4966–4971.

Park, B. K. *et al.* (2004). Intergenic enhancers with distinct activities regulate *Dlx* gene expression in the mesenchyme of the branchial arches. *Dev. Biol.* **268**, 532–545.

Parker, H. G. *et al.* (2004). Genetic structure of the purebred domestic dog. *Science* **304**, 1160–1164.

Patrinos, G. P. *et al.* (2004). Multiple interactions between regulatory regions are required to stabilize an active chromatin hub. *Genes Dev.* **18**, 1495–1509.

Paul, C. R. C. and Smith, A. B. (Eds.) (1988). "Echinoderm Phylogeny and Evolutionary Biology." Clarendon, Oxford.

Pearse, J. S. and Cameron, R. A. (1991). Echinodermata: Echinoidea. *In* "Reproduction of Marine Invertebrates" (A. C. Giese, J. S. Pearse and V. B. Pearse, Eds.), Vol. VI, pp. 514–662. The Boxwood Press, Pacific Grove, CA.

Pearson, J. C., Lemons, D., and McGinnis, W. (2005). Modulating Hox gene functions during animal body patterning. *Nature Rev. Genet.* **6**, 893–904.

Pennacchio, L. A., Baroukh, N., and Rubin, E. M. (2003). Human-mouse comparative genomics: Successes and failures to reveal functional regions of the human genome. *Cold Spring Harbor Symp. Quant. Biol.* **68**, 303–309.

Peterson, K. J. and Butterfield, N. J. (2005). Origin of the Eumetazoa: Testing ecological predictions of molecular clocks against the Proterozoic fossil record. *Proc. Natl. Acad. Sci. USA* **102**, 9547–9552.

Peterson, K. J., Cameron, R. A., and Davidson, E. H. (1997). Set-aside cells in maximal indirect development: Evolutionary and developmental significance. *BioEssays* **19**, 623–631.

Peterson, K. J., Harada, Y., Cameron, R. A., and Davidson, E. H. (1999). Expression pattern of *Brachyury* and *Not* in the sea urchin: Comparative implications for the origins of mesoderm in the basal deuterostomes. *Dev. Biol.* **207**, 419–431.

Peterson, K. J., Cameron, R. A., and Davidson, E. H. (2000). Bilaterian origins: Significance of new experimental observations. *Dev. Biol.* **219**, 1–17.

Peterson, K. J. *et al.* (2004). Estimating metazoan divergence times with a molecular clock. *Proc. Natl. Acad. Sci. USA* **101**, 6536–6541.

Peterson, R. E., and McClay, D. R. (2003). Primary mesenchyme cell patterning during the early stages following ingression. *Dev. Biol.* **254**, 68–78.

Pfeffer, P. L., Bouchard, M., and Busslinger, M. (2000). Pax2 and homeodomain proteins cooperatively regulate a 435 bp enhancer of the mouse *Pax5* gene at the midbrain-hindbrain boundary. *Development* **127**, 1017–1028.

Pfeffer, P. L., Payer, B., Reim, G., di Magliano, M. P., and Busslinger, M. (2002). The activation and maintenance of *Pax2* expression at the mid-hindbrain boundary is controlled by separate enhancers. *Development* **129**, 307–318.

Phan, D. *et al.* (2005). BOP, a regulator of right ventricular heart development, is a direct transcriptional target of MEF2C in the developing heart. *Development* **132**, 2669–2678.

Pirrotta, V. (1997). Chromatin-silencing mechanisms in *Drosophila* maintain patterns of gene expression. *Trends Genet.* **13**, 314–318.

Polli, M. and Amaya, E. (2002). A study of mesoderm patterning through the analysis of the regulation of *Xmyf-5* expression. *Development* **129**, 2917–2927.

Posakony, J. W. (1994). Nature versus nurture: Asymmetric cell divisions in *Drosophila* bristle development. *Cell* **76**, 415–418.

Poustelnikova, E., Pisarev, A., Blagov, M., Samsonova, M., and Reinitz, J. (2004). A database for management of gene expression data *in situ*. *Bioinformatics* **20**, 2212–2221.

Punzo, C., Seimiya, M., Flister, S., Gehring, W. J., and Plaza, S. (2002). Differential inter-actions of *eyeless* and *twin of eyeless* with the *sine oculis* enhancer. *Development* **129**, 625–634.

Qian, L., Liu, J., and Bodmer, R. (2005). Neuromancer Tb20-related genes (H15/midline) promote cell fate specification and morphogenesis of the *Drosophila* heart. *Dev. Biol.* **279**, 509–524.

Quiring, R., Walldorf, U., Kloter, U., and Gehring, W. J. (1994). Homology of the *eyeless* gene of *Drosophila* to the *Small eye* gene in mice and Aniridia in humans. *Science* **265**, 785–789.

Raff, R. A. (1996). "The Shape of Life. Genes, Development, and the Evolution of Animal Form," p. 208. The University of Chicago Press, Chicago.

Raible, F. and Brand, M. (2004). *Divide et Impera* – the midbrain-hindbrain boundary and its organizer. *Trends Neurosci.* **27**, 727–734.

Rajewsky, N., Vergassola, M., Gaul, U., and Siggia, E. D. (2002). Computational detection of genomic *cis*-regulatory modules applied to body patterning in the early *Drosophila* embryo. *BMC Bioinformatics* **3**, 30.

Ramos, E., Price, M., Rohrbaugh, M., and Lai, Z.-C. (2003). Identifying functional *cis*-acting regulatory modules of the *yan* gene in *Drosophila melanogaster*. *Dev. Genes Evol.* **213**, 83–89.

Ramsey, S., Orrell, D., and Bolouri, H. (2005). Dizzy: Stochastic simulation of large-scale genetic regulatory networks. *J. Bioinformatics Computational Biol.* **3**, 437–454.

Ransick, A. and Davidson, E. H. (1993). A complete second gut induced by transplanted micromeres in the sea urchin embryo. *Science* **259**, 1134–1138.

Ransick, A. and Davidson, E. H. (1995). Micromeres are required for normal vegetal plate specification in sea urchin embryos. *Development* **121**, 3215–3222.

Ransick, A. and Davidson, E. H. (1998). Late specification of veg$_1$ lineages to endodermal fate in the sea urchin embryo. *Dev. Biol.* **195**, 38–48.

Ransick, A. and Davidson, E. H. (2006). *cis*-Regulation of Notch signaling input in the sea urchin *gcm* gene. *Dev. Biol.*, submitted.

Ransick, A., Ernst, S., Britten, R. J., and Davidson, E. H. (1993). Whole mount *in situ* hybridization shows *Endo-16* to be a marker for the vegetal plate territory in sea urchin embryos. *Mech. Dev.* **42**, 117–124.

Ransick, A., Rast, J. P., Minokawa, T., Calestani, C., and Davidson, E. H. (2002). New early zygotic regulators of endomesoderm specification in sea urchin embryos discovered by differential array hybridization. *Dev. Biol.* **246**, 132–147.

Rao, M. V., Donoghue, M. J., Merlie, J. P., and Sanes, J. R. (1996). Distinct regulatory elements control muscle-specific, fiber-type-selective, and axially graded expression of a myosin light-chain gene in transgenic mice. *Mol. Cell. Biol.* **16**, 3909–3922.

Rast, J. P., Cameron, R. A., Poustka, A. J., and Davidson, E. H. (2002). *brachyury* target genes in the early sea urchin embryo isolated by differential macroarray screening. *Dev. Biol.* **246**, 191–208.

Rat Genome Sequencing Project Consortium (2004). Genome sequence of the brown Norway rat yields insights into mammalian evolution. *Nature* **428**, 493–521.

Rebeiz, M., Reeves, N. L., and Posakony, J. W. (2002). SCORE: A computational approach to the identification of *cis*-regulatory modules and target genes in whole-genome sequence data. *Proc. Natl. Acad. Sci. USA* **99**, 9888–9893.

Rebeiz, M., Stone, T., and Posakony, J. W. (2005). An ancient transcriptional regulatory linkage. *Dev. Biol.* **281**, 299–308.

Reeves, G. T., Kalifa, R., Klein, D. E., Lemmon, M. A., and Shvartsman, S. Y. (2005). Computational analysis of EGFR inhibition by Argos. *Dev. Biol.* **284**, 523–535.

Rehorn, K.-P., Thelen, H., Michelson, A. M., and Reuter, R. (1996). A molecular aspect of hematopoiesis and endoderm development common to vertebrates and *Drosophila*. *Development* **122**, 4023–4031.

Reichert, H. and Simeone, A. (2001). Developmental genetic evidence for a monophyletic origin of the bilaterian brain. *Phil. Trans. R. Soc. Lond.* B **356**, 1533–1544.

Reijnen, M. J. *et al.* (1995). Polycomb and BMI-1 homologs are expressed in overlapping patterns in *Xenopus* embryos and are able to interact with each other. *Mech. Dev.* **53**, 35–46.

Reim, I., Lee, H.-H., and Frasch, M. (2003). The T-box-encoding Dorsocross genes function in amnioserosa development and the patterning of the dorsolateral germ band downstream of Dpp. *Development* **130**, 3187–3204.

Reim, I., Mohler, J. P., and Frasch, M. (2005). *Tbx20*-related genes, *mid* and *H15*, are required for *tinman* expression, proper patterning, and normal differentiation of cardioblasts in *Drosophila*. *Mech. Dev.* **122**, 1056–1069.

Reiter, J. F., Kikuchi, Y., and Stainier, D. Y. R. (2001). Multiple roles for Gata5 in zebrafish endoderm formation. *Development* **128**, 125–135.

Revilla-i-Domingo, R. and Davidson, E. H. (2003). Developmental gene network analysis. *Int. J. Dev. Biol.* **47**, 695–703.

Revilla-i-Domingo, R., Minokawa, T., and Davidson, E. H. (2004). R11: A *cis*-regulatory element that controls expression of *SpDelta* and responds to the Pmar1 repression system. *Dev. Biol.* **274**, 438–451.

Rhinn, M. and Brand, M. (2001). The midbrain-hindbrain boundary organizer. *Curr. Opin. Neurobiol.* **11**, 34–42.

Ring, C. *et al.* (2002). The role of a Williams-Beuren syndrome-associated helix-loop-helix domain-containing transcription factor in activin/nodal signaling. *Genes Dev.* **16**, 820–835.

Ringrose, L., Rehmsmeier, M., Dura, J.-M., and Paro, R. (2003). Genome-side prediction of Polycomb/Trithorax response elements in *Drosophila melanogaster*. *Dev. Cell* **5**, 759–771.

Rodriguez, T. A., Casey, E. S., Harland, R. M., Smith, J. C., and Beddington, R. S. P. (2001). Distinct enhancer elements control *Hex* expression during gastrulation and early organogenesis. *Dev. Biol.* **234**, 304–316.

Rodriguez-Esteban, C. *et al.* (1999). The T-box genes *Tbx4* and *Tbx5* regulate limb outgrowth and identity. *Nature* **398**, 814–818.

Ronshaugen, M. and Levine, M. (2005). Visualization of *trans*-homolog enhancer-promoter interactions at the *Abd-B* Hox locus in the *Drosophila* embryo. *Dev. Cell* **7**, 925–932.

Ronshaugen, M., McGinnis, N., and McGinnis, W. (2002). Hox protein mutation and macroevolution of the insect body plan. *Nature* **415**, 914–917.

Rosenthal, N. *et al.* (1992). The myosin light-chain 1/3 locus: A model for developmental control of skeletal muscle differentiation. *In* "Neuromuscular Development and Disease" (A. M. Kelly and H. M. Blau, Eds.), pp. 131–144. Raven Press, Ltd., New York.

Rossant, J. and Tam, P. P. L. (2004). Emerging asymmetry and embryonic patterning in early development. *Dev. Cell* **7**, 155–164.

Rossel, M. and Capecchi, M. R. (1999). Mice mutant for both *Hoxa1* and *Hoxb1* show extensive remodeling of the hindbrain and defects in craniofacial development. *Development* **126**, 5027–5040.

Roth, S. and Schüpbach, T. (1994). The relationship between ovarian and embryonic dorsoventral patterning in *Drosophila*. *Development* **120**, 2245–2257.

Rothenberg, E. V. and Davidson, E. H. (2003). Regulatory co-options in the evolution of deuterostome immune systems. In *Innate Immunity* (R. A. B. Ezekowitz and J. A. Hoffman, eds.), pp. 61–88. Humana Press, Totowa, NJ.

Rozowski, M. and Akam, M. (2002). Hox gene control of segment-specific bristle patterns in *Drosophila Genes Dev.* **16**, 1150–1162.

Rubin, G. M. *et al.* (2000). Comparative genomics of the eukaryotes. *Science* **287**, 2204–2215.

Ruiz-Trillo, I. *et al.* (2002). A phylogenetic analysis of myosin heavy chain type II sequences corroborates that Acoela and Nemertodermatida are basal bilaterians. *Proc. Natl. Acad. Sci. USA* **99**, 11246–11251.

Ruiz-Trillo, I., Riutort, M., Fourcade, H. M., Baguña, J., and Boore, J. L. (2004). Mitochondrial genome data support the basal position of Acoelomorpha and the polyphyly of the Platyhelminthes. *Mol. Phylogenet. Evol.* **33**, 321–332.

Rusch, J. and Levine, M. (1996). Threshold responses to the dorsal regulatory gradient and the subdivision of primary tissue territories in the *Drosophila* embryo. *Curr. Opin. Genet. Dev.* **6**, 416–423.

Rushlow, C. A., Han, K., Manley, J. L., and Levine, M. (1989). The graded distribution of the dorsal morphogen involves selective nuclear transport in *Drosophila*. *Cell* **59**, 1165–1177.

Ruvkun, G. and Hobert, O. (1998). The taxonomy of developmental control in *Caenorhabditis elegans*. *Science* **282**, 2033–2041.

Ryan, K. M., Hoshizaki, D. K., and Cripps, R. M. (2005). Homeotic selector genes control the patterning of *seven-up* expressing cells in the *Drosophila* dorsal vessel. *Mech. Dev.* **122**, 1023–1033.

St. Johnston, D. and Nüsslein-Volhard, C. (1992). The origin of pattern and polarity in the *Drosophila* embryo. *Cell* **68**, 201–219.

Sanchez, M. J. *et al.* (1999). An SCL 3′ enhancer targets developing endothelium together with embryonic and adult haematopoietic progenitors. *Development* **126**, 3891–3904.

Sander, M. *et al.* (2000). Homeobox gene Nkx6.1 lies downstream of Nkx2.2 in the major pathway of beta-cell formation in the pancreas. *Development* **127**, 5533–5540.

Sardet, C., Nishida, H., Prodon, F., and Sawada, K. (2003). Maternal mRNAs of *PEM* and *macho 1,* the ascidian muscle determinant, associate and move with a rough endoplasmic reticulum network in the egg cortex. *Development* **130**, 5839–5849.

Satoh, N., Makabe, K. W., Katsuyama, Y., Wada, S., and Saiga, H. (1996). The ascidian embryo: An experimental system for studying genetic circuitry for embryonic cell specification and morphogenesis. *Dev. Growth Differ.* **38**, 325–340.

Satou, Y., Kusakabe, T., Araki, I., and Satoh, N. (1995). Timing of initiation of muscle-specific gene expression in the ascidian embryo precedes that of developmental fate restriction in lineage cells. *Dev. Growth Differ.* **37**, 319–327.

Scemama, J.-L., Hunter, M., McCallum, J., Prince, V., and Stellwag, E. (2002). Evolutionary divergence of vertebrate *Hoxb2* expression patterns and transcriptional regulatory loci. *J. Exp. Zool. (Mol. Dev. Evol.)* **294**, 285–299.

Schier, A. F. (2001). Axis formation and patterning in zebrafish. *Curr. Opin. Genet. Dev.* **11**, 393–404.

Schnabel, R. and Preiss, J. R. (1997). Specification of cell fates in the early embryo. *In* "*C. Elegans* II" (D. L. Riddle, T. Blumenthal, B. J. Meyer and J. R. Preiss, Eds.), pp. 361–382. Cold Spring Harbor Laboratory Press, Cold Spring Harbor, NY.

Schuler-Metz, A., Knöchel, S., Kaufmann, E., and Knöchel, W. (2000). The homeodomain transcription factor Xvent-2 mediates autocatalytic regulation of BMP-4 expression in *Xenopus* embryos. *J. Biol. Chem.* **275**, 34365–34374.

Schwartz, R. J., and Olson, E. N. (1999). Building the heart piece by piece: Modularity of *cis*-elements regulating *Nkx2–5* transcription. *Development* **126**, 4187–4192.

Scott, M. P. (1994). Intimations of a creature. *Cell* **79**, 1121–1124.

Scully, K. M., and Rosenfeld, M. G. (2002). Pituitary development: Regulatory codes in mammalian organogenesis. *Science* **295**, 2231–2235.

Sea Urchin Sequencing Consortium (2006). The *Strongylocentrotus purpuratus* genome. Submitted.

Seipel, K., and Schmid, V. (2005). Evolution of striated muscle: Jellyfish and the origin of triploblasty *Dev. Biol.* **282**, 14–26.

Seo, H.-C. *et al.* (2001). Miniature genome in the marine chordate *Oikopleura dioica*. *Science* **294**, 2506.

Servitja, J. M., and Ferrer, J. (2004). Transcriptional networks controlling pancreatic development and beta cell function. *Diabetologia* **47**, 597–613.

Shao, X. *et al.* (2002). Regulatory DNA required for *vnd/NK-2* homeobox gene expression pattern in neuroblasts. *Proc. Natl. Acad. Sci. USA* **99**, 113–117.

Shapiro, M. D. *et al.* (2004). Genetic and developmental basis of evolutionary pelvic reduction in threespine sticklebacks. *Nature* **428**, 717–723.

Sharon-Friling, R. *et al.* (1998). Lens-specific gene recruitment of ζ-crystallin through Pax6, Nrl-Maf, and brain suppressor sites. *Mol. Cell. Biol.* **18**, 2067–2076.

Sherwood, D. R., and McClay, D. R. (1997). Identification and localization of a sea urchin Notch homologue: Insights into vegetal plate regionalization and Notch receptor regulation. *Development* **124**, 3363–3374.

Sherwood, D. R., and McClay, D. R. (1999). LvNotch signaling mediates secondary mesenchyme specification in the sea urchin embryo. *Development* **126**, 1703–1713.

Sherwood, D. R., and McClay, D. R. (2001). LvNotch signaling plays a dual role in regulating the position of the ectoderm-endoderm boundary in the sea urchin embryo. *Development* **128**, 2221–2232.

Shilo, B. Z. (2003). Signaling by the *Drosophila* epidermal growth factor receptor pathway during development. *Exp. Cell Res.* **284**, 140–149.

Shimada, N., Aya-Murata, T., Reza, H. M., and Yasuda, K. (2003). Cooperative action between L-Maf and Sox2 on δ-*crystallin* gene expression during chick lens development. *Mech. Dev.* **120**, 455–465.

Shiojima, I. *et al.* (1999). Context-dependent transcriptional cooperation mediated by cardiac transcription factors Csx/Nkx-2.5 and GATA-4. *J. Biol. Chem.* **274**, 8231–8239.

Shivdasani, R. A. (2002). Molecular regulation of vertebrate early endoderm development. *Dev. Biol.* **249**, 191–203.

Shu, D. *et al.* (1999). A pipiscid-like fossil from the Lower Cambrian of south China. *Nature* **400**, 746–749.

Silver, S. J., and Rebay, I. (2005). Signaling circuitries in development: Insights from the retinal determination gene network. *Development* **132**, 3–13.

Simeone, A. (2000). Positioning the isthmic organizer where *Otx* and *Gbx2* meet. *Trends Genetics* **16**, 237–240.

Simon, J. A., and Tamkun, J. W. (2002). Programming off and on states in chromatin: Mechanisms of Polycomb and trithorax group complexes. *Curr. Opin. Genet. Dev.* **12**, 210–218.

Simon, J., Peifer, M., Bender, W., and O'Connor, M. (1990). Regulatory elements of the bithorax complex that control expression along the anterior-posterior axis. *EMBO J.* **9**, 3945–3956.

Simon, J., Chiang, A., Bender, W., Shimell, M. J., and O'Connor, M. (1993). Elements of the *Drosophila* bithorax complex that mediate repression by *Polycomb* group products. *Dev. Biol.* **158**, 131–144.

Sinclair, A. M. *et al.* (1999). Distinct 5′ SCL enhancers direct transcription to developing brain, spinal chord and endothelium; conserved neural expression is GATA factor dependent. *Dev. Biol.* **209**, 128–142.

Singh, M. K. *et al.* (2005). *Tbx20* is essential for cardiac chamber differentiation and repression of *Tbx2 Development* **132**, 2697–2707.

Smale, S. T. (2001). Core promoters: Active contributors to combinatorial gene regulation. *Genes Dev.* **15**, 2503–2508.

Small, S., Kraut, R., Hoey, T., Warrior, R., and Levine, M. (1991). Transcriptional regulation of a pair-rule stripe in *Drosophila*. *Genes Dev.* **5**, 827–839.

Small, S., Blair, A., and Levine, M. (1992). Regulation of *even-skipped* stripe 2 in the *Drosophila* embryo. *EMBO J.* **11**, 4047–4057.

Small, S., Blair, A., and Levine, M. (1996). Regulation of two pair-rule stripes by a single enhancer in the *Drosophila* embryo. *Dev. Biol.* **175**, 314–324.

Sokol, N. S., and Ambros, V. (2005). Mesodermally expressed *Drosophila* microRNA-1 is regulated by Twist and is required in muscle during larval growth. *Genes Dev.* **19**, 2343–2354.

Soltysik-Espanola, M., Klinzing, D. C., Pfarr, K., Burke, R. D., and Ernst, S. G. (1994). Endo16, a large multidomain protein found on the surface and ECM of endodermal cells during sea urchin gastrulation, binds calcium. *Dev. Biol.* **165**, 73–85.

Spieler, D. *et al.* (2004). Involvement of *Pax6* and *Otx2* in the forebrain-specific regulation of the vertebrate homeobox gene *ANF/Hesx1*. *Dev. Biol.* **269**, 567–579.

Spitz, F., Gonzalez, F., and Duboule, D. (2003). A global control region defines a chromosomal regulatory landscape containing the *HoxD* cluster. *Cell* **113**, 405–417.

Stamatoyannopoulos, G., and Grosveld, F. (2001). Hemoglobin switching. *In* "The Molecular Basis of Blood Diseases," 3rd Ed. (G. Stamatoyannopoulos *et al.*, Eds.), pp. 135–182. W. B. Saunders, Philadelphia.

Stathopoulos, A., and Levine, M. (2005). Localized repressors delineate the neurogenic ectoderm in the early *Drosophila* embryo. *Dev. Biol.* **280**, 482–493.

Stathopoulos, A., Van Drenth, M., Erives, A., Markstein, M., and Levine, M. (2002). Whole-genome analysis of dorsal-ventral patterning in the *Drosophila* embryo. *Cell* **111**, 687–701.

Stathopoulos, A., Tam, B., Ronshaugen, M., Frasch, M., and Levine, M. (2004). *pyramus* and *thisbe:* FGF genes that pattern the mesoderm of *Drosophila* embryos. *Genes Dev.* **18**, 687–699.

Stein, D., and Nüsslein-Volhard, C. (1992). Multiple extracellular activities in *Drosophila* egg perivitelline fluid are required for establishment of embryonic dorsal-ventral polarity. *Cell* **68**, 429–440.

Stein, L., Sternberg, P., Durbin, R., Thierry-Mieg, J., and Spieth, J. (2001). WormBase: Network access to the genome and biology of *Caenorhabditis elegans*. *Nucl. Acids Res.* **29**, 82–86.

Stennard, F., Carnac, G., and Gurdon, J. B. (1996). The *Xenopus* T-box gene, *Antipodean*, encodes a vegetally localised maternal mRNA and can trigger mesoderm formation. *Development* **122**, 4179–4188.

Stennard, F. A. *et al.* (2005). Murine T-box transcription factor Tbx20 acts as a repressor during heart development, and is essential for adult heart integrity, function and adaptation. *Development* **132**, 2451–2462.

Stern, C. D. (2004). Neural induction. *In* "Gastrulation. From Cells to Embryo" (C. D. Stern, Ed.). Cold Spring Harbor Laboratory Press, Cold Spring Harbor, NY.

Steward, R. (1989). Relocalization of the dorsal protein from the cytoplasm to the nucleus correlates with its function. *Cell* **59**, 1179–1188.

Steward, R. and Govind, S. (1993). Dorsal-ventral polarity in the *Drosophila* embryo. *Curr. Opin. Genet. Dev.* **3**, 556–561.

Su, W., Jackson, S., Tjian, R., and Echols, H. (1991). DNA looping between sites for transcriptional activation—self-association of DNA-bound Sp1. *Genes Dev.* **5**, 820–826.

Sulston, J. E., Schierenberg, E., White, J. G., and Thomson, J. N. (1983). The embryonic cell lineage of the nematode *Caenorhabditis elegans*. *Dev. Biol.* **100**, 64–119.

Sumiyama, K. *et al.* (2002). Genomic structure and functional control of the *Dlx3–7* bigene cluster. *Proc. Natl. Acad. Sci. USA* **99**, 780–785.

Sun, G. F. *et al.* (2004). TBX5, a gene mutated in holt-oram syndrome, is regulated through a GC box and T-box binding elements (TBEs). *J. Cell. Biochem.* **92**, 189–199.

Sweet, H. C., Hodor, P. G., and Ettensohn, C. A. (1999). The role of micromere signaling in Notch activation and mesoderm specification during sea urchin embryogenesis. *Development* **126**, 5255–5265.

Sweet, H. C., Gehring, M., and Ettensohn, C. A. (2002). LvDelta is a mesoderm-inducing signal in the sea urchin embryo and can endow blastomeres with organizer-like properties. *Development* **129**, 1945–1955.

Szebenyi, G. and Fallon, J. F. (1999). Fibroblast growth factors as multifunctional signaling factors. *Int. Rev. Cytol. – Survey Cell Biol.* **185**, 45–106.

Szeto, D. P., Griffin, K. J. P., and Kimelman, D. (2002). hrT is required for cardiovascular development in zebrafish. *Development* **129**, 5093–5101.

Tada, M., Casey, E. S., Fairclough, L., and Smith, J. C. (1998). Bix1, a direct target of *Xenopus* T-box genes, causes formation of ventral mesoderm and endoderm. *Development* **125**, 3997–4006.

Takahashi, T. and Holland, P. W. H. (2004). Amphioxus and ascidian Dmbx homeobox genes give clues to the vertebrate origins of midbrain development. *Development* **131**, 3285–3294.

Takeuchi, J. K. *et al.* (2003). *Tbx5* and *Tbx4* trigger limb initiation through activation of the Wnt/Fgf signaling cascade. *Development* **130**, 2729–2739.

Tanaka, M. *et al.* (1999). Complex modular *cis*-acting elements regulate expression of the cardiac specifying homeobox gene *Csx/Nkx2.5*. *Development* **126**, 1439–1450.

Teboul, L. *et al.* (2002). The early epaxial enhancer is essential for the initial expression of the skeletal muscle determination gene *Myf5* but not for subsequent, multiple phases of somitic myogenesis. *Development* **129**, 4571–4580.

Technau, U. (2001). *Brachyury*, the blastopore and the evolution of the mesoderm. *BioEssays* **23**, 788–794.

Teng, Y., Girard, L., Ferreira, H. B., Sternberg, P. W., and Emmons, S. W. (2004). Dissection of *cis*-regulatory elements in the *C. elegans* Hox gene *egl-5* promoter. *Dev. Biol.* **276**, 476–492.

Tickle, C. (2004). The contribution of chicken embryology to the understanding of vertebrate limb development. *Mech. Dev.* **121**, 1019–1029.

Tidyman, W. E. *et al.* (2003). *In vivo* regulation of the chicken cardiac troponin T gene promoter in zebrafish embryos. *Dev. Dyn.* **227**, 484–496.

Tokuoka, M., Imai, K. S., Satou, Y., and Satoh, N. (2004). Three distinct lineages of mesenchymal cells in *Ciona intestinalis* embryos demonstrated by specific gene expression. *Dev. Biol.* **274**, 211–224.

Tomioka, M., Miya, T., and Nishida, H. (2002). Repression of zygotic gene expression in the putative germline cells in ascidian embryos. *Zool. Sci.* **19**, 49–55.

Tomoyasu, Y., Wheller, S. R., and Denell, R. E. (2005). *Ultrabithorax* is required for membranous wing identity in the beetle *Tribolium castaneum*. *Nature* **433**, 643–647.

Tour, E., Pillemer, G., Gruenbaum. Y., and Fainsod, A. (2002a). *Otx2* can activate the isthmic organizer genetic network in the *Xenopus* embryo. *Mech. Dev.* **110**, 3–13.

Tour, E., Pillemer, G., Gruenbaum, Y., and Fainsod, A. (2002b). *Gbx2* interacts with *Otx2* and patterns the anterior-posterior axis during gastrulation in *Xenopus*. *Mech. Dev.* **112**, 141–151.

Trindade, M. Tada, M., and Smith, J. C. (1999). DNA-binding specificity and embryological function of XOM (Xvent-2). *Dev. Biol.* **216**, 442–456.

Tümpel, S., Maconochie, M., Wiedemann, L. M., and Krumlauf, R. (2002). Conservation and diversity in the *cis*-regulatory networks that integrate information controlling expression of *Hoxa2* in hindbrain and cranial neural crest cells in vertebrates. *Dev. Biol.* **246**, 45–56.

Uchikawa, M., Ishida, Y., Takemoto, T., Kamachi, Y,., and Kondoh, H. (2003). Functional analysis of chicken *Sox2* enhancers highlights an array of diverse regulatory elements that are conserved in mammals. *Dev. Cell* **4**, 509–519.

Uemur, O. *et al.* (2005). Comparative functional genomics revealed conservation and diversification of three enhancers of the *isl1* gene for motor and sensory neuron-specific expression. *Dev. Biol.* **278**, 587–606.

Vachon, G. *et al.* (1992). Homeotic genes of the bithorax complex repress limb development in the abdomen of the *Drosophila* embryo through the target gene *Distal-less*. *Cell* **71**, 437–450.

Valentine, J. W. (2004). *On the Origin of Phyla*. University of Chicago Press, Chicago.

Valentine, S. A. *et al.* (1998). Dorsal-mediated repression requires the formation of a multiprotein repression complex at the ventral silencer. *Mol. Cell. Biol.* **18**, 6584–6594.

Vaudin, P., Delanoue, R., Davidson, I., Silber, J., and Zider, A. (1999). TONDU (TDU), a novel human protein related to the product of *vestigial* (*vg*) gene of *Drosophila melanogaster* interacts with vertebrate TEF factors and substitutes for Vg function in wing formation. *Development* **126**, 4807–4816.

Venter, J. C. *et al.* (2001). The sequence of the human genome. *Science* **291**, 1304–1351.

Vilimas, T., Abraham, A., and Okkema, P. G. (2004). An early pharyngeal muscle enhancer from the *Caenorhabditis elegans ceh-22* gene is targeted by the Forkhead factor PHA-4. *Dev. Biol.* **266**, 388–398.

von Both, I. *et al.* (2004). Foxh1 is essential for development of the anterior heart field. *Dev. Cell* **7**, 331–345.

von Hippel, P. H. and Berg, O. G. (1986). On the specificity of DNA-protein interactions. *Proc. Natl. Acad. Sci. USA* **83**, 1608–1612.

Von Ohlen, T. and Doe, C. Q. (2000). Convergence of Dorsal, Dpp, and Egfr signaling pathways subdivides the *Drosophila* neuroectoderm into three dorsal-ventral columns. *Dev. Biol.* **224**, 362–372.

Walossek, D. (1999). On the Cambrian diversity of Crustacea. Proc. Fourth Int. Crustacean Congress. Koninklijke Brill, Leiden, The Netherlands.

Walton, K., Croce, J., Wu, S.-Y., LePage, T., and McClay, D. R. (2006). Hedgehog, Notch, and the TGF superfamily: conserved developmental signaling pathways in the sea urchin. Submitted.

Wang, D.-Z. *et al.* (2001). Activation of cardiac gene expression by myocardin, a transcriptional cofactor for serum response factor. *Cell* **105**, 851–862.

Wang, J. *et al.* (2004). The concerted activities of Pax4 and Nkx2.2 are essential to initiate pancreatic β-cell differentiation. *Dev. Biol.* **266**, 178–189.

Wang, X. *et al.* (2002). Control of megakaryocyte-specific gene expression by GATA-1 and FOG-1: Role of Ets transcription factors. *EMBO J.* **21**, 5225–5234.

Wassarman, K. M., Lewandoski. M., Campbell, K., Joyner, A. L., Rubenstein, J. L. R., Martinez, S., and Martin, G. R. (1997). Specification of the anterior hindbrain and establishment of a normal mid/hindbrain organizer is dependent on *Gbx2* gene function. *Development* **124**, 2923–2934.

Watabe, T. *et al.* (1995). Molecular mechanisms of Spemann's organizer formation: Conserved growth factor synergy between *Xenopus* and mouse. *Genes Dev.* **9**, 3038–3050.

Weatherbee, S. D., Halder, G., Kim, J., Hudson, A., and Carroll, S. (1998). Ultrabithorax regulates genes at several levels of the wing-patterning hierarchy to shape the development of the *Drosophila* haltere. *Genes Dev.* **12**, 1474–1482.

Weaver, C. and Kimelman, D. (2004). Move it or lose it: Axis specification in *Xenopus*. *Development* **131**, 3491–3499.

Weiss, J. B. *et al.* (1998). Dorsoventral patterning in the *Drosophila* central nervous system: The *intermediate neuroblasts defective* homeobox gene specifies intermediate column identity. *Genes Dev.* **12**, 3591–3602.

Weitzel, H. E. *et al.* (2004). Differential stability of β-catenin along the animal-vegetal axis of the sea urchin embryo mediated by disheveled. *Development* **131**, 2947–2956.

Wellik, D. M. and Capecchi, M. R. (2003). *Hox10* and *Hox11* genes are required to globally pattern the mammalian skeleton. *Science* **301**, 363–367.

Wenick, A. S. and Hobert, O. (2004). Genomic *cis*-regulatory architecture and *trans*-acting regulators of a single interneuron-specific gene battery in *C. elegans*. *Dev. Cell*. **6**, 757–770.

West, A. G., Gaszner, M., and Felsenfeld, G. (2002). Insulators: Many functions, many mechanisms. *Genes Dev.* **16**, 271–288.

Wienholds, E. *et al.* (2005). MicroRNA expression in zebrafish embryonic development. *Science* **309**, 310–311.

Wikramanayake, A. H., Huang, L., and Klein, W. H. (1998). β-catenin is essential for patterning the maternally specified animal-vegetal axis in the sea urchin embryo. *Proc. Natl. Acad. Sci. USA* **95**, 9343–9348.

Wikramanayake, A. H. *et al.* (2003). An ancient role for nuclear β-catenin in the evolution of axial polarity and germ layer segregation. *Nature* **426**, 446–450.

Wikramanayake, A. H. *et al.* (2004). Nuclear β-catenin-dependent Wnt8 signaling in vegetal cells of the early sea urchin embryo regulates gastrulation and differentiation of endoderm and mesoderm cell lineages. *Genesis* **39**, 194–205.

Wijgerde, M., Grosveld, F., and Fraser, P. (1995). Transcription complex stability and chromatin dynamics *in vivo*. *Nature* **377**, 209–213.

Williams, J. A., Paddock, S. W., Vorwerk, K., and Carroll, S. B. (1994). Organization of wing formation and induction of a wing-patterning gene at the dorsal/ventral compartment boundary. *Nature* **368**, 299–305.

Williams, T. M. *et al.* (2005). Candidate downstream regulated genes of HOX group 13 transcription factors with and without monomeric DNA binding capability. *Dev. Biol.* **279**, 462–480.

Wilson, E. B. (1896). "The Cell in Development and Inheritance." Johnson Reprint Corporation, New York, London, 1966.

Wilson, E. B. (1925). "The Cell in Development and Heredity." The Macmillan Company, New York.

Witze, E. S., Field, E. D., Hunt, D. F., and Rothman, J. H. (2006). Asymmetrically localized pur alpha transcription factor regulates endoderm development in response to MAP kinase signaling in *C. elegans*. Submitted.

Wu, J. and Cohen, S. M. (1999). Proximodistal axis formation in the *Drosophila* leg. Subdivision into proximal and distal domains by Homothorax and Distalless. *Development* **126**, 109–117.

Wu, L. H. and Lengyel, J. A. (1998). Role of *caudal* in hindgut specification and gastrulation suggests homology between *Drosophila* aminoproctodeal invagination and vertebrate blastopore. *Development* **125**, 2433–2442.

Wu, P. *et al.* (2004). *Evo-Devo* of amniote integuments and appendages. *Int. J. Dev. Biol.* **48**, 249–270.

Wu, X., Vakani, R., and Small, S. (1998). Two distinct mechanisms for differential positioning of gene expression borders involving the *Drosophila* gap protein giant. *Development* **125**, 3765–3774.

Wurst, W. and Bally-Cuif, L. (2001). Neural plate patterning: Upstream and downstream of the isthmic organizer. *Nature Rev. Neurosci.* **2**, 99–108.

Xanthos, J. B., Kofron, M., Wylie, C., and Heasman, J. (2001). Maternal VegT is the initiator of a molecular network specifying endoderm in *Xenopus laevis*. *Development* **128**, 167–180.

Yagi, K., Satoh, N., and Satou, Y. (2004a). Identification of downstream genes of the ascidian muscle determinant gene *Ci-macho1*. *Dev. Biol.* **274**, 478–489.

Yagi, K., Satou, Y., and Satoh, N. (2004b). A zinc finger transcription factor, ZicL, is a direct activator of *Brachyury* in the notochord specification of *Ciona intestinalis*. *Development* **131**, 1279–1288.

Yagi, K., Takatori, N., Satou, Y., and Satoh, N. (2005). *Ci-Tbx6b* and *Ci-Tbx6c* are key mediators of the maternal effect gene *Ci-macho1* in muscle cell differentiation in *Ciona intestinalis* embryos. *Dev. Biol.* **282**, 535–549.

Yamada, L., Kobayashi, K., Satou, Y., and Satoh, N. (2005). Microarray analysis of localization of maternal transcripts in eggs and early embryos of the ascidian, *Ciona intestinalis*. *Dev. Biol.* **284**, 536–550.

Yamashita, M. *et al.* (2002). Identification of a conserved GATA3 response element upstream proximal from the interleukin-13 gene locus. *J. Biol. Chem.* **277**, 42399–42408.

Yan, S.-J., Gu, Y., Li, W. X., and Fleming, R. J. (2004). Multiple signaling pathways and a selector protein sequentially regulate *Drosophila* wing development. *Development* **131**, 285–298.

Yin, Z., Xu, X.-L., and Frasch, M. (1997). Regulation of the Twist target gene *tinman* by modular *cis*-regulatory elements during early mesoderm development. *Development* **124**, 4971–4982.

Yoshida, T. and Yasuda, K. (2002). Characterization of the chicken L-Maf, MafB and c-Maf in crystalline gene regulation and lens differentiation. *Genes to Cells* **7**, 693–706.

Yuh, C.-H. and Davidson, E. H. (1996). Modular *cis*-regulatory organization of *Endo16*, a gut-specific gene of the sea urchin embryo. *Development* **122**, 1069–1082.

Yuh, C.-H., Ransick, A., Martinez, P., Britten, R. J., and Davidson, E. H. (1994). Complexity and organization of DNA-protein interactions in the 5′ regulatory region of an endoderm-specific marker gene in the sea urchin embryo. *Mech. Dev.* **47**, 165–186.

Yuh, C.-H., Moore, J. G., and Davidson, E. H. (1996). Quantitative functional interrelations within the *cis*-regulatory system of the *S. purpuratus Endo16* gene. *Development* **122**, 4045–4056.

Yuh, C.-H., Bolouri, H., and Davidson, E. H. (1998). Genomic *cis*-regulatory logic: Functional analysis and computational model of a sea urchin gene control system. *Science* **279**, 1896–1902.

Yuh, C.-H., Bolouri, H., and Davidson, E. H. (2001). *cis*-Regulatory logic in the *endo16* gene: Switching from a specification to a differentiation mode of control. *Development* **128**, 617–628.

Yuh, C.-H. *et al.* (2002). Patchy interspecific sequence similarities efficiently identify positive *cis*-regulatory elements in the sea urchin. *Dev. Biol.* **246**, 148–161.

Yuh, C.-H., Dorman, E. R., Howard, M. L., and Davidson, E. H. (2004). An *otx cis*-regulatory module: A key node in the sea urchin endomesoderm gene regulatory network. *Dev. Biol.* **269**, 536–551.

Yuh, C.-H., Dorman, E. R., and Davidson, E. H. (2005). Brn1/2/4, the predicted midgut regulator of the *endo16* gene of the sea urchin embryo. *Dev. Biol.* **281**, 286–298.

Yusufzai, T. M., Tagami, H., Nakatani, Y., and Felsenfeld, G. (2004). CTCF tethers an insulator to subnuclear sites, suggesting shared insulator mechanisms across species. *Mol. Cell* **13**, 291–298.

Zaffran, S. and Frash, M. (2002). Early signals in cardiac development. *Circ. Res.* **91**, 457–469.

Zaffran, S., Xu, X., Lo, P. C. H., Lee, H.-H., and Frasch, M. (2002). Cardiogenesis in the *Drosophila* model: Control mechanisms during early induction and diversification of cardiac progenitors. *Cold Spring Harbor Symp. Quant. Biol.* **67**, 1–12.

Zeller, R. W. *et al.* (1995). A multimerizing transcription factor of sea urchin embryos capable of looping DNA. *Proc. Natl. Acad. Sci. USA* **92**, 2989–2993.

Zhang, C., Basta, T., Fawcett, S. R., and Klymkowsky, M. W. (2005). SOX7 is an immediate-early target of VegT and regulates Nodal-related gene expression in *Xenopus*. *Dev. Biol.* **278**, 526–541.

Zhang, D., Penttila, T. L., Morris, P. L., Teichmann, M., and Roeder, G. (2001). Spermatogenesis deficiency in mice lacking the *Trf2* gene. *Science* **292**, 1153–1155.

Zhang, J. and King, M. L. (1996). *Xenopus VegT* RNA is localized to the vegetal cortex during oogenesis and encodes a novel T-box transcription factor involved in mesodermal patterning. *Development* **122**, 4119–4129.

Zhang, J. and Williams, T. (2003). Identification and regulation of tissue-specific *cis*-acting elements associated with the human AP-2α gene. *Dev. Dyn.* **228**, 194–207.

Zhang, Y., Yin, L., Xiao, S., and Knoll, A. H. (1998). Permineralized fossils from the terminal Proterozoic Doushantuo Formation, South China. *J. Paleontol.* **72**, 1–52.

Zhou, B., Bagri, A., and Beckendorf, S. K. (2001). Salivary gland determination in *Drosophila:* A salivary-specific, *fork head* enhancer integrates spatial pattern and allows *fork head* autoregulation. *Dev. Biol.* **237**, 54–67.

Zhou, Y.-H., Zheng, J. B., Gu, X., Saunders, G. F., and Yung, W.-K. A. (2002). Novel PAX6 binding sites in the human genome and the role of repetitive elements in the evolution of gene regulation. *Genome Res.* **12**, 1716–1722.

Zhu, X. and Rosenfeld, M. G. (2004). Transcriptional control of precursor proliferation in the early phases of pituitary development. *Curr. Opin Genet. Dev.* **14**, 567–574

Zilberman, A., Dave, V., Miano, J., Olson, E. N., and Periasamy, M. (1998). Evolutionarily conserved promoter region containing CArG*-like elements is crucial for smooth muscle myosin heavy chain gene expression. *Circ. Res.* **82**, 566–575.

Zuber, M. E., Gestri, G., Viczian, A. S., Barsacchi, G., and Harris, W. A. (2003). Specification of the vertebrate eye by a network of eye field transcription factors. *Development* **130**, 5155–5167.

Zuniga, A. *et al.* (2004). Mouse *limb deformity* mutations disrupt a global control region within the large regulatory landscape required for *Gremlin* expression. *Genes Dev.* **18**, 1553–1564.

FIGURE CITATIONS

FIGURE 1.4. Accurate expression of *evenskipped (eve)* stripe 2, and *eve* stripes 3+7, generated by individual *cis*-regulatory elements. Small *et al.* (1996). *Dev. Biol.* **175**, 314–324; copyright Elsevier Inc; with additional material kindly provided by M. Levine.

FIGURE 2.1. Examples of modular *cis*-regulatory organization from diverse bilaterians. (A1) Rodriguez *et al.* (2001). *Dev. Biol.* **234**, 304–316; copyright Elsevier Inc. (A2) Kwan *et al.* (2001). *Dev. Biol.* **232**, 176–190; copyright Elsevier Inc. (A3) Schwartz and Olsen (1999). *Development* **126**, 4187–4192 and The Company of Biologists Ltd.; Tanaka *et al.* (1999). *Development* **126**, 1439–1450 and The Company of Biologists Ltd. (A4) Carvajal *et al.* (2001). *Development* **128**, 1857–1868 and The Company of Biologists Ltd. Photographs from Hadchouel *et al.* (2003). *Development* **130**, 3415–3426 and The Company of Biologists Ltd. (B1) Landemann *et al.* (2004). *Dev. Biol.* **265**, 478–490; copyright Elsevier Inc. (B2) Wenick and Hobert (2004). *Dev. Cell.* **6**, 757–770; copyright Elsevier Inc. (C1) Lehman *et al.* (1999). *Development* **126**, 1793–1803 and The Company of Biologists Ltd. (C2) Simon *et al.* (1990). *EMBO J.* **9**, 3945–3956. Reprinted by permission from Macmillan Publishers Ltd. (C3) Jones *et al.* (2004). *Dev. Biol.* **266**, 374–387; copyright Elsevier Inc. (C4) Gómez-Skarmeta *et al.* (1995). *Genes Dev.* **9**, 1869–1882. Copyright (1995) Cold Spring Harbor Laboratory Press.

FIGURE 2.2. Identification of multiple *cis*-regulatory modules by interspecific sequence conservation. (A1) Revilla-i-Domingo *et al.* (2004). *Dev. Biol.* **274**, 438–451; copyright Elsevier Inc.; and unpublished data of R. Revilla-i-Domingo and Davidson. (A2) Yuh *et al.* (2002). *Dev. Biol.* **246**, 148–161; copyright Elsevier Inc. (B1) Kleinjan *et al.* (2004). *Dev. Biol.* **265**, 462–477; copyright Elsevier Inc. Griffin *et al.* (2002). *Mech. Dev.* **112**, 89–100; copyright Elsevier Inc. (B2) Uchikawa *et al.* (2003). *Dev. Cell.* **4**, 509–519; copyright Elsevier Inc. (B3) Kurokawa *et al.* (2004). *Development* **131**, 3307–3317 and The Company of Biologists Ltd. Kimura-Yoshida *et al.* (2004). *Development* **131**, 57–71 and The Company of Biologists Ltd.

FIGURE 2.3. *cis*-Regulatory logic in the *endo16* gene. (A1)–(A6) Ransick *et al.* (1993). *Mech. Dev.* **42**, 117–124; copyright Elsevier Inc. (B) Yuh *et al.* (1994). *Mech. Dev.* **47**, 165–186; copyright Elsevier Inc. (C) Yuh *et al.* (2005). *Dev. Biol.* **281**, 286–298; copyright Elsevier Inc. (D) Yuh *et al.* (2001). *Development* **128**, 617–628 and The Company of Biologists Ltd. (E) and (F) Yuh *et al.* (1998). *Science* **279**, 1896–1902. Copyright (1998) AAAS. (G) and (H) Yuh *et al.* (2001). *Development* **128**, 617–628 and The Company of Biologists Ltd.

FIGURE 2.5. *cis*-Regulatory integration of diverse spatial inputs. (A1) Barrow *et al.* (2000). *Development* **127**, 933–944 and The Company of Biologists Ltd. (A2)–(A6) Manzanares *et al.* (2002). *EMBO J.* **21**, 365–376. Reprinted by permission from Macmillan Publishers Ltd. (B1) and (B2) Kim *et al.* (1996). *Nature* **382**, 133–138. Reprinted by permission from Macmillan Publishers Ltd. (B3) Kim *et al.* (1997).

Nature **388**, 304–308. Reprinted by permission from Macmillan Publishers Ltd. (B4) Halder *et al.* (1998). *Genes Dev.* **12**, 3900–3909. Copyright (1998) Cold Spring Harbor Laboratory Press. (B5)–(B9) Guss *et al.* (2001). *Science* **292**, 1164–1167. Copyright (2001) AAAS. (C1)–(C4) Han *et al.* (2002). *Dev. Biol.* **252**, 225–240; copyright Elsevier Inc.; incorporating information discovered by Halfon *et al.*, 2000; Knirr and Frasch, 2001, and prior studies referred to therein. (D1)–(D4) Gebelein *et al.* (2004). *Nature* **431**, 653–659. Reprinted by permission from Macmillan Publishers Ltd. (D5) and (D6) Gebelein *et al.* (2002). *Dev. Cell.* **3**, 487–498; copyright Elsevier Inc. (D7) Gebelein *et al.* (2004). *Nature* **431**, 653–659. Reprinted by permission from Macmillan Publishers Ltd.

FIGURE 2.6. *cis*-**Regulatory repression in the definition of spatial boundaries of pair rule gene expression in *Drosophila*.** (A1) Small *et al.*, (1992). *EMBO J.* **11**, 4047–4057. Reprinted by permission from Macmillan Publishers Ltd. (A2) Image kindly provided by J. Reinitz. (A3) and (A4) Wu *et al.* (1998). *Development* **125**, 3765–3774 and The Company of Biologists Ltd. (B1)–(B3) Clyde *et al.* (2003). *Nature* **426**, 849–853. Reprinted by permission from Macmillan Publishers Ltd. (C1)–(C4) Andrioli *et al.* (2004). *Dev. Biol.* **276**, 541–551; copyright Elsevier Inc.

FIGURE 2.7. *cis*-**Regulatory design in specification of ventral to dorsal development domains in *Drosophila*.** (A1) and (A2) Rushlow *et al.* (1989). *Cell* **59**, 1165–1177; copyright Elsevier Inc. (A3)–(A5) Huang *et al.* (1997). *Genes Dev.* **11**, 1963–1973. Copyright (1997) Cold Spring Harbor Laboratory Press. (B1) and (B2) Gray *et al.* (1994). *Genes Dev.* **8**, 1829–1838. Copyright (1994) Cold Spring Harbor Laboratory Press. (B3) Gray and Levine (1996). *Genes Dev.* **10**, 700–710. Copyright (1996) Cold Spring Harbor Laboratory Press. (C) Diagram kindly provided by A. Stathopoulos and M. Levine, from experimental *cis*-regulatory and computational data of Stathopoulos *et al.*, 2002, 2004; Markstein *et al.*, 2004; and references therein.

FIGURE 2.8. External interactions of *cis*-regulatory modules by genomic looping. (A1)–(A4) Zeller *et al.* (1995). *Proc. Natl. Acad. Sci. USA* **92**, 2989–2993. Copyright (1995) National Academy of *Sciences*, U.S.A. (B1) Calhoun and Levine (2003). *Proc. Natl. Acad. Sci. USA* **100**, 9878–9883. Copyright (2003) National Academy of *Sciences*, U.S.A. (B2) and (B3) Calhoun *et al.* (2002). *Proc. Natl. Acad. Sci. USA* **99**, 9243–9247. Copyright (2002) National Academy of *Sciences*, U.S.A. (B4) and (B5) Calhoun and Levine (2003). *Proc. Natl. Acad. Sci. USA* **100**, 9878–9883. Copyright (2003) National Academy of *Sciences*, U.S.A. (C1) and (C2) Patrinos *et al.* (2004). *Genes Dev.* **18**, 1495–1509. Copyright (2004) Cold Spring Harbor Laboratory Press. (C3) Wijgerde *et al.* (1995). *Nature* **377**, 209–213. Reprinted by permission from Macmillan Publishers Ltd. (D1)–(D6) Ronshaugen and Levine (2005). *Dev. Cell.* **7**, 925–932; copyright Elsevier Inc.

FIGURE 3.1. Anisotropies in animal eggs and the differential establishment of early regulatory states. (A) Davidson (1990). *Development* **108**, 365–389 and The Company of Biologists Ltd. (B) Ochoa-Espinosa *et al.* (2005). *Proc. Natl. Acad. Sci. USA* **102**, 4960–4965. Copyright (2005) National Academy of *Sciences*, U.S.A. (C) Bowerman *et al.* (1993). *Cell* **74**, 443–452; copyright Elsevier Inc. (D1)–(D3) Nishida (2002a). *BioEssays* **24**, 613–624. Reprinted with permission of Wiley-Liss Inc., a subsidiary of John Wiley & Sons, Inc. (E1)–(E2) Zhang and King (1996). *Development* **122**, 4119–4129 and The Company of Biologists Ltd. (F1) Image kindly provided by C. A. Ettensohn. (F2)–(F3) Logan *et al.* (1999). *Development* **126**, 345–357 and The Company of Biologists Ltd.

FIGURE 3.2. Lineage and fate maps in sea urchin, nematode and ascidian embryos. (A) A. Ransick and E. Davidson. (B1) and (B2 upper) Goldstein (1995). *Development* **121**, 1227–1236 and The Company of Biologists Ltd. (B2 lower) and (B3) Maduro and Rothman (2002). *Dev. Biol.* **246**, 68–85; copyright Elsevier Inc. (C1) Satoh *et al.* (1996). *Develop. Growth Differ.* **38**, 325–340. Blackwell Publishing. (C2) Nishida (2002a). *BioEssays* **24**, 613–624. Reprinted with permission of Wiley-Liss Inc., a subsidiary of John Wiley & Sons, Inc. (C3)–(C4) Satoh *et al.* (1996). *Develop. Growth Differ.* **38**, 325–340. Blackwell Publishing. (C5) Nishida (2002a). *BioEssays* **24**, 613–624. Reprinted with permission of Wiley-Liss Inc., a subsidiary of John Wiley & Sons, Inc.

FIGURE 3.4. Syncytial and early cellular development in *Drosophila*. (A) Image kindly provided by D. Kosman and W. McGinnis. (B)–(C) Hartenstein (1993). "Atlas of *Drosophila* Development." Copyright (1993) Cold Spring Harbor Laboratory Press. (D) Stathopoulos and Levine (2005). *Dev. Cell.* **9**, 449–462; copyright Elsevier Inc.

FIGURE 3.5. Progenitor fields and their subdivision in the formation of various adult bilaterian body parts. (A1) Lecuit and Cohen (1997). *Nature* **388**, 139–145. Reprinted by permission from Macmillan Publishers Ltd. (A2) and (A3) Wu and Cohen (1999). *Development* **126**, 109–117 and The Company of Biologists Ltd. (B1) and (B2) Harvey (1996). *Dev. Biol.* **178**, 203–216; copyright Elsevier Inc. (C1) and (C2) Logan *et al.* (1998). *Development* **125**, 2825–2835 and The Company of Biologists Ltd. (D) Grapin-Botton and Melton (2000). *Trends Genet.* **16**, 124–130; copyright Elsevier Inc.; see also Beck *et al.*, 2000. (E) Baylies *et al.* (1998). *Cell* **93**, 921–927; copyright Elsevier Inc. (F) Isshiki *et al.* (2001). *Cell* **106**, 511–521; copyright Elsevier Inc. (G1) Gray *et al.* (2004). *Science* **306**, 2255–2257; Supporting Online Material.

FIGURE 4.2. Network for specification of sea urchin embryo endomesoderm: process diagram, view from the genome, and *cis*-regulatory evidence of selected subcircuit functions. (A) Revilla-i-Domingo and Davidson (2003). *Int. J. Dev. Biol.* **47**, 695–703; copyright UBC Press. (C2)–(C5) Minokawa *et al.* (2005). *Dev. Biol.* **288**, 545–558; copyright Elsevier Inc. (D2) Oliveri *et al.* (2002). *Dev. Biol.* **246**, 209–228; copyright Elsevier Inc. (D3) Revilla-i-Domingo *et al.* (2004). *Dev. Biol.* **274**, 438–451; copyright Elsevier Inc. (F2) Ransick *et al.* (2002). *Dev. Biol.* **246**, 132–147; copyright Elsevier Inc. (F3) Ransick and Davidson (2006). *Dev. Biol.*, submitted. (G2) Yuh *et al.* (2004). *Dev. Biol.* **269**, 536–551; copyright Elsevier Inc.

FIGURE 4.3. Detailed and extensive conservation of a kernel of the endomesoderm network. Davidson and Erwin (2006). *Science* **311**, 796–800. Copyright (2006) AAAS.

FIGURE 4.4. Gene regulatory network for specification of mesoderm in *Xenopus*, subcircuits, and examples of supporting *cis*-regulatory evidence. (A) After Koide *et al.* (2005). *Proc. Natl. Acad. Sci. USA* **102**, 4943–4948. Copyright (2005) National Academy of Sciences, U.S.A. (B2) Hilton *et al.* (2003). *Mech. Dev.* **120**, 1127–1138; copyright Elsevier Inc. (C2) Koide *et al.* (2005). *Proc. Natl. Acad. Sci. USA* **102**, 4943–4948. Copyright (2005) National Academy of Sciences, U.S.A. (C3) and (C4) Laurent *et al.* (1997). *Development* **124**, 4905–4916 and The Company of Biologists Ltd. (D2) Polli and Amaya (2002). *Development* **129**, 2917–2927 and The Company of Biologists Ltd. (D3) Trindade *et al.* (1999). *Dev. Biol.* **216**, 442–456; copyright Elsevier Inc. (E2) Casey *et al.* (1998). *Development* **125**, 3887–3894 and The Company of Biologists Ltd. (E3) Lin *et al.* (2003). *Dev. Dyn.* **226**, 51–58. Reprinted with permission of Wiley-Liss Inc., a subsidiary of John Wiley & Sons, Inc.

FIGURE 4.5. Gene regulatory network underlying dorsoventral specification in *Drosophila*; subcircuits and their functions. (A) Levine and Davidson (2005). *Proc. Natl. Acad. Sci. USA* **102**, 4936–4942. Copyright (2005) National Academy of Sciences, U.S.A. (B2) Cowden and Levine (2002). *Development* **129**, 1785–1793 and The Company of Biologists Ltd. (B3) Nambu *et al.* (1991). *Cell* **67**, 1157–1167; copyright Elsevier Inc. (C2) and (C3) Chang *et al.* (2001). *Int. J. Dev. Biol.* **45**, 715–724; copyright UBC Press.

FIGURE 4.7. Gene regulatory network underlying differential gene expression in right and left ASE taste neurons in *C. elegans*. Figure kindly provided by O. Hobert, after Johnston *et al.*, 2005.

FIGURE 5.2. Changes in the different parts of a gene regulatory network and their qualitatively diverse evolutionary consequences. Davidson and Erwin (2006). *Science* **311**, 796–800. Copyright (2006) AAAS.

FIGURE 5.3. Homologies in nervous system gene expression patterns across the Bilateria. (A) and (B) Lichtneckert and Reichert (2005). *Heredity* **94**, 465–477. Reprinted by permission from Macmillan Publishers Ltd. (C1)–(C5) Lowe *et al.* (2003). *Cell* **113**, 853–865; copyright Elsevier Inc. (D1) and (D2) Cornell and Von Ohlen (2000). *Curr. Opin. Neurobiol.* **10**, 63–71; copyright Elsevier Inc.

FIGURE 5.4. Chordate hindbrain gene expression patterns and evidence for vertebrate rhombomere specification kernel. (A) Takahashi and Holland (2004). *Development* **131**, 3285–3294 and The Company of Biologists Ltd. (B) Raible and Brand (2004). *Trends Neurosci.* **27**, 727–734; copyright Elsevier Inc. (C1) Davidson (2001). "Genomic Regulatory Systems. Development and Evolution." Academic Press/Elsevier, pp. 136–139. (C2) and (C3) Tümpel *et al.* (2002). *Dev. Biol.* **246**, 45–56; copyright Elsevier Inc.

FIGURE 5.5. Endoderm specification in vertebrates. (A1)–(A3) Reiter *et al.* (2001). *Development* **128**, 125–135 and The Company of Biologists Ltd. (B1) and (B2) Shivdasani (2002). *Dev. Biol.* **249**, 191–203; copyright Elsevier Inc. (C1) and (C2) Afouda *et al.* (2005). *Development* **132**, 763–774 and The Company of Biologists Ltd.

FIGURE 5.6. A putative pan-bilaterian kernel for heart specification. Davidson and Erwin (2006). *Science* **311**, 796–800. Copyright (2006) AAAS.

FIGURE 5.7. *hox* genes as regional off or on switches, applied to developmental patterning functions (A1) and (A2) Cohn and Tickle (1999). *Nature* **399**, 474–479. Reprinted by permission from Macmillan Publishers Ltd. (B1) Cover image from *Science*, vol. **221**, no. 4605, 1 July 1983. Image: E. B. Lewis, Caltech, Pasadena, CA 91125. Bender *et al.* (1983). *Science* **221**, 23–29. Copyright (1983) AAAS. (B2) Weatherbee *et al.* (1998). *Genes Dev.* **12**, 1474–1482. Copyright (1998) Cold Spring Harbor Laboratory Press; modified by inclusion of data from Galant *et al.*, 2002. (B3) Galant *et al.* (2002). *Development* **129**, 3115–3126 and The Company of Biologists Ltd. (B4) Hersh and Carroll (2005). *Development* **132**, 1567–1577 and The Company of Biologists Ltd. (C1) and (C2) Wellik and Capecchi (2003). *Science* **301**, 363–367. Copyright (2003) AAAS. (D1)–(D3) Anand *et al.* (2003). *Proc. Natl. Acad. Sci. USA* **100**, 15666–15669. Copyright (2003) National Academy of Sciences, U.S.A. (E1) and (E2) Wellik and Capecchi (2003). *Science* **301**, 363–367. Copyright (2003) AAAS (F) Lo *et al.* (2002). *Dev. Biol.* **251**, 307–319; copyright Elsevier Inc.

INDEX

Page numbers in italics refer to the pages on which the figure appears.

Printed and bound by CPI Group (UK) Ltd, Croydon, CR0 4YY

03/10/2024

01040312-0007